# POLLUTANTS
# IN A MULTIMEDIA
# ENVIRONMENT

# POLLUTANTS IN A MULTIMEDIA ENVIRONMENT

Edited by

## Yoram Cohen

University of California, Los Angeles
Los Angeles, California

PLENUM PRESS • NEW YORK AND LONDON

Library of Congress Cataloging in Publication Data

Pollutants in a multimedia envornment.

Proceedings of a workshop held in Santa Monica, Calif., Jan. 21–24, 1986, sponsored
by the National Center for Intermedia Transport Research and the Engineering and
Systems Analysis for the Control of Toxics Program.
Includes bibliographical references and index.
1. Environmental impact analysis—Mathematical models—Congresses. 2. Environmental protection—Mathematical models—Congresses. 3. Transport theory—Mathematical
models—Congresses. I. Cohen, Yoram. II. National Center for Intermedia Transport
Research (U.S.) III. University of California, Los Angeles. Engineering and Systems
Analysis for Control of Toxics Program.

| TD194.6.P65   1986 | 628.5 | 86-22704 |
|---|---|---|

ISBN-13: 978-1-4612-9314-9     e-ISBN-13: 978-1-4613-2243-6
DOI: 10.1007/978-1-4613-2243-6

Proceedings of a workshop on Pollutant Transport and Accumulation in a
Multimedia Environment, sponsored by the National Center for Intermedia
Transport Research and the University of California, Los Angeles Engineering
and Systems Analysis for the Control of Toxics Program,
held January 21–24, 1986, in Santa Monica, California

© 1986 Plenum Press, New York
Softcover reprint of the hardcover 1st edition 1986
A Division of Plenum Publishing Corporation
233 Spring Street, New York, N.Y. 10013

## PREFACE

Pollutants released to the environment are distributed among the many environmental media such as air, water, soil, and vegetation, as the result of complex physical, chemical and biological processes. The possible environmental impact associated with chemical pollutants is related to their concentration levels and persistence in the various environmental compartments. Therefore, information regarding the migration of pollutants across environmental phase boundaries (eg., air-water, soil-water) and their accumulation in the environment is essential if we are to assess the potential environmental impact and the associated risks.

In recent years it has become apparent that environmental pollution is a multimedia problem. Risk assessment and the design of appropriate pollution control measures require that we carefully consider the transport and accumulation of pollutants in the environment. We are now recognizing that the environment must be considered as a whole, and the scientific and regulatory approaches must consider the interactions of environmental media. It is also becoming apparent that single-medium approaches are partial and often counter-productive. On the other hand any multimedia program must carefully consider the rate of each environmental medium in the overall multimedia scheme.

In recognizing the need to evaluate our current ability to assess the risk that is imposed by pollutant migration in the environment, the National Center for Intermedia Transport Research (NCITR) and the Engineering and Systems Analysis for the Control of Toxics Program (ESACT), supported by the University of California Toxic Substance Research and Training Program sponsored the first workshop on "Pollutant Transport and Accumulation in a Multimedia Environment" on January 21-24, 1986 in Santa Monica, California. The workshop signifies the first time that scientists working on problems of multimedia transport and accumulation have focused their attention on the applicability of multimedia models of pollutant transport to exposure and risk assessment.

"Pollutants in a Multimedia Environment" is based in part on the papers presented in the workshop on "Pollutant Transport and Accumulation in a Multimedia Environment". The book is comprised of 14 papers arranged in the following three parts: multimedia transport modeling; multimedia analyses of exposure and risk; and field studies of pollutant transport and exposure assessment. The book presents a state-of-the-art review of multimedia models of pollutant transport, exposure and risk assessment, as well as experimental field studies of pollutant transport and human exposure.

Much effort has gone into the preparation of this book and the workshop which served as the focal point for formulating and refining many of the ideas expressed in this volume. I am confident that the collective contribution of the distinguished scientists who participated in the workshop will mark the beginning of a new multidisciplinary approach to environmental research.

I am particularly grateful for the generous financial support for the workshop and this book provided by the National Center for Intermedia Transport Research, and the Engineering and Systems Analysis for the Control of Toxics Program. I am indebted to Professor S.K. Friedlander, Director of both the NCITR and ESACT, for his encouragement, cooperation and assistance throughout the development of the multimedia program at UCLA. This multimedia program was the impetus for both the workshop and this book. I want to thank the members of the workshop science advisory committee, Dr. G. Hidy, Dr. R. Stephens, Dr. B. Fischback, and Professors V.L. Vilker and S.K. Friedlander who contributed their time and knowledge. I also benefited from the valuable discussions with Professors D. Mackay and L. Thibodeaux. Lastly, I want to especially acknowledge Sandy Malone, Administrative Analyst for the National Center for Intermedia Transport Research, for her assistance in the organization of the workshop and the preparation of this volume.

June 1986                                                    Yoram Cohen

# CONTENTS

## FIELD STUDIES OF POLLUTANT TRANSPORT
## AND EXPOSURE ASSESSMENT

A MULTIMEDIA APPROACH TO POLLUTION CONTROL:  CROSS-MEDIA PROBLEMS

J. Clarence Davies

The Conservation Foundation
Washington, D.C.

The conference on pollutant transport and accumulation in a multimedia environment is of great importance because, to my knowledge, it is the first time that the most distinguished scientists working on transport and accumulation have been brought together to focus specifically on cross-media transfer of pollutants.  The conference is one, but by no means the only sign that a new view of environmental problems is beginning to take hold.  We are recognizing that the environment must be considered as a whole, that scientific and regulatory approaches alike must take into account the complex interactions that make any one-medium approach at best partial and at worst counter-productive.

## Nature of the Cross-Media Problem

There are many aspects to the cross-media problem and many definitions of both the problem and its solutions.  I have found it useful to divide consideration of the problem into four stages:  pollution sources, pollution control, environmental transport and transformation, and human and environmental exposure.

Pollution Sources.  The initial stage of cross-media problems can take three forms:

1)  A single pollutant can be emitted by a source into several different media.  The emissions into different media can be simultaneous. For example, the New Jersey industrial survey and the analysis of the data by Inform, showed that a number of chemical plants emitted toluene both into the air and into the water.  Some plants emitted it into the air and water and at the same time disposed of some toluene on land.  A different form of the cross-media problem occurs when a source has a choice of which medium it discharges a pollutant into.  Thus a plant may dispose of a toxic chemical into water, but when water disposal becomes more expensive or regulated it may switch to air discharge of the substance.  Regulation of air and water discharge may result in increased land disposal.  There is a natural tendency for pollutants to be discharged into the environmental medium that is least regulated because the form of disposal that is least regulated is also likely to be the form that is least expensive.

2)  A second type of initiating factor is when discharge in one medium is only a problem when transferred to another medium.  This obviously involves environmental transport, which is discussed below.  However, in

1

some situations, notably those involving non-point pollution sources, the connection between discharge to one medium and impact on another is so direct that it may be considered more as a source problem than a transport problem. (All of these classifications are descriptive and heuristic, so it is not worth arguing about which problems belong in which categories.) A good example is urban street run-off, which is a water pollution problem created by wastes disposed on land.

3) A different type of problem is multiple pollutants emitted from a single source. This is a cross-media problem in that it may be possible or necessary to trade-off emissions of different pollutants from the same plant, even though the pollutants may be emitted into different media. Thus, for example, the type of fuel, the temperature, and the combustion process used by a steam electric generating plant will have an important influence on how much of which types of pollutants are emitted into different parts of the environment. The Electric Power Research Institute has done extensive work on this type of problem and is convinced that a more integrated approach to pollution control at power plants could greatly increase the efficiency of such plants without any loss in environmental quality.

Pollution Control. Pollution controls can themselves be an important source of environmental contamination, particularly by transferring pollutants from one medium of the environment to another. This aspect of the cross-media problem is both the most troubling and perhaps, the most challenging, because it calls into question so much of what we are now doing to improve the environment.

The cross-media aspects of pollution control appear no matter where we look. The most startling recent discovery has been the contribution of municipal waste water treatment plants to volatile organic chemical air pollution. The transfer from water control to land problems through sludge disposal has been a longstanding problem. The residues from air pollution control devices similarly have been a disposal problem in other media, and the dilution approach to air pollution control has clearly not worked well with respect to sulfur and nitrogen oxides. The cross-media impacts of land disposal have been starkly revealed by the Superfund program. Incineration is likely to pose more subtle problems, but we are just beginning to analyze the incineration by products that will have to be disposed of somewhere.

The point of reciting these control problems is not to show that all pollution control is futile. Rather, it is to show that we tend to ask the wrong questions about how to control pollution. What we ask is how to remove pollutant X from some particular part of the environment or how to prevent it from getting into that medium. What we should be asking is what is the least risky way of dealing with pollutant X, taking into account transport, degradation, and transformation. Waste reduction and recycling are obviously good answers to this question in many circumstances. Where these approaches are not appropriate, either because of cost or physical constraints, a crude way of looking at the remaining options is which controls will result in chemical transformation of the pollutant to a less risky form before exposure to target organisms occurs. In this crude sense, real pollution control is structuring the race between degradation of the pollutant and exposure in such a way that degradation wins. The degradation products cannot, of course, be ignored. Some of them may be more toxic and, even if they are less toxic the significance of the residual risk must be considered.

<u>Environmental Transport and Transformation</u>. Because all cross-media problems involve transport and transformation, this aspect of the problem cuts across all the others. There are, however, several aspects of this stage that are worth noting.

A whole school of literature and study is devoted to global cycles, such as those for nitrogen, sulfur, or carbon. But these studies have not adequately been related to the studies of the impacts of pollution in particular geographical areas. It is as if global transport and sub-global transport were two independent phenomena, which clearly they are not. The modeling of the greenhouse effect is beginning to bridge this gap, but it is only a beginning.

A second tendency to err, which I am delighted to see that this conference has mostly avoided, is to talk about the "fate" of a pollutant. The word fate has a finality to it that is quite misleading when it comes to analyzing the behavior of pollutants. What is the fate of a pollutant under some conditions (burial in the bottom sediments of lakes, for example) becomes a source of the pollutant when temperature, or time of year, or other conditions change. Use of the term "fate" too often encourages the partial non-systematic view of environmental problems that has led us into the cross-media problem.

Finally, I would make a plea that the biota not be ignored when considering transport and transformation. The scientists at this conference are too smart to commit such an error, but some other scientists and analysts are not. In most cases, the relevant biota should be considered a medium in the same sense as the air, water, or land.

<u>Human and Environmental Exposure</u>. The exposure stage of the cross-media problem arises from the fact that the same target organism, humans for example, may be exposed to the same pollutant through several different routes. The policy corollary of this fact is that regulatory standards should take multiple exposure into account. They rarely do.

The scientific corollary is that there are a number of unanswered questions that urgently need to be explored. For example, what types of chemicals have the same toxicological effects on humans regardless of route of exposure and what types have different effects? Lead tends to accumulate in the body in the same places regardless of how exposure occurs. Therefore the toxicological burden of lead is the same regardless of how much of the exposure comes from breathing, eating, drinking, or touching. Asbestos, however, appears to be quite different. We know that breathing asbestos can cause cancer. However, based on animal studies, it appears that drinking asbestos fibres in water does not cause cancer. How many of what kinds of chemicals are like lead and how many like asbestos? We are just beginning to explore this type of question.

The interactive effects of simultaneous exposure to many different kinds of chemicals is the parallel problem to the source stage question of multiple pollutants emitted from a single source. Whether it is part of the cross-media problem is open to argument. That it is an important problem is not open to argument.

<u>Extent and Seriousness of Cross-Media Problems</u>

The above brief description is, I hope, sufficient to demonstrate that cross-media problems are important and pervasive. But measuring or delineating the extent and seriousness of the problems is a difficult

task.  It is made difficult in part by the great lack of relevant data.
This is a problem for all environmental studies, but cross-media data are
particularly scarce because there are few control programs that want such
data.  Since research and data collection efforts tend to be driven by the
needs of control programs, lack of data and information is even greater for
cross-media problems than for most aspects of environmental science and
policy.

We must of course ask what kind of data we would want if we could get
it.  I will deal with this below in discussing tools and data.  However,
even the answers to this question are not completely satisfactory, and it
is only through workshops such as this that the answers will become more
clear.

From a policy standpoint, the central question is whether the extent
and seriousness of the cross-media problems are so great that radical
policy changes need to be made, or can the problems be dealt with simply by
tinkering and fine-tuning existing environmental policies.  My own opinion
is that radical changes are probably needed.  But more than anecdotal
evidence is needed to justify such changes.

## Cross-Media Policy Approaches

The policies needed to deal with cross-media problems can be described
through three-different approaches--coordination, tools and data, and
legislation.  Coordination is the approach to be used if one believes that
only fine-tuning of the policy system is needed.  Tools and data need to be
developed and improved for any policy changes.  Legislation is the major
approach of those who think that coordinating mechanisms will not be
adequate to deal with the seriousness of cross-media problems.

Coordination.  Coordination can take a variety of forms.  The earliest
major policy move justified by concerns about cross-media transfers was the
creation, in 1970, of the Environmental Protection Agency.  The ostensible
purpose of creating the agency was to allow the air, water, and other
pollution control programs to be better coordinated.  That this goal was
not completely fulfilled is evidenced by the latest policy move justified
on the same basis, EPA Administrator Lee Thomas' shuffling of personnel in
the agency so as to give the programs a more coordinated and integrated
perspective.

EPA has taken a number of coordinating steps in recent years that have
addressed, directly or indirectly, cross-media problems.  The Integrated
Environmental Management Program (IEMP), directed by Dan Beardsley, has
focused largely on coordination of pollution control programs in particular
geographical areas.  Much of the data and many of the tools relevant to
cross-media analysis have come from this program.  The IEMP has also done
cross-media analyses of two industrial sectors--steel and chemicals.

Efforts to deal with specific chemical pollutants on a coordinated
basis have not been very successful, perhaps because the chemical-specific
approach to coordination represents more of a threat to the established
pollution control programs.  Efforts have been made to deal on a
coordinated basis with asbestos, solvents, PCBs, and cadmium among other
substances.  Some of these efforts were just within EPA, others involved
several different federal agencies.  Some, such as the effort to deal with
formaldehyde, resulted in a more integrated approach to risk analysis, but
still foundered on the rocks of differing programs based on differing
statutes implemented by narrowly focused personnel.

4

<u>Tools and Data</u>. There are certain tools of cross-media analysis that are so important to understanding and dealing with cross-media problems that they need to be an integral part of any policy approach to such problems. Time does not allow a full discussion of these tools, but they include:

o  Integrated risk analysis and integrated exposure analysis. The path to cross-media policy is likely to be laid on the foundation of risk assessment. True analysis of risks will likely demonstrate the necessity for a cross-media approach.

o  Materials balance analysis. Materials balances--accounting for the total input and output of a substance--can be done around a plant, a local area, or the universe. New Jersey and Maryland have implemented a materials balance reporting form for their industrial plants, and a Superfund amendment would require EPA to do this nationally. The New Jersey experience clearly demonstrates the utility of reporting on a materials balance basis and the uselessness of the current fragmented reporting system. Integrated permits are the next logical step after integrated reporting.

o  Environmental transport analysis. We do not know enough about transport and transformation in the environment to construct a fully national pollution control system. The more we learn the better able we will be to judge the adequacy of the existing system and the firmer the basis for any improvements in pollution control policies.

o  Data. As has already been noted, research and data collection on cross-media problems are caught in a vicious circle. Research and data collection are funded and motivated by regulatory programs, there is no cross-media regulatory program of any magnitude, therefore cross-media research is not conducted and data are not collected on a cross-media basis, therefore the evidence to support a cross-media regulatory program is not available.

<u>Legislation</u>. It can be argued that no matter how much coordination takes place, and no matter how well we develop the tools and data to deal with cross-media problems, a truly integrated environment program is impossible as long as there are separate statutes dealing with air, water, and land disposal. It is not clear what an integrated environmental statute would look like since, to my knowledge, no one has tried to draft one. But the stumbling blocks are not wholly or even primarily intellectual. Integrated legislation, at least in the United States, runs counter to the structure of legislative committees, bureaucratic programs, and lobbying groups.

## Future of the Cross-Media Approach

Cross-media is not a motherhood issue. If taken seriously it will bring changes to government agencies, legislative committees, business firms, and even environmental groups. But the logic of a more integrated approach tends to be irresistible. As data accumulate about the need to deal with cross-media problems and the inadequacy of existing approaches to meet the need, action will follow.

A number of institutions are now focusing on the cross-media problem. Lee Thomas, in a recent speeech devoted to the cross-media problem, stated: "If the Environmental Protection Agency is ever going to live up to its name in the fullest sense, if it is ever going to become more than a holding company for single medium programs, we are going to have to re-examine the roots of environmental policy." (Speech to the Natural Resources Council of America, May 30, 1985)

A committee of the National Academy of Sciences, reviewing what studies needed to be done with respect to pollution control, gave highest priority to a study of the cross-media problem. The Academy has now received funds from EPA to initiate the study, and is in the process of forming the study committee.

The Electric Power Research Institute (EPRI) has been concerned with a more integrated pollution control approach for the utility industry for several years. Next February in Pittsburgh it will hold a two-day conference on the subject. Two previous conferences on the subject were also sponsored by EPRI and several other national organizations.

The Organization for Economic Cooperation and Development (OECD), an influential international organization composed of the governments of Western Europe, Canada, Japan, and the United States, has placed cross-media problems and approaches on its agenda. It will shortly canvass the member nations to find out what each one has done about the problem.

The Conservation Foundation will continue its research and policy analysis on cross-media issues. Within the next few months we will issue a report that examines state-level initiatives for dealing with cross-media problems. We are convinced that the importance of such problems increasingly will be recognized; that cross-media approaches are the wave of the future; and that therefore workshops such as this one will have a significant impact on public policy.

# INTERMEDIA TRANSPORT MODELING IN MULTIMEDIA SYSTEMS

Yoram Cohen

University of California, Los Angeles
Department of Chemical Engineering
Los Angeles, California 90024

## INTRODUCTION

The hazardous potential of chemical pollutants in the environment
depends upon the degree of multimedia exposure of human and ecological
receptors to these chemicals and the associated health effects. Therefore,
a multimedia understanding of pollutant fate and transport in the
environment is essential for the early assessment of the associated
environmental impact.

It is clear however, that the large number of current and future
pollutants precludes detailed experimental multimedia evaluation of the
potential impact of each chemical on the environment. Therefore,
mathematical models of pollutant fate and transport in a multimedia
environment have become extremely attractive, because they offer a
relatively rapid and inexpensive assessment of potential hazards.
Furthermore, such theoretical models are often the only source of guidance
for predicting complex environmental events.

Multimedia models share a common feature of linking various
environmental compartments (well-mixed or non-uniform) through intermedia
transport to processes which lead to pollutant transport across media
boundaries (see Table I). Such models may be formulated at different
levels of detail with regard to the model structure. Yet, irrespective of
the mathematical complexity of the model, a true forecasting capability can
only be achieved through an accurate description of the physical, chemical
and biological-intermedia and transformation processes. Unfortunately,
there is still a serious deficiency in our understanding of various

TABLE I

---

**Summary of Major Intermedia Transport Routes**

---

A.  <u>Transport form the Atmosphere to Land and Water</u>
     1.  Dry deposition of particulate and gaseous pollutants
     2.  Precipitation scavenging of gases and aerosols
     3.  Adsorption onto particulate matter and subsequent dry and wet deposition

B.  <u>Transport from Water to Atmosphere, Sediment and Organisms</u>
     1.  Volatilization
     2.  Aerosol formation at the air/water interface
     3.  Sorption by sediment and suspended solids
     4.  Sedimentation and resuspension of solids
     5.  Uptake and release by biota

C.  <u>Transport from Soil to Water, Sediment, Atmosphere or Biota</u>
     1.  Dissolution in rain water
     2.  Adsorption on soil particles and transport by runoff or wind erosion
     3.  Volatilization from soil and vegetation
     4.  Leaching into groundwater
     5.  Resuspension of contaminated soil particles by wind
     6.  Uptake by microorganisms, plants, and animals

intermedia transport processes. Consequently, many of the existing multimedia models employ highly approximate treatment of intermedia transport processes (Cohen, 1986). Therefore, there are uncertainties in the predictions of these models due to parameter uncertainties associated with various intermedia transport processes. Another problem which arises is the question of accuracy of measured field concentrations that is required for the evaluation of model parameters (i.e., model calibration).

The evaluation of model sensitivity due to parametric uncertainty, associated with intermedia transport processes, can often be assessed with the use of simple multimedia compartmental models. Such models are particularly useful for studying the relative sensitivity of various transport pathways and the resulting effect on pollutant distribution in the environment. Although there is a loss of spatial resolution when uniform compartments are employed, much useful information can be obtained regarding the macro-behavior of the environmental system, exposure estimates and risk analysis (Mackay and Paterson, 1986; McKone and Kastenberg, 1986).

It is beyond the scope of this paper to review all of the pertinent intermedia transport processes, and the interested reader is referred to other pertinent references (Cohen, 1984, 1986; Cohen and Ryan, 1985; Ryan and Cohen, 1986; Thibodeaux, 1977; Swann and Eshenroeder, 1983). In this paper we will review several theoretical models of predicting intermedia transport parameters associated with pollutant transport across the air-water interface. These include gaseous transport across the air-water interface, dry deposition and rain-scavenging of gaseous and particle-bound pollutants. Subsequently, we demonstrate the applicability of a sensitivity analysis for evaluating the response of a simple multimedia-compartmental models subject to parameter uncertainties.


## GASEOUS MASS TRANSFER ACROSS THE AIR-WATER INTERFACE

The mass flux, N, across the air-water interface can be described by the usual model of resistances in series (Lewis and Whitman, 1924):

$$N = K_{OL} (C_w - C_w^*) = K_{OG} (C_a^* - C_a) \tag{2.1}$$

$$C_a^* = HC_w \tag{2.2}$$

$$\frac{1}{K_{OL}} = \frac{1}{K_L} + \frac{RT}{HK_g} \tag{2.3}$$

9

$$\frac{1}{K_{OG}} = \frac{H}{RTK_L} + \frac{1}{K_g} \tag{2.4}$$

in which $K_{OL}$, $K_{OG}$, $K_L$ and $K_g$ are the overall liquid phase, overall gas phase, and liquid and gas phase mass transfer coefficients respectively. $C_w$ is the chemical concentration in the liquid, $C_w^*$ is the concentration in the liquid phase in equilibrium with the chemical partial pressure in the air phase, and H is the Henry's law constant. $C_a$ is the concentration in the air and $C_a^*$ is the air concentration in equilibrium with the liquid phase.

The important parameters that are needed in order to predict the flux N are $K_L$, $K_g$ and H. Detailed discussions of various prediction methods and experimental techniques of evaluating H have been described by Mackay (1982), Mackay et al. (1979) and Mackay and Shiu (1975, 1984).

### The Liquid Phase Mass Transfer Coefficient, $K_L$

Until recently, predictions of the liquid side mass transfer coefficient for deep water bodies were accomplished via empirical correlations. Most of the empirical correlations did not consider the role of the wind-induced water drift current and the wind-generated water waves. Consequently, frequent misuse of the various proposed correlations led to underprediction or overprediction of $K_L$ (Cohen, 1983). A few theoretical models for $K_L$ were proposed by Hasse and Liss (1980), and Deacon (1977) on rather weak momentum-mass transfer analogies for solid walls. Brkto and Kabel (1979) proposed an analogy based on the eddy-cell (roll) models. All of the above models failed to yield an adequate prediction of $K_L$ at both low and high surface shear velocities.

Recently, Cohen (1983) and Cohen and Ryan (1985) presented a theoretical model for $K_L$ which was shown to accurately predict both laboratory and field data of $K_L$. The analysis is based on the realization that wave-induced turbulence controls mass transfer at low wind speeds (or low shear velocities), while the wind-induced water drift current dominates at high shear velocities. Accordingly, a composite description of $K_L$ based on a surface renewal model and a theoretical mass transfer analogy were proposed.

According to the surface renewal model $K_L$ can be predicted from the following expression:

$$K_L = 0.4 Sc^{-1/2} (\nu_w \varepsilon)^{1/4} \tag{2.5}$$

in which $\nu_w$ and $\varepsilon$ are the kinematic viscosity and rate of turbulent energy dissipation near the surface, respectively. The Schmidt number, Sc, is given by the ratio $\nu_w/D$, where D is molecular mass diffusivity. The rate of energy dissipation is calculated from:

$$\varepsilon = \varepsilon_w + \varepsilon_d + \varepsilon_s \qquad (2.6)$$

where $\varepsilon_w$, $\varepsilon_d$ and $\varepsilon_s$ are the contributions associated with the wave field, water drift current and mechanical mixing, respectively, to the total rate of turbulent energy dissipation near the surface. Those are defined by (Cohen, 1983):

$$\varepsilon_w = \nu_w A^4 K^4 \sigma^2 \qquad (2.7)$$

$$\varepsilon_d = b_o (U_w^*)^4 / \nu_w \qquad (2.0)$$

where A, K and $\sigma$ are the wave amplitude, and radian wave number and frequency, respectively. The surface shear velocity $U_w^*$ is given by:

$$U_w^* = \sqrt{\tau_s / \rho_w} \qquad (2.9)$$

in which $\tau_s$ is the shear stress imported by the wind on the water surface, and $\rho_w$ is the water density. Finally, the scaling constant $b_o$ was determined empirically to be $3.25 \times 10^{-4}$.

At low shear velocity ($U_w^* \lesssim 1.5$ cm/s) the wave field dominates the turbulent field. This is shown in Figure 1 where $K_L Sc^{1/2}$ is plotted versus the contribution of the wave field to the term $(\nu_w \varepsilon)^{1/4}$ in Equation 2.5. Accordingly, it is apparent that Equation 2.5 with $\varepsilon$ determined from Equation 2.7 accurately predicts the mass transfer coefficients. At high wind speeds $\varepsilon_d$ (Eq. 2.8) becomes significant and must be incorporated into the calculation of $\varepsilon$. The above model for $K_L$ is theoretical for $U_w^* \lesssim 1.5$ cm/s and semi-empirical at $U_w^* > 1.5$ cm/s. The average error of the composite model in predicting $K_L$ data was about 5.5 per cent. Since wave properties are not readily available the following empirical correlation was proposed:

$$K_L Sc^{1/2} = A_o + 0.048 \, U_w^* \, , \, \text{cm/s} \qquad (2.10)$$

where $A_o$ has the value of 0.0148 for a well mixed water body and for $U_w^* > 1.5$ cm/s. For non flowing water bodies a value of 0.0029 for $A_o$ is recommended for $U_w^* \lesssim 1.5$ cm/s.

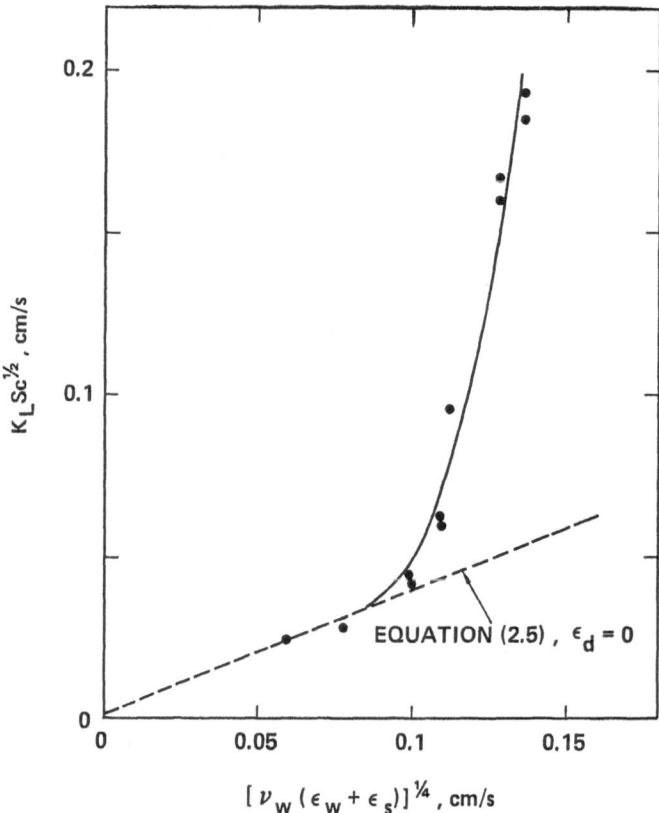

Figure 1.  The correlation of the liquid side mass
transfer coefficient with the rate of
turbulent energy dissipation associated
with the wave field.

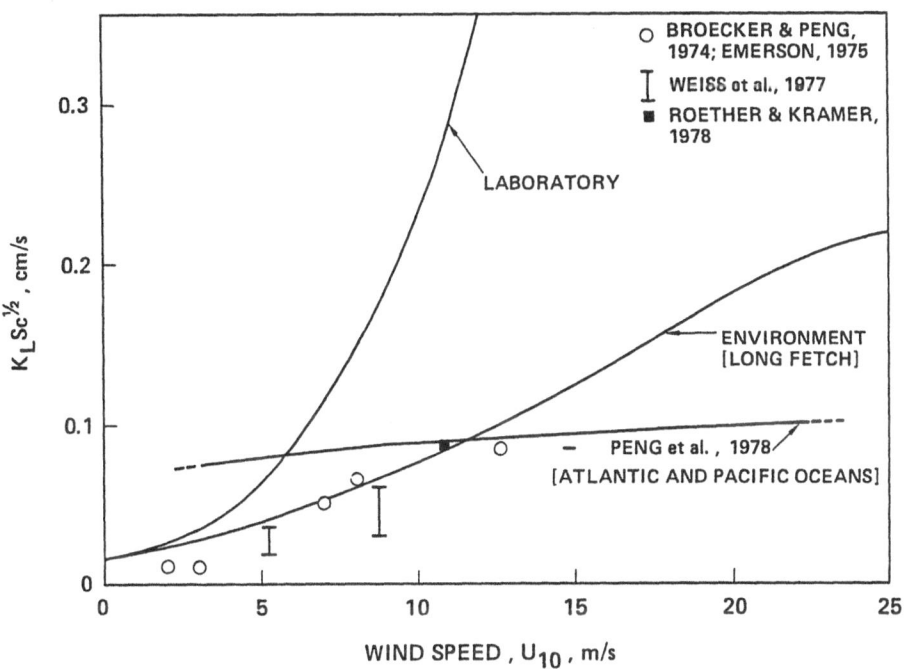

Figure 2.  The prediction of the water side mass transfer coefficient, $K_L$ in laboratory and environmental water bodies (Y. Cohen, 1983).

More recently, Cohen and Ryan (1985) proposed the following theoretical analogy, which was found to be in excellent agreement with available data for a water-side friction velocity range of 0.5-6 cm/s,

$$K_L/U_w^* = a Sc^{-n} \tag{2.11}$$

in which the constants a and n are weak functions of the dimensionless water surface velocity $U_s^+$ ($U_s^+ = U_s/U_w^*$, in which $U_s$ is the surface water velocity), given by:

$$a = a_o - a_1 \ln U_s^+ \tag{2.12a}$$
$$n = n_o - n_1 \ln U_s^+ \tag{2.12b}$$

where

$$a_o = 0.09691; \quad a_1 = 0.010535;$$
$$\tag{2.12c}$$
$$n_o = 0.5778; \quad n_1 = 0.01771$$

Equation 2.11 was found to be in excellent agreement with laboratory data from wind-wave facilities with an average error of about 16 per cent.

In order to apply the various predictive models for $K_L$ to the field, it is necessary to estimate $U_w^*$ from either field data or from an appropriate correlation. A useful correlation, for a fully developed wave field, was proposed by Wu (1980):

$$
\begin{aligned}
C_D &= 8.5 \times 10^{-4}, & U_{10} &< 5 \text{ m/s} \\
C_D &= \left[0.85 + 0.11 \, (U_{10}-5)\right] \times 10^{-3}, & 5 \text{ m/s} &< U_{10} < 20 \text{ m/s} \\
C_D &= 2.5 \times 10^{-3}, & U_{10} &< 20 \text{ m/s}
\end{aligned}
\tag{2.13}
$$

where $C_D$ is the wind-stress coefficient defined as:

$$C_D = \left(\frac{U_w^*}{U_{10}}\right)^2 \left(\frac{\rho_w}{\rho_a}\right) \tag{2.14}$$

in which $\rho_a$ is the air density and $U_{10}$ is the wind speed measured at a height of 10 m above the water surface.

A comparison of the prediction of Equation 2.11 with field data is shown in Figure 2. The least square fit of the data of Peng et al. (1979) is most intriguing since it does not reveal a definite trend of $K_L$ versus $U_{10}$. The data however, reveal lower $K_L$ values compared with laboratory measurements which is consistent with the theory. It is worth noting that one of the key problems in comparing field data with available theories is

14

that in situ results are based on measurements over long periods of time during which environmental conditions vary considerably.

The prediction of $K_L$ in flowing water bodies (i.e. rivers) requires a consideration of the river current. Additionally, the river depth must be considered since significant turbulence in flowing rivers originates at the river bottom. Several recent papers by Smith et al. (1979, 1980), Mackay and Yeun (1983) propose empirical approximations to predict $K_L$ for organic chemicals in rivers. Also, Rathbun (1972), Eloubaidy and Plate (1972) and Munnich et al. (1983) considered the additional effect of wind-shear in flowing water bodies on reaeration and carbon dioxide exchange.

There are numerous studies on the reaeration coefficient in flowing streams (O'Connor, 1983; and references therein). The reaeration coefficient $k_v$ $(hr^{-1})$ can be related to $K_L$ by the following expression:

$$k_L = \left( \frac{D}{D_o} \right)^{1/2} h \ k_v \qquad (2.15)$$

in which D and $D_o$ are the mass diffusivities of the compound of interest and of oxygen in water, and h is the river depth. Lyman et al. (1982) recommended the following empirical correlations:

$$k_v = 1.08 \ (1 + 0.17F^2)(V_{curr}S)^{0.375}, \ hr^{-1} \qquad (2.16a)$$

$$k_v = 0.00102 \ V_{curr}^{2.695}h^{-3.085}S^{-0.823}, \ hr^{-1} \qquad (2.16b)$$

$$k_v = 638 \ V_{curr}S, \ hr^{-1} \qquad (2.16c)$$

in which S is the river bed slope (m drop/m run), and F is the Froude number ($F = V_{curr}/\sqrt{gh}$). The use of an average value of $k_v$ determined from the above three equations was recommended.

A theoretical expression for $K_L$ can be formulated by using the surface renewal model (Eq. 2.5) with $\varepsilon$ estimated from the study of Jirka and Brutsaert (1984):

$$K_L = 0.3181 \ Sc^{-1/2} \left[ \frac{\nu_w U_b^{*3}}{h} \right]^{1/4} \qquad (2.17)$$

in which $U_b^*$ is the bottom shear velocity. Equation 2.17 should only be regarded as an estimate since it does not consider bottom roughness.

Two recent studies on wind effects on the liquid side mass transfer coefficient by Jirka and Brutsaert (1984) and Plate and Friedrich (1984) have pointed out the important effect of wind-generated turbulence on the rate of mass transfer across the air-water interface in flowing streams. When wind blows parallel to the water flow, above a water surface shear velocity of about 1.5 cm/s, the rate of mass transfer is controlled by the wind-induced current. Below a water surface shear velocity of 1.5 cm/s the turbulence generated by the natural river flow controls the rate of mass transfer (Cohen, 1983). It is difficult to predict the wind effect on $K_L$ since the wind direction changes continuously. Nonetheless, it is possible to construct an approximate composite formula using the surface renewal model (Eq. 2.5) with the rate of turbulent energy dissipation due to the mean flow ($\varepsilon_m$) calculated as (Plate and Friedrich, 1983):

$$\varepsilon_m = U_b^{*2} \cdot \frac{\overline{U}_w}{h} \tag{2.18}$$

in which $\overline{U}_w$ is the mean flow velocity. The total rate of energy dissipation is then taken to be the linear sum of $\varepsilon_m$ and $\varepsilon_d$ (Eq. 2.8), accordingly, $K_L$ becomes:

$$K_L = 0.4 \, Sc^{-1/2} \left[ \nu \left( U_b^{*2} \frac{\overline{U}_w}{h} + b_o U_w^{*4} \right) \right]^{1/4} \tag{2.19}$$

Alternatively, one could approximate $\varepsilon_m$ by $\varepsilon_m = 0.4 U_b^{*3}/h$ (Jirka and Brutsaert, 1984).

It is worth noting that all of the above equations for $K_L$ are based on local theories and/or laboratory experiments of air/water mass transfer. Therefore, one must ensure that the associated parameters such as $C_D$, $\overline{U}_w$, $U_w^*$, and $U_b^*$ are determined at the appropriate fetch. Also, it is emphasized that for shallow water bodies of depth less than about 30 cm, the contribution of the drift-current to the rate of turbulent energy dissipation near the surface is affected by the depth of the water body. This behavior is the result of a strong interaction between the turbulent field associated with the wind-induced water drift-current at the surface region and the turbulent field generated by the natural current or return current at the bottom for confined water bodies (e.g. surface water impoundments). For shallow but unconfined water bodies (i.e. rivers) the effect of depth arises naturally as seen from Equations 2.15-2.19 although non-linear bottom-surface interactions are not considered.

Recently, Lunney and Thibodeaux (1985) have argued that for confined water bodies, $K_L$ should also be a function of the fetch/depth (F/D) ratio. Their experiments in which the depth of a wave tank was varied for a constant fetch clearly show that $K_L$ varies with depth. Their conclusion however, that F/D is the proper correlating parameter is yet to be verified as they did not vary the fetch in their experiments. Nonetheless, the correlations of Lunny and Thibodeaux (1985) should provide useful estimates of $K_L$ for confined shallowed water bodies of a depth and fetch similar to that used to derive their correlations.

## The Gas Phase Mass Transfer Coefficient, $K_g$

As the result of numerous studies on water evaporation and heat transfer from rough and smooth surfaces, there is a significant body of knowledge regarding the gas phase mass transfer coefficient. The studies of Dipprey and Sabersky (1962) and Owen and Thompson (1963) have provided the early theoretical framework for predicting $K_g$ and an excellent review of the early studies of $K_g$ was given by Brutsaert (1982). Later studies by Brutsaert (1975), Mangarella et al. (1973), Street (1979), Deacon (1977), have attempted to account for the observed increase in $k_g$ as a consequence of wind-wave interactions. Experimental studies of $k_g$ in laboratory wind-wave tanks include the work of Mangarella (1972), Liss and Slater (1973), Cohen (1976), and Mackay and Yuen (1983). All of the above studies have demonstrated that $k_g \gg k_L$, and hence $k_g$ would be of importance for gases with a low Henry's law constant or chemicals that are reactive in the aqueous phase. As mentioned above, numerous theories and empirical equations have been proposed to predict $K_g$. Cohen and Ryan (1985) have recommended the theoretical expressions advanced by Brutsaert (1975). For rough surface, $K_g$ is given by:

$$\frac{K_g}{U_a^*} = C_D^{1/2} \left[ \varepsilon_D^+ (C_D^{1/2} - 5) + 7.3 \, Re_o \, Sc^{1/2} \right]^{-1}, \quad Re_o > 2 \qquad (2.20)$$

and for smooth surfaces, $k_g$ is given by:

$$\frac{K_g}{U_a^*} = C_D^{1/2} \left[ \varepsilon_D^+ (C_D^{1/2} - 13.5) + 13.6 \, Sc^{2/3} \right]^{-1}, \quad Re_o < 0.13 \qquad (2.21)$$

in which $\varepsilon_D^+$ is the ratio of the eddy momentum diffusivity ($\varepsilon_M$) to the eddy mass diffusivity ($\varepsilon_D$), and $Re_o$ is the roughness Reynolds number defined by:

$$Re_o = U_a^* \, Z_o / \nu_a \qquad (2.22)$$

where $Z_o$ is the effective roughness height. For smooth surfaces $Z_o$ is given by $Z_o \approx 0.135 \, (U_a^*/\nu_a)$, while for rough surfaces it can be estimated from either Equations 2.29 or 2.30.

An important consideration in the calculation of $k_g$ is the effect of buoyancy-induced changes on the flux-gradient relationships. These effects are characterized by the stability parameter, $Z/L$, where $Z$ is a reference height above ground level and $L$ is the Obukonin-Obukhov length scale defined by

$$L = \frac{-U_a^{*3}\,\rho_a}{k_g g} \cdot \frac{1}{(H/\overline{T}C_p) + 0.61E} \tag{2.23}$$

in which $H$ is the sensible heat flux, $E$ is the rate of evaporation, $\overline{T}$ is the mean air temperature, $g$ is the acceleration due to gravity, and $C_p$ is the air heat capacity. According to the extended similarity hypothesis of Monin and Obukhov (Monin and Yaglom, 1971); Wesely and Hicks, 1977) the mass transfer coefficient $k_g$ can be written as:

$$\frac{U_a^*}{k_g} = \frac{1}{k}\left[\ln(Z/Z_{oc}) - \psi_c\right] \tag{2.24}$$

in which $k$ is the von Karman constant (about 0.41). $Z_{oc}$ is the effective surface roughness length and $\psi_c$ is a function of $Z/L$. Several expressions which relate $\psi_c$ to $Z/L$ have been proposed in the literature. A particularly useful set of equations was recommended by Brutsaert (1982):

$$\psi_c = 2\ln\left[(1 + \xi^2)/2\right] \tag{2.25}$$

where

$$\xi = (1 - 16\tfrac{Z}{L})^{1/4} \tag{2.26}$$

The roughness length $Z_{oc}$ can be estimated by the following relations (Jirka and Brutsaert, 1982):

$$Z_{oc} = 0.395\nu a \quad , \quad U_a^* < 6.89 \text{ cm/s} \tag{2.27}$$

$$Z_{oc} = Z_o \exp\left[-k\,(7.3\,Z_o^{*1/4} + Pr^{1/2} - 5)\right] \quad , \quad U_a^* > 6.89 \text{ cm/s} \tag{2.28}$$

in which $Z_o^+ = U_a^* Z_o/\nu$ is the roughness Reynolds number and $Pr$ is the prandtl number for air. A simple empirical correlation for the momentum roughness length ($Z_o$), based on the data of Kondo (1975), was proposed by Yasuda (1975):

$$Z_o = a\,U_a^{*b} \quad , \quad \text{cm} \tag{2.29}$$

where $a = 1.69 \times 10^{-2}$ and $b = -1$ for $U_a^* < 6.89$ cm/s, and $a = 1.65 \times 10^{-4}$,

$b = 1.4$ for $U_a^* > 6.89$ cm/s. For an open sea at full development Wu (1980) suggested that the following relation holds:

$$\frac{z_o g}{U_a^{*2}} = 0.0144 \tag{2.30}$$

More recently, Hwang and Thibodeaux (1983) proposed a mass transfer model which accounts for atmospheric stability in terms of the Richardson's number. Accordingly, the eddy diffusivity was expressed by:

$$\frac{\varepsilon_m}{\nu} = \frac{k z^+}{U_a^* \phi_m} - 1 \tag{2.31}$$

$$(\phi_m^2 \, Sc_{H_2O}^t)^{-1} = (1 \pm 50 \, R_i)^{\pm 1/2} \tag{2.32}$$

in which $D_i$, $D_{H_2O}$ are the diffusivities of the pollutant and water, respectively, $Sc_{H_2O}^t$ is the turbulent Schmidt number for water. The left hand side of the equation is the stability correction and $R_i$ is the Richardson number defined by (Eskinazi, 1972):

$$R_i = \frac{g}{T} \frac{(T - T_o) (Z - Z_o)}{U} \tag{2.33}$$

T and $T_o$ are the temperatures at reference heights Z and $Z_o$, T is the arithmetic average of $T_o$ and T, U is the velocity at $Z_o$. The range of applicability of Equation 2.32 includes $|R_i| \leq 4$ for unstable conditions ($R_i$ negative) and $|R_i| \leq 1$ for unstable conditions ($R_i$ negative). In Equation 2.32 −50 and +1/2 apply for unstable conditions and +50 and −1/2 apply for stable conditions. Although the model proposed by Hwang and Thibodeaux (1983) is simple to use, it should be noted that it overestimates the mass transfer coefficient since it only considers the logarithmic portion of the wind velocity profile.

In contrast with the previous corrections for non-neutral atmospheric conditions, the theory of Brutsaert (1975) considers both the logarithmic region and the surface region. Accordingly, the expressions for $K_g$ is given by:

$$\frac{K_g}{U_a^*} = C_D^{1/2} \left[ \alpha_1 Sc^{2/3} - \alpha_2 \, \varepsilon_D^+ - \frac{1}{k} \left( \psi\left(\frac{Z}{L}\right) - \varepsilon_D^+ \ln \frac{z_o}{L} \right) \right]^{-1} \tag{2.34}$$

in which $\alpha_1 = 7.3$, $\alpha_2 = 5$ for rough surfaces and $\alpha_1 = 13.6$, $\alpha_2 = 13.5$ for smooth surfaces.

# RAIN SCAVENGING OF GASEOUS POLLUTANTS

The scavenging of gaseous pollutants from the atmosphere by rain is an efficient mechanism of atmospheric cleansing and it has been the subject of numerous theoretical studies (Hales, 1972, 1978; Engelman, 1968, 1970; Semonin and Beadle, 1974; Walcek et al., 1984; Walcek and Pruppacher, 1984a, 1984b; Topalian et al., 1984). Advances that have been made by Pruppacher and coworkers in particular have enabled the formulation of a simple parameterized model (for non-reacting drops) as presented below.

The pollutant flux from the atmosphere to the underlying either land or water surfaces is written as:

$$N = \Lambda^* \left( \frac{C_a}{H'} \right) R \tag{3.1}$$

in which $C_a$ is the concentration of the gaseous pollutant in the atmosphere, $H'$ is the dimensionless air to water partition coefficient for the given pollutant, and $\Lambda^*$ is the scavenging ratio given as:

$$\Lambda^* = C_w / \left[ C_a / H' \right] \tag{3.2}$$

The theoretical expression for $\Lambda^*$ can be obtained by performing a component mass balance on a single raindrop and integrating over the spectrum of raindrop sizes. The resulting expression for $\Lambda^*$ (Cohen, 1984)

$$\Lambda^* = \frac{1}{V_r} \int_0^\infty \left[ 1 - \left( 1 - \frac{C_{wo} H'}{C_a} \right) \exp \left( \frac{-6K_{OL} L_c}{V_t D} \right) \right] \frac{\pi D^3}{6} N_D dD \tag{3.3}$$

in which $C_{wo}$ is the initial concentration in the drop, $V_r$ is the volume of rain per unit volume of air, $V_t$ is the terminal falling velocity of the drop, and $K_{OL}$ is the overall mass transfer coefficient (Eq. 2.3).

It is common practice in the atmospheric pollution literature to define a gaseous scavenging coefficient that describes the pollutant removal from the atmosphere (in the absence of other removal processes) as

$$\frac{dC_a}{dt} = - \bar{\Lambda} C_a \tag{3.4}$$

By comparing Equations 3.4 and 4.7 (with $\Lambda'$ replaced by $\Lambda^*$) we obtain the relation

$$\bar{\Lambda} = \frac{\Lambda^* R}{(1 - V_r) L_c H'} \tag{3.5}$$

which clearly illustrates that the rain scavenging coefficient is not a universal constant as has been often implied in various multimedia transport models.

In order to calculate $\Lambda^*$ it is necessary to specify both $K_L$ and $K_g$. Recently Walcek (1984) carried out numerical simulations of mass transfer into raindrops by solving the convective diffusion equation, accounting for the velocity field inside the drops. His results were parameterized by the following set of equations (Ryan, 1984):

$$\frac{K_L R_d}{D_L} = 14.5 \qquad\qquad (3.6)$$

in which

$$D_2/D_o = 1 \qquad\qquad R < 0.05 \text{ cm} \qquad\qquad (3.7a)$$

$$\frac{D_L}{D_o} = 21.34 \ (R_d/R_{d_{0.05}}) - 20.34 \ ; \quad 0.05 \text{ cm} < R_d < 0.09 \text{ cm} \qquad (3.7b)$$

$$D_L/D_o = 18.07 \qquad\qquad R_d > 0.09 \text{ cm} \qquad\qquad (3.7c)$$

where $D_o$ and $D_L$ are the molecular and effective diffusivities (units of $\text{cm}^2/\text{s}$) in the water phase, respectively, $R_d$ is the raindrop radius (cm) and $R_{d_{0.05}}$ is the limiting reference radius of 0.05 cm.

The gas phase mass transfer coefficient, $K_g$ can be calculated from (Bird et al., 1966),

$$K_g = \frac{D_g}{2R_d} \ (2 + 0.6 \ Sc^{1/3} \ Re^{1/2}) \qquad\qquad (3.8)$$

in which $D_g$ is the diffusivity in the gas phase, Sc is the gas phase Schmidt number and Re is the raindrop's Reynolds number.

Equation 3.3 with the above expressions for $K_L$, $K_g$, Equation 4.3 for $N_D$, and the appropriate expressions for $V_t$ based on the data of Gunn and Kinzer (1949) can be solved to give $\Lambda^*$ for selected chemicals, as a function of the height of the cloud base, the rate of rainfall, and the initial concentration in the raindrop relative to the equilibrium concentration.

For the purpose of illustration, the prediction of Equation 3.3 for $\Lambda^*$ is shown in Figure 3 for trichloroethylene (TCE). The scavenging ratio, $\Lambda^*$, increases with the rate of rainfall and rapidly rises towards unity with increasing distance below the cloud base. For the depicted example, at a cloud base distance greater than 100 meters, the concentration of TCE in the falling rain is practically equal to the equilibrium concentration.

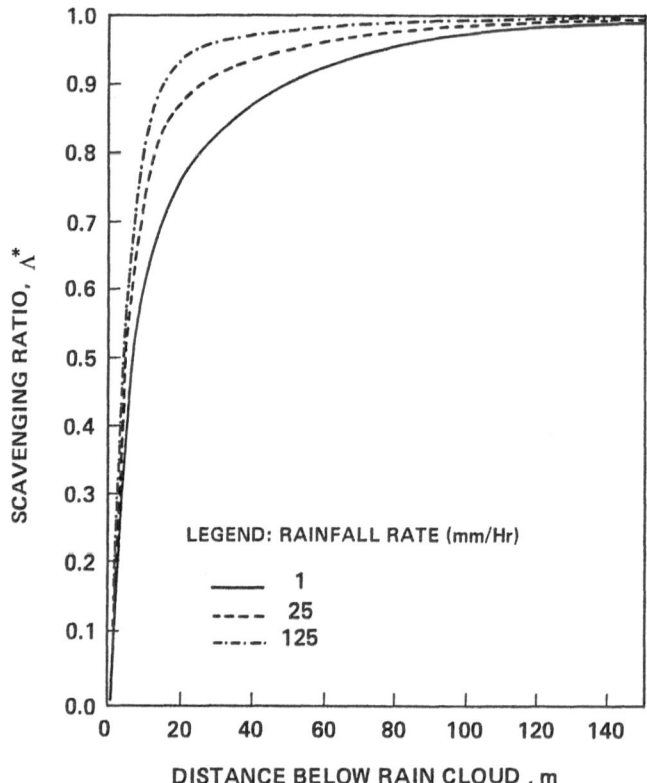

Figure 3. The scavenging ratio for trichloroethylene.

This means that, for most situations of interest, $\Lambda^*$ for unreactive organics may be assumed to be unity.

A parametric model for the rain scavenging ratio of chemicals that undergo chemical reactions in the liquid phase (i.e., raindrop) is possible, but it is specific to the chemical kinetics of the chemical under consideration. Therefore, it is not feasible to develop a general model for $\Lambda^*$. A discussion of the above is beyond the scope of this paper. An extensive treatment of the prediction of rain scavenging for reactive organics can be found in the papers of Walcek and Pruppacher (1984a, 1984b).

## RAIN SCAVENGING OF PARTICLE-BOUND POLLUTANTS

Precipitation scavenging is an important process for the transport of particulate pollutants from the atmosphere to both land and water bodies. Rain scavenging is prevelant in most parts of the country and has been the subject of numerous studies. Unfortunately, most multimedia models incorporate rain scavenging in a rather approximate manner without regard to particle and raindrop size distributions. Moreover, it is often that rain scavenging coefficients or scavenging ratios are utilized, in error, as universal parameters. In the analysis that follows, we present the proper format of the rain scavenging parameters that are to be included in multimedia transport models.

The pollutant flux from the atmosphere to either water or land surfaces can be expressed by:

$$N = \Lambda' R C_a \qquad (4.1)$$

in which R is the rate of rainfall (i.e., units of length/time), $C_a$ is the pollutant concentration in the atmosphere (mass/volume), and $\Lambda'$ is the scavenging ratio defined as the ratio of the average pollutant concentration in the falling rain to the pollutant concentration in the air phase.

The theoretical expression of $\Lambda'$ can be derived by performing a material balance on a single raindrop and then integrating over the spectrum of raindrop and particle sizes (Ryan and Cohen, 1986). Following such an analysis, it can be shown that $\Lambda'$ is given by:

$$\Lambda' = \frac{\overline{C_w}}{C_a} = \frac{\overline{C_{wo}}}{C_a} + \frac{1}{V_r} \int_0^\infty \frac{3}{2} \frac{L_c}{D} \left[ \int_0^\infty E(a,D) f(a) \, da \right] \frac{\pi D^3}{6} N_D \, dD \qquad (4.2)$$

where $\overline{C_w}$ and $\overline{C_{wo}}$ are the pollutant concentrations in the falling rain at ground level and at the cloud-base, respectively. $V_r$ is the volume of rain per unit volume of air at gound level, $L_c$ is the height of the cloud base, D is the raindrop diameter, and a is the particle diameter. f(a) is the mass distribution of the particle-bound pollutant in the particle phase per unit volume of air normalized with respect to the pollutant concentration, $C_a$. The distribution of raindrop sizes is denoted by $N_D$, and the collection efficiency of a particle of diameter a by a raindrop of diameter D is denoted by the function E(a,D). In order to calculate $\Lambda'$, one must specify the functions $N_D$, f(a) and E(a,D). The raindrop size distribution can be approximated by available measured distributions, such as the Marshall and Palmer (1948) distribution

$$N_D = N_o e^{-cD}, \qquad N_o = 0.08 \; Cm^{-4}, \qquad c = 41R^{-.21} \tag{4.3}$$

in which the rate of rainfall, R is in units of mm/hr and D is the drop diameter in mm. The chemical distribution f(a) can only be obtained from experimental data which are unfortunately scarce.

One of the difficulties in calculating $\Lambda'$ lies in the lack of suitable models for the collision efficiency E(a,D). The existing theoretical models underpredict the collision efficiency of submicron particles by one to two orders of magnitude (Radke et al., 1980). Although some recent models (Grover, 1976; and references therein) have suggested that electrical charges carried by the drops may lead to a significant enhancement of E(a,D), the application of such models is not feasible due to the lack of correlations of the electrical fields associated with macroscopic meteorological variables such as rainfall rates, temperature, season of the year, etc. The field studies of Radke et al. (1974, 1980) revealed that the collision efficiency did not correlate with the rate of rainfall, although a definite dependence on particle size was observed. The above observation suggests that as a first order approximation the collision efficiency may be taken to be independent of the raindrop size. This assumption, with the Marshall and Palmer raindrop size distribution leads to the following equation for $\Lambda'$:

$$\Lambda' = \frac{\overline{C_{wo}}}{C_a} + L_c (5.69 \times 10^{-4} R^{-0.21}) \int_0^\infty E(a) f(a) da \tag{4.4}$$

The parameters $L_c$ (height of cloud base) and the rainfall rate R can be easily determined for the geographical region of interest. The collision efficiency may be approximated from available models (Slinn, 1977) or as

24

suggested by Ryan and Cohen (1986) from the empirical fit to the extensive field data of Radke et al. (1974, 1980). According to the latter approach $E(a)$ is given by:

$$E(a) = \begin{cases} (S - 1/12)\ (S + 7/12) & ,\ a > 1.35\ \mu m \\ 0.005 & ,\ 0.9\ \mu m \le a \le 1.35\ \mu m \\ 0.125/(0.5 + a^2) & ,\ a < 0.9\ \mu m \end{cases} \tag{4.5}$$

in which the parameter $S$ is equal to $0.1038\ a^2$, and the scavenging gap, $0.9\ \mu m < a < 1.35\ \mu m$ represents the region where only a negligible fraction of particles are scavenged by rain.

Most of the studies on rain scavenging normally report the rain scavenging coefficient rather than the rain scavenging ratio. It is usually assumed that in the absence of all other atmospheric removal processes but rain, the change in the atmospheric concentration of the particle-bound pollutant, $C_a$ can be expressed by:

$$\frac{dC_a}{dt} = -\Lambda\ C_a \tag{4.6}$$

in which $\Lambda$ is the rain scavenging coefficient (in units of $time^{-1}$). A pollutant mass balance on the atmosphere leads to the following equation:

$$-(1 - V_r)\ V_a\ \frac{dC_a}{dt} = A_i\ \Lambda'\ C_a \tag{4.7}$$

in which $A_i$ is the interfacial area between the atmosphere and the land or water surfaces, in the region under consideration, and $V_a$ is the volume of the air compartment (i.e., $V_a = A_i\ L_c$). Upon comparing Equations 4.6 and 4.7 it is clear that $\Lambda$ is related to $\Lambda'$ by:

$$\Lambda = \frac{A_i\ R\ C_a\ \Lambda'}{(1 - V_r)\ V_a} \tag{4.8}$$

Subject to the usual assumption that $V_r \ll 1$ and $\overline{C_{wo}} = 0$, the expression for $\Lambda$ can be written with the aid of Equation 4.4 as

$$\Lambda = 5.69 \times 10^{-4}\ R^{0.79} \int_0^\infty E(a)f(a)da \tag{4.9}$$

Equation 4.9 is in qualitative agreement with the study of Dana and Wolf (1974) who reported $\Lambda \propto R^{0.75}$, and with the studies of both Vali (1974) and Scott (1983) who reported $\Lambda \propto R^{0.8}$.

# TABLE II

## Predicted and Measured Scavenging Coefficients

| Rainfall (mm/hr) | Scavenging Coefficient ($\Lambda$) ($\times 10^4$ s$^{-1}$) | | |
|---|---|---|---|
| | Lead | | Benzo(a)Pyrene |
| | Experimental[†] | Predicted | Predicted |
| 0.3 | 2.0 | 4.7 | 0.73 |
| 0.7 | 1.2 | 4.6 | 1.42 |
| 1.7 | 4.4 | 3.2 | 2.87 |
| 2.7 | 1.2 | 1.6 | 4.14 |
| 2.8 | 3.0 | 1.0 | 4.26 |

[†] Calculated from the data of Lusis et al. (1983)

The important feature of Equations 4.4 and 4.9 is the clear indication that both $\Lambda'$ and $\Lambda$ (in the absence of in-cloud scavenging) will be different for different chemical species if the chemical mass distributions f(a), are different. The prediction of Equation 4.9 for the rain scavenging of lead, from a smelter plume (Lusis et al., 1983), using f(a) obtained from the data of Knuth et al. (1983), and for Benzo(a)Pyrene, using the distribution f(a) obtained from the data of Miguel and Friedlander (1978), are shown in Table II. It is evident that rain scavenging is a sensitive function of both the rate of rainfall and the chemical distribution in the particle phase.

## DRY DEPOSITION TO WATER SURFACES

The flux of particle-bound pollutants from the atmosphere to water surfaces is expressed by the following flux equation:

$$N_A = V_d C_a$$

in which $C_a$ is the mass of the pollutant in the particle phase per unit volume of air, and $V_d$ is the overall chemical deposition velocity defined by:

$$V_d = \int_0^\infty V_d(a)\ f(a)\ da \qquad (5.1)$$

in which $V_d(a)$ is the deposition velocity for a particle of diameter a, and f(a) is the normalized chemical mass distribution of the chemical in the particle phase given by the ratio $C(a)/C_a$, C(a) being the mass of distribution of the chemical in the particle phase per unit volume of air.

The determination of $V_d(a)$ for water surfaces requires a consideration of the water surface roughness as a function of wind speed, and the effect of wave-breaking as well as humidity. A model which considers the various resistances to particle deposition onto natural water surfaces was proposed by Williams (1982). The model for $V_d(a)$ as simplified by Ryan (1984) is given by:

$$V_d(a) = \left(\frac{A}{B}\right)\left[(1 - \theta)(k_{ss} + V_{gw}) + \frac{k_m\,\theta(k_{bs} + V_{gw})}{k_m + \theta(k_t + k_{bs} + V_{gw})}\right]$$
$$+ \frac{\theta^2(k_{bs} + V_{gw})(k_t + V_{gd})}{k_t + \theta(k_t + k_{bs} + V_{gw})} \qquad (5.2)$$

$$A = k_t\left[(1 - \theta)k_t + \theta k_t + V_{gd})\right]$$
$$+ \theta\,(1 - \theta)(k_t + V_{gd})(k_t + k_{bs} + V_{gw}) \qquad (5.3)$$

$$B = k_t\left[(1 - \theta)(k_t + k_{ss}) + \theta(k_t + k_{bs}) + V_{gw}\right]$$
$$+ \theta\,(1 - \theta)(k_t + k_{ss} + V_{gw})(k_t + k_{bs} + V_{gw}) \qquad (5.4)$$

in which the different parameters are defined by:

$\theta$ - fraction of total area occupied by broken surface area

$k_t$ - turbulent transfer coefficient

$k_{bs}$ - broken surface transfer coefficient

$k_{ss}$ - surface transfer coefficient

$V_{gd}$ - gravitational settling velocity based on dry particle diameter

$V_{gw}$ - gravitational settling velocity based on wet particle diameter

The fraction of broken water surface, $\theta$, can be calculated based on the correlation of Wu (1979):

$$\theta = 1.7 \times 10^{-6}\ U_{10}{}^{3.75} \qquad (U_{10} \text{ in m/s}) \qquad (5.5)$$

and $V_{gw}$ and $V_{gd}$ are given by Stoke's law:

$$V_{gx} = \rho_p g D_x^2 / 18\nu a \qquad\qquad (5.6)$$

where the subscript x designates either d (dry particle) or w (wet particle). The wet particle diameter can be estimated from the relationship given by Fitzgerald (1975) who related the dry to wet particle diameter at different levels of relative humidity.

The turbulent transport coefficient $k_t$, can be calculated using Equations 2.20 or 2.22 with the appropriate estimate for $Z_o$. The transport coefficient, $k_{ss}$, for a smooth surface is assumed to be controlled by diffusion and impaction. It is given by

$$k_{ss} = \frac{U_a^{*2}}{0.4U_{10}} \left[ 10^{-3/St} + Sc^{-1/2} \right] \qquad\qquad (5.7)$$

in which St is the Stokes number defined as:

$$St = U_{10}^{*2} \, V_{gw} / g\mu_a \qquad\qquad (5.8)$$

and Sc is the particle Schmidt number, $Sc = \nu_a/D_c$, in which the diffusivity can be calculated from (Davies, 1966):

$$D_c = \left[ (2.38 \times 10^{-7}/D_x)\, (1 + 0.163/D_x + \frac{0.548}{D_x} \exp\,(-\,6.66\,D_x) \right] \quad (5.9)$$

where $D_x$ is the particle diameter in microns. A prediction method for the broken surface transport coefficient, $k_{bs}$, is unavailable at the present time, and an average value of 10 cm/s was recommended by Ryan (1984). Alternatively, one could set $k_{ss} = k_{bs}$ which corresponds to neglecting the broken surface effect on the tranport coefficient in the deposition layer.

The above prediction method is currently the most complete description available of particle deposition to wavy water surfaces. It must be noted that the prediction method is approximate as uncertainties in the various model parameters may lead to up to an order of magnitude uncertainty in the predicted rates of $V_d$.


## SENSITIVITY ANALYSIS OF COMPARTMENTAL MULTIMEDIA MODELS

The reliability of multimedia transport models is undoubtedly a function of the accuracy to which the various model parameters are known. Therefore, it is useful to quantify the effects of parametric uncertainty and variability of the model output. Such information should be useful

when regulatory decisions are to be based on information from a particular
model, and to determine the parameters to which the model is most
sensitive. The latter piece of information could suggest the degree of
parameter accuracy that is needed for a particular model application.

Although various methods of studying parameter sensitivity are now
available (Tomovic and Vukobratovic, 1970; Frank, 1978), their applications
to multimedia models are scarce. Most sensitivity studies are done by a
"brute force" method of systematically varying different parameters and
correlating the model results with the parameter variation to obtain model
sensitivity correlations. Such a procedure is time consuming and may mask
the detailed effect of coupled parameter uncertainties when the number of
model parameters is large. Therefore, there is a need to employ more
systematic mathematical methods to quantify the effect of parametric
sensitivity.

In this chapter a simple parametric sensitivity analysis is
illustrated for two recent multimedia-compartmental case studies. The
first study is on the distribution of trichloroethylene in the San Deigo
region (Cohen and Ryan, 1985) and the second is on the distribution of
benzo(a)pyrene in the Southeast Ohio region (Ryan and Cohen, 1986). We
illustrate, for the above cases, the sensitivity of the models to the
uncertainty and/or variability of the air/water mass transfer coefficient
and the dry deposition velocity.

## Multimedia-Compartmental (MCM) Model

The dynamic distribution of a given pollutant in a compartmental
system of uniform compartments can be described by the following set of
equations:

$$
V_i \frac{dC_i}{dt} = \sum_{j=1}^{N} K_{ij} a_{ij} (C_{ij}^* - C_i) + V_i k_i \xi_i C_i + \sum_{j=1}^{N} Q_{ji} C_i -
$$

$$
\sum_{j=1}^{N} Q_{ij} C_i + s_i \quad , \quad i = 1,\ldots.N \quad , \quad i \neq j \tag{6.1}
$$

$$
C_i = C_i(0) \quad \text{at } t = 0 \tag{6.2}
$$

The concentration of $(mol/m^3)$ of the species of interest in
compartment i is designated by $C_i$ and $s_i$ is the source strength $(mol/h)$ in

the ith compartment. $K_{ij}$ values are the overall mass-transfer coefficients in units of meters per hour, based on compartment i, for the exchange of mass between compartments i and j, $a_{ij}$ is the corresponding interfacial area ($m^2$), and $V_i$ is the compartmental volume ($m^3$). The term ($C_{ij}^* - C_i$) is the concentration driving force for interfacial mass transport constrained by equilibrium. The variable $C_{ij}^*$ is defined as the concentration of the pollutant in compartment i in equilibrium with compartment j. The equilibrium relationship is assumed to have the linear form $C_{ij}^* = C_j H_{ji}$, in which $H_{ji}$ is the dimensionless i to j partition coefficient. For particulate pollutants, $C_{ij}^*$ vanishes and $K_{ij}$ is identified with the dry deposition velocity.

For simplicity, in the above model, we assume that the pollutant undergoes a first-order transformation reaction (chemical or biochemical). The reaction rate constant is $k_i$, and the corresponding coefficient $\xi_i$ equals -1 for a degradation reaction and +1 for a production reaction. The terms $Q_{ij}C_j$ and $Q_{ij}C_i$ represent the pollutant mass flow rate (mol/h) from compartments j to compartment i and from compartment i to j, respectively, where $Q_{ji}$ and $Q_{ij}$ are the corresponding flow rates ($m^3$/h) or convection terms. For example, in a flowing water body such as a river, the flow rate into and out from the water compartment is readily identified with the average river flow rate. The convection term for the atmospheric compartment, however, must be adjusted to account for the extensive degree of air recirculation as proposed by Cohen and Ryan (1985). Finally, we note that the convection terms can also be identified with the rain scavenging of gaseous or particle-bound pollutants. In this latter case the convection terms QC are replaced by the scavenging flux N (Eq. 3.1 or 4.1) multiplied by the appropriate interfacial area (Cohen and Ryan, 1985; Ryan and Cohen, 1986).

In order to simplify the subsequent analysis, the system of linear equations (Eq. 6.1) can be written more conveniently in the following matrix form:

$$\frac{d\underset{\sim}{C}}{dt} = \underset{\approx}{A}\, \underset{\sim}{C} + \underset{\sim}{S}$$

$$\underset{\sim}{C} = \underset{\sim}{C}(0) \qquad at \qquad t = 0$$

(6.3)

in which $\underset{\sim}{C}$ and $\underset{\sim}{S}$ are the concentration and source vectors and $\underset{\approx}{A}$ is the parameter matrix whose coefficients are defined as:

$$A_{ij} = a_{ij}\, K_{ij}\, H_{ji}/V_i + Q_{ji}/V_i \;, \qquad i \neq j$$

(6.4)

$$A_{ii} = \sum_{\substack{j=1 \\ i \neq j}}^{N} \left[ -a_{ij}K_{ij}/V_i - Q_{ij}/V_i + \xi_i k_i \right] \qquad (6.5)$$

The solution of the above model system can be obtained analytically where $A_{ij}$ and $S_i$ are time invariant. When $A_{ij}$ and $S_i$ are time variant and/or when the number of compartments increases (say above N = 4) then a numerical solution is the feasible approach.

## Sensitivity Model

In the particular multimedia model described by Equation 6.1 variation may occur in the parameters of the matrix $\underset{\sim}{A}$ the source vector $\underset{\sim}{S}$. The variations in these parameters may be due to meteorological factors, measurement errors or approximations of the parameters. Although uncertainties may also exist in the initial conditions these are usually less severe and will not be treated in this paper. The interested reader is referred to other excellent references that deal with model sensitivity associated with initial conditions (Frank, 1978; Tomovic and Vukobratovic, 1970).

In order to quantify the significance of the parameter uncertainties a sensitivity model for the multimedia model is first constructed. We consider a parameter $\alpha$ which is subjected to a small variation $\Delta\alpha$ such that

$$\alpha = \alpha_o + \Delta\alpha \qquad (6.6)$$

where $\alpha_o$ is the value of the parameter for the original multimedia system whose solution is $\underset{\sim}{C}^o$. The change in the solution of Equation 6.3 due to the variation $\Delta\alpha$ can be expressed via a Taylor series expansion

$$\underset{\sim}{C} = \underset{\sim}{C}^o + \left. \frac{\partial C}{\partial \alpha} \right|_{\alpha_o} \Delta\alpha + \left. \frac{\partial^2 \underset{\sim}{C}}{\partial \alpha^2} \right|_{\alpha_o} \frac{\Delta\alpha^2}{2} + \ldots \qquad (6.7)$$

If we limit our interest to the effect of small variations and relative sensitivities, then second order terms are neglected and Equation 6.7 reduces to:

$$\underset{\sim}{C} = \underset{\sim}{C}^o + \underset{\sim}{\sigma}_\alpha \Delta\alpha \qquad (6.8)$$

in which $\underset{\sim}{\sigma}_\alpha = \underset{\sim}{\sigma} (t, \alpha_o) = \left. \frac{\partial C}{\partial \alpha} \right|_{\alpha_o}$ is termed the output sensitivity function. The inclusion of the second order coefficients is possible but it requires complex algorithms to solve the sensitivity model (Kramer et al., 1982).

After differentiating Equation 6.3 with respect to the $\alpha$ parameter, the following sensitivity model is obtained:

$$\dot{\sigma}_\alpha = \underset{\sim}{A}^o \underset{\sim}{\sigma}_\alpha + \underset{\sim}{\psi}_\alpha \, \underset{\sim}{c}^o + \underset{\sim}{\tau}_\alpha \, , \qquad \alpha = 1 \ldots M = N \times N \qquad (6.9)$$

in which $\dot{\sigma}_\alpha$ is the time derivative of $\sigma_\alpha$ and

$$\underset{\sim}{\tau}_\alpha = \left. \frac{\partial \underset{\sim}{S}}{\partial \alpha} \right|_{\alpha_o} \, ; \qquad \underset{\sim}{\psi}_\alpha = \left. \frac{\partial \underset{\sim}{A}}{\partial \alpha} \right|_{\alpha_o} \qquad (6.10)$$

Equation 6.9 represents a set of N x N equations for each parameter $\alpha$. Since $\alpha$ represents any one of the parameters of $\underset{\sim}{A}$, a total of N x N x N equations need to be solved for a complete description of the model sensitivity. Since the source vector $\underset{\sim}{S}$ is independent of the parameter $\alpha$ for $\alpha = A_{ij}$, it is useful to introduce a parameter variation that is identified with variation of the source such that

$$\beta = \beta_o + \Delta\beta \qquad (6.11)$$

in which $\beta$ designate any one of the components of the source vector $\underset{\sim}{S}$. Furthermore, in the remainder of the analysis $\alpha$ will be understood to represent any of the components of the parameter matrix $\underset{\sim}{A}$. According to the above simplification we write Equation 6.9 as two separable sensitivity models, namely

$$\dot{\sigma}_\alpha = \underset{\sim}{A}^o \underset{\sim}{\sigma}_\alpha + \underset{\sim}{\psi}_\alpha \, \underset{\sim}{c}^o \qquad (6.12)$$

and

$$\dot{\sigma}_\beta = \underset{\sim}{A}^o \underset{\sim}{\sigma}_\beta + \underset{\sim}{\phi}_\beta \, , \qquad \beta = 1 \ldots N \qquad (6.13)$$

in which

$$\phi_\beta = \left. \frac{\partial \underset{\sim}{S}}{\partial \beta} \right|_{\beta_o} \qquad (6.14)$$

Since the initial conditions for $\underset{\sim}{c}^o$ and $\underset{\sim}{c}$ are unchanged, the initial conditions for Equations 6.12 and 6.13 are:

$$\underset{\sim}{g}_\alpha = 0 \qquad \text{and} \quad \underset{\sim}{g}_\beta = 0 \qquad \text{at } t = 0 \qquad\qquad (6.15)$$

Once the set of output sensitivity functions $\underset{\sim}{g}$ is known, the response of the system to a variation in any of the parameters $A_{ij}$ or $S_i$ can be calculated from the linear approximation of Equation 6.8. This approximation should hold for up to approximately 30% variation in the model parameters. When several parameters are varied simultaneously, the new concentration vector can be determined from:

$$\underset{\sim}{C} = \underset{\sim}{C}^o + \sum_{\alpha=1}^{M} \underset{\sim}{g}_\alpha \, \Delta\alpha + \sum_{\beta=1}^{N} \underset{\sim}{g}_\beta \, \Delta\beta \qquad\qquad (6.16)$$

The important simplification that is introduced by the sensitivity model is that we need to solve Equation 6.9 only once rather than obtain a new solution of the compartmental system for each parameter variation. Unlike the MCM model, the sensitivity model has to be solved numerically since $\underset{\sim}{C}^o$ is a function of time. Simple solutions are possible for the case of steady state if a steady state solution for $\underset{\sim}{C}^o$ exists.

The recent studies of Mackay and Paterson (1981, 1982, 1986), Mackay et al. (1986), Wiersma (1979), McKone (1986), Cohen and Ryan (1985, 1986), demonstrated that the steady-state predictions of MCM models often yield results that are useful for screening analysis in risk assessment. In such cases, the output sensitivity function can be easily determined for the steady-state solution.

At steady state Equations 6.11 and 6.12 are solved to yield the solution:

$$\underset{\sim}{g}_\alpha = - \underset{\approx}{A}^{o^{-1}} \underset{\approx}{\psi}_\alpha \, \underset{\sim}{C}^o \qquad\qquad (6.17)$$

$$\underset{\sim}{g}_\beta = - \underset{\approx}{A}^{o^{-1}} \underset{\sim}{\phi}_\beta \qquad\qquad (6.18)$$

The above solution can then be applied to the steady state case of any compartmental system whose parameter matrix $\underset{\approx}{A}$ and source vector $\underset{\sim}{S}$ are time invariant and independent of $\underset{\sim}{C}$.

The steady-state sensitivity analysis should be applied with caution since pollutant partitioning in the environment is a dynamic process. In general, the sensitivity output function $\underset{\sim}{g}_\alpha$ is a function of time even for a time invariant parameter matrix $\underset{\approx}{A}$, and output vector $\underset{\sim}{S}$.

33

**Steady-State Sensitivity Analysis for the Multimedia Distribution of**
**Benzo(a)Pyrene.** The environmental distribution of benzo(a)pyrene
(B(a)P) in the Southeast Ohio region was recently studied by Ryan and Cohen
(1986) utilizing a dynamic three-compartment MCM model.  Their model system
can be represented by Equation 3 with the parameter matrix represented by:

$$\underset{\sim}{A} = \begin{bmatrix} A_{11} & 0 & 0 \\ A_{21} & A_{22} & 0 \\ A_{31} & 0 & A_{33} \end{bmatrix} \tag{6.19}$$

and the source vector by:

$$\underset{\sim}{S} = \begin{bmatrix} S_1 \\ S_2 \\ 0 \end{bmatrix} \tag{6.20}$$

in which the subscripts 1, 2 and 3 refer to the air, water and soil
compartments, repsectively.  The above multimedia system is inherently
unsteady due to rain scavenging of particle-bound B(a)P.  After a
simulation of twelve years, however, the fluctuations in the B(a)P
concentrations in the soil and water compartments were found to be
negligible.  The atmospheric concentration decreased significantly during
rain events but reached a steady concentration level very rapidly during
non-rainy periods.  Consequently, it is possible to investigate the model
sensitivity during such intermittent "pseudo-steady state" periods.  It is
emphasized however, that the parametric sensitivity of the rain scavenging
ratio cannot be determined from the steady-state solution.

Although the above steady-state example is a trivial case, it
illustrates the power of the sensitivity model approach.  The sensitivity
model for the above system can be solved analytically for the steady state
solution.  Accordingly, the complete sensitivity of the above MCM model can
be represented by:

$$\sigma_{11} = -\frac{c_1^0}{\Delta} \begin{pmatrix} A_{22}A_{33} \\ -A_{21}A_{33} \\ -A_{22}A_{31} \end{pmatrix} \qquad \sigma_{21} = -\frac{c_1^0}{\Delta} \begin{pmatrix} 0 \\ A_{11}A_{33} \\ 0 \end{pmatrix}$$

$$\sigma_{22} = -\frac{c_2^0}{\Delta} \begin{pmatrix} 0 \\ A_{11}A_{33} \\ 0 \end{pmatrix} \qquad \sigma_{31} = -\frac{c_1^0}{\Delta} \begin{pmatrix} 0 \\ 0 \\ A_{11}A_{22} \end{pmatrix} \qquad (6.21)$$

$$\sigma_{33} = -\frac{c_3^0}{\Delta} \begin{pmatrix} 0 \\ 0 \\ A_{11}A_{22} \end{pmatrix}$$

in which $\Delta = |A_{22}A_{22}A_{33}|$ and the subscripts of $\sigma$ refer to the sensitivity of the corresponding $A_{ij}$ parameter.

The output sensitivity vectors associated with the source terms $S_1$ and $S_2$ can be shown to be:

$$\sigma_{S_1} = -\frac{1}{\Delta} \begin{pmatrix} A_{22}A_{33} \\ -A_{21}A_{33} \\ -A_{22}A_{31} \end{pmatrix} \qquad \sigma_{S_2} = -\frac{1}{\Delta} \begin{pmatrix} 0 \\ A_{11}A_{33} \\ 0 \end{pmatrix} \qquad (6.22)$$

in which $\sigma_{S_1}$ and $\sigma_{S_2}$ are the output sensitivity vectors corresponding to the source terms, $S_1$ and $S_2$, respectively.

The structure of the sensitivity vectors provides useful information regarding the parametric sensitivity of the model. For example, from $\sigma_{S_2}$ it is seen that the MCM model output for the air and soil compartments is not sensitive to the B(a)P source in the water compartment, while variations in source term of the air compartment affect all three compartments. Variations in the degradation rate constant in the soil and water compartments affect the model output only for the soil and water compartments, respectively. The sensitivity of the MCM model with respect to the dry deposition velocity to the soil surface, $V_{ds}$ (i.e., $K_{13}$) is reflected by the variation in the parameters $A_{11}$ and $A_{31}$ (see Eqs. 6.4, 6.5). Consequently, from both $\sigma_{11}$ and $\sigma_{31}$ it is apparent that <u>all three</u> compartments will respond to variations in the dry deposition velocity, $V_{ds}$. This conclusion is not immediately obvious from the structure of the MCM model, since the $V_{ds}$ does not appear in the compartmental balance for the water compartment. The above simple examples clearly illustrate that the output sensitivity vectors represent the parametric environmental interactions. Finally, as a simple illustration, the MCM model sensitivity to the dry deposition velocity $V_{ds}$ can be written as:

$$\frac{C_1 - C_1^o}{C_1^o} = \left[ + \frac{A_{22} \, A_{33}}{\Delta} \right] \left( \frac{A_{as}}{V_a} \right) \Delta V_{ds} \qquad (6.23a)$$

$$\frac{C_2 - C_2^o}{C_1^o} = \left[ - \frac{A_{21} \, A_{33}}{\Delta} \right] \left( \frac{A_{as}}{V_a} \right) \Delta V_{ds} \qquad (6.23b)$$

$$\frac{C_3 - C_3^o}{C_1^o} = \left[ - \frac{A_{22} \, A_{31}}{\Delta} \right] \left( \frac{A_{as}}{V_a} \right) \Delta V_{ds} \qquad (6.23c)$$

$$- \left[ \frac{A_{11} \, A_{22}}{\Delta} \right] \left( \frac{A_{as}}{V_s} \right) \Delta V_{ds}$$

The above expressions can also be viewed as the accuracy in the measurements of $\underset{\sim}{C}$ which are needed for a determination of $\underset{\approx}{A}$ within a specified uncertainty level.

**Multimedia Distribution of Trichloroethylene.** In a recent multimedia study Cohen and Ryan (1985) utilized a six compartment system to simulate the distribution of trichloroethylene (TCE) in the San Diego region in California. The mathematical description of the above six compartment system is given by Equation 6.3 with the matrix $\underset{\approx}{A}$ given by:

$$\underset{\approx}{A} = \begin{bmatrix} A_{11} & A_{12} & A_{13} & 0 & 0 & 0 \\ A_{21} & A_{22} & 0 & 0 & 0 & 0 \\ A_{31} & A_{32} & A_{33} & A_{34} & A_{35} & A_{36} \\ 0 & 0 & A_{43} & A_{44} & 0 & 0 \\ 0 & 0 & A_{53} & 0 & A_{55} & 0 \\ 0 & 0 & A_{63} & 0 & 0 & A_{66} \end{bmatrix}$$

in which the components $A_{ij}$ are defined in Equations 6.4 and 6.5. The subscripts i, j=1,...,6 refer to the compartments which are numbered as follows: 1 - air; 2 - soil; 3 - water; 4 - biota; 5 - suspended solids; and 6 - sediment. In this model sources were assumed to be present only in the air compartment. Therefore the source vector S is defined by:

$$\underset{\sim}{S} = \begin{bmatrix} S_1 \\ 0 \\ 0 \\ 0 \\ 0 \\ 0 \end{bmatrix}$$

The numerical values of the coefficients calculated from the data of Cohen and Ryan (1985) are given in Table III. The solution for the above compartmental system (Equation 6.3 with $\underset{\approx}{A}$ and $\underset{\sim}{S}$ given by Equations 6.19 and 6.20, respectively, was solved analytically by Cohen and Ryan (1984).

The sensitivity model as represented by Equations 6.12 and 6.13 and the compartmental model (Eq. 6.3) was solved numerically. As an illustration, the sensitivity of the model output to the overall water-side air/water mass transfer coefficient, $K_{OL}$, can be evaluated by observing that a variation in $K_{31}$ leads to a variation of the coefficients $A_{11}$, $A_{13}$, $A_{31}$ and $A_{33}$. Therefore, the sensitivity of the MCM model is represented by the sensitivity vector $\sigma_{11}$, $\sigma_{13}$, $\sigma_{31}$ and $\sigma_{33}$ in which the subscripts indicate the model sensitivity with respect to the corresponding parameters.

TABLE III

Components of $\underset{\approx}{A}$ (hr$^{-1}$)

| | | | | | |
|---|---|---|---|---|---|
| A(1,1) | = | $-\ 2.00 \times 10^{-2}$ | A(3,5) | = | 0.392 |
| A(1,2) | = | $2.82 \times 10^{-9}$ | A(3,6) | = | $2.138 \times 10^{-8}$ |
| A(1,3) | = | $4.00 \times 10^{-5}$ | A(4,3) | = | 1.755 |
| A(2,1) | = | $4.839 \times 10^{-3}$ | A(4,4) | = | $-\ 9.159 \times 10^{-2}$ |
| A(2,2) | = | $-\ 6.580 \times 10^{-4}$ | A(5,3) | = | $4.2 \times 10^{-5}$ |
| A(3,1) | = | $1.28 \times 10^{-2}$ | A(5,5) | = | $-\ 7.838 \times 10^{4}$ |
| A(3,3) | = | $-\ 2.104$ | A(6,3) | = | $6.8 \times 10^{-6}$ |
| A(3,4) | = | $4.579 \times 10^{-8}$ | A(6,6) | = | $-\ 2.138 \times 10^{-6}$ |

The parametric sensitivity can be conveniently expressed as a relative sensitivity

$$F_\alpha = \underset{\sim}{\sigma}_\alpha / \underset{\sim}{c}^o$$

Therefore, $F_\alpha$ multiplied by the parameter variation $\Delta\alpha$ yields the fractional uncertainty in the predicted concentration $\underset{\sim}{c}$. The relative parameter sensitivities for $A_{11}$, $A_{13}$, $A_{31}$ and $A_{33}$ are shown in Figures 4-7. The relative sensitivities $F_{A_{11}}$ and $F_{A_{13}}$ (Figs. 4,5) increase with time for all the compartments towards their respective steady state values. The sensitivity to the degradation rate constant of TCE in the air, $k_1$, is reflected by $F_{A_{11}}$. As expected, the concentration of TCE in the air is most sensitive to $k_1$ relative to the sensitivity of all other compartments on $k_1$. The important observation, however, is that all of the compartments are sensitive to $k_1$ within the same order of magnitude. For example, an order of magnitude overestimate of the value of $k_1$, given a base value of 0.01 h$^{-1}$ for $k_1$ (Yung et al., 1975), would lead to an absolute uncertainty in the range of 450-500 percent for the compartmental concentrations at steady state. An order of magnitude underestimate of $k_1$ would lead to a 45-50 percent uncertainty (relative to the base value of $k_1$) in the predicted steady state concentrations. Similarly, an order of magnitude variation in the degradation rate constant in the water phase ($k_3$), would lead to an absolute uncertainty of approximately 0.25-25 percent (see Fig. 7). From the above it is clear that the compartmental system is more sensitive to $k_1$ compared to $k_3$, and that the resulting uncertainties in $\underset{\sim}{c}$ reach their maximum at steady state.

The sensitivity model also reveals that the sensitivity of the air and soil compartments to the $A_{31}$ and $A_{33}$ parameters is negligible. Additionally, although the TCE model is dynamic, it is apparent that the relative sensitivity of the water, suspended solids, biota and sediment is time invariant.

When the physicochemical parameter (i.e. $K_L$, $K_g$, $k_i$) appear in more than one of the components of the parameter matrix $\underset{\approx}{A}$, it is useful to express the sensitivity in the following way

$$\frac{\underset{\sim}{c} - \underset{\sim}{c}^o}{\underset{\sim}{c}_o} = \sum_{i=1}^{L} \sum_{\alpha=1}^{M} K_i \left( \frac{\underset{\sim}{\sigma}_\alpha}{\underset{\sim}{c}} \right) \left( \frac{\partial \alpha}{\partial K_i} \right) \left( \frac{\Delta K}{K_i} \right)$$

or

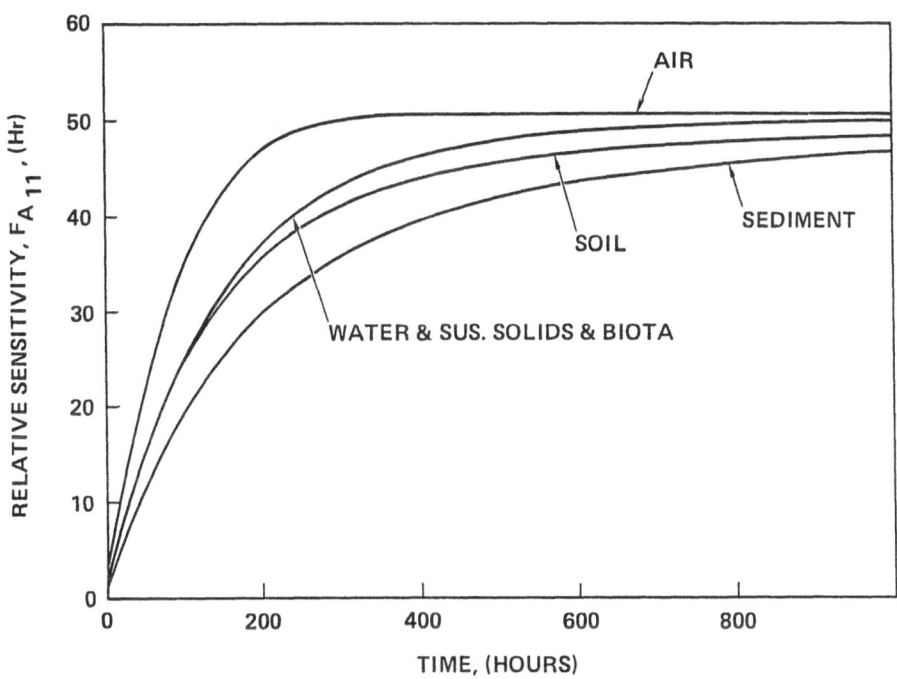

Figure 4. Relative sensitivity of the TCE model to $A_{11}$.

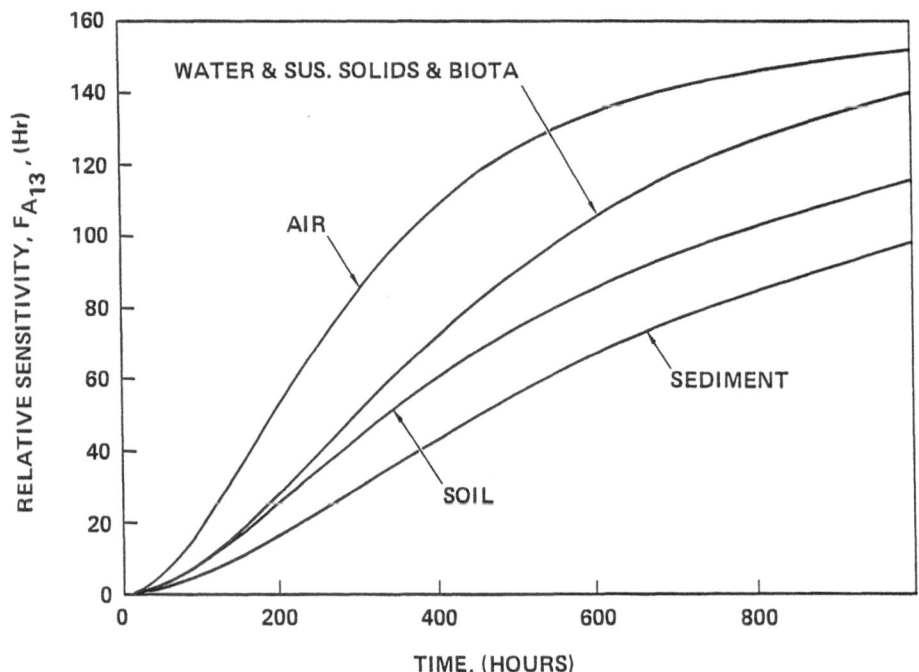

Figure 5.   Relative sensitivity of the TCE model to $A_{13}$.

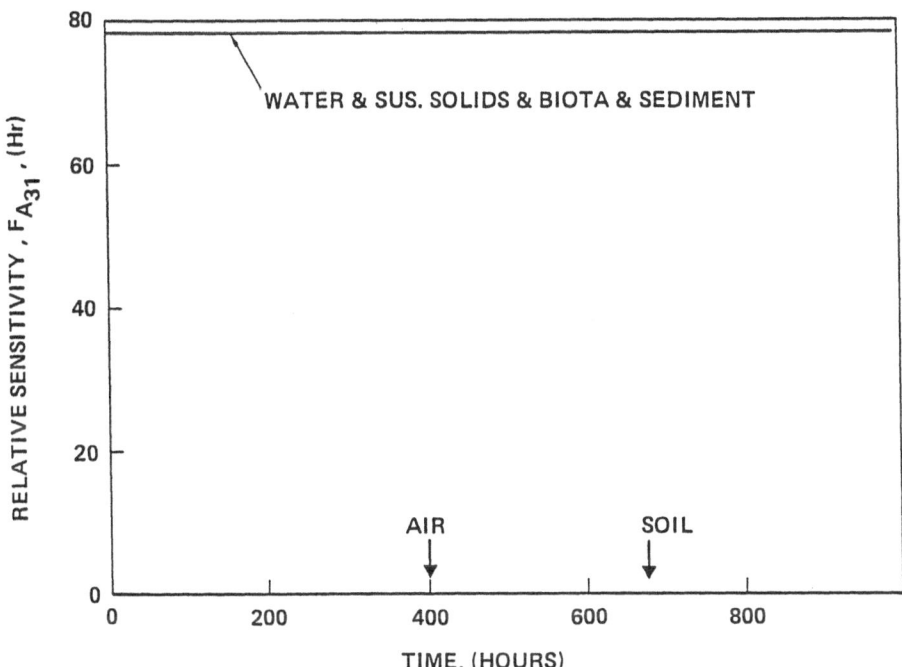

Figure 6. Relative sensitivity of the TCE model to $A_{31}$.

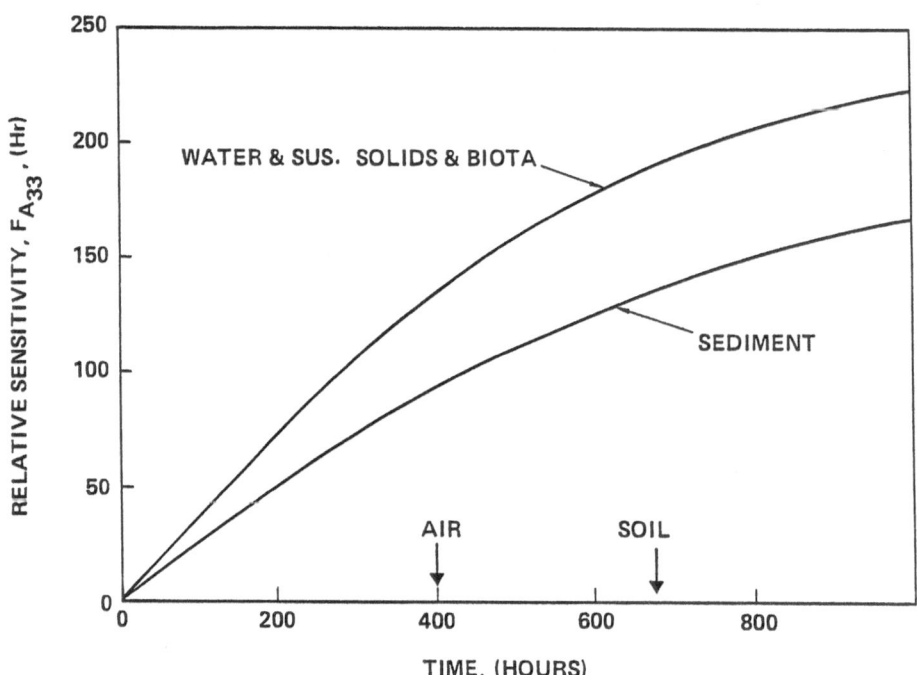

Figure 7.   Relative sensitivity of the TCE model to $A_{33}$.

$$\frac{\underline{C} - \underline{C}^O}{\underline{C}_o} = \sum_{i=1}^{L} E_{K_i} \, (\Delta K_i/K_i)$$

in which $E_{K_i} = \sum_{\alpha=1}^{M} K_i \left(\frac{\sigma_\alpha}{\underline{C}}\right) \left(\frac{\partial \alpha}{\partial K_i}\right)$

where $K_i$ is designate any given physicochemical parameter, L is the number of physicochemical parameters, and $\Delta K/K$ is the fractionsl parameter uncertainty. As an illustration the $K_L$ relative sensitivity for the TCE model is shown in Figure 8. It is clear that the concentrations of the air and soil compartments are not sensitive to $K_L$. Also, the relative sensitivity $E_{K_L}$ for the water, suspended solids, biota and sediment compartments decreases with time. This latter result, which is not immediately obvious, suggests that the per cent uncertainty in the concentration prediction for the above compartments is in fact smaller than the per cent uncertainty in the value of $K_L$. For example, an uncertainty of 50 percent in $K_L$ results in an uncertainty of 18 per cent in the TCE concentration in the water compartment at t = 800 hrs. The highest uncertainty in $C_3$ occurs at short times and it is of the same magnitude as the uncertainty in $K_L$. The above results suggest that in order to estimate $K_L$ from field data it is necessary to obtain concentration data during initial uptake (or release) of the chemical of interest. It should also be apparent that the uncertainty of the estimated $K_L$ value will be about the same as the uncertainty in the measured $C_3$ data.


**SUMMARY**


Intermedia transport models play a central role in dynamic multimedia fate and transport models. Theoretical evaluation of intermedia transport parameters, backed by laboratory and/or field data, should be used whenever available. In this manner it will be possible to clearly evaluate the conditions over which the parameter estimation techniques are valid. Additionally, theoretical intermedia models reveal the required field data for parameter estimation (e.g. f(a) needed to calculate both the dry deposition velocity and the scavenging ratio of particle-bound pollutants). The use of empirical laboratory correlations should be preferably restricted to correlations that are based on sound theoretical principles.

Finally, the effect of intermedia transport parameter uncertainties on multimedia fate and transport can be readily assessed via the use of compartmental multimedia transport models coupled with appropriate

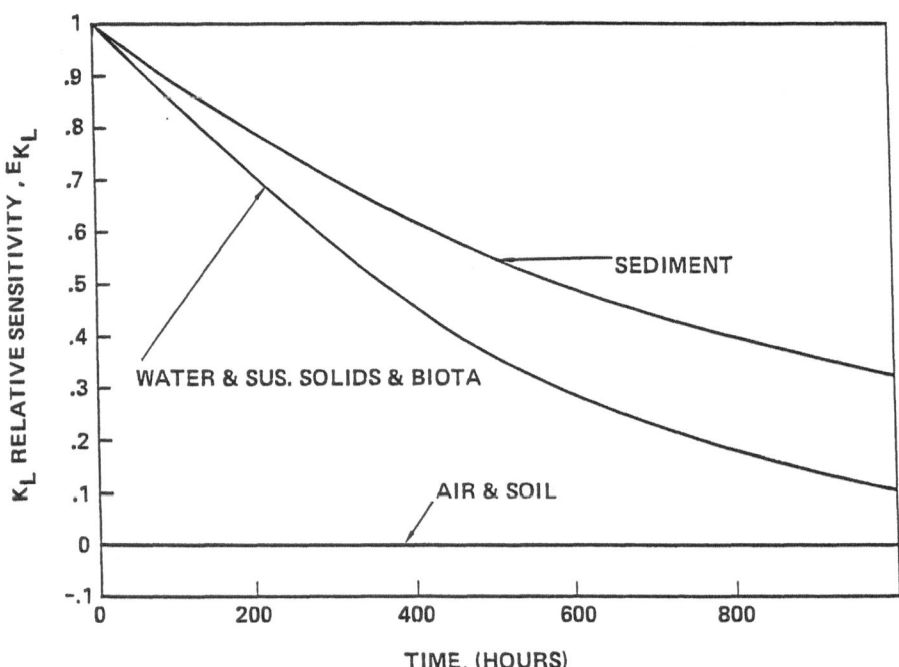

Figure 8.  Relative sensitivity, $E_K$, of the TCE model to the liquid-side
air/water mass transfer coefficient.

sensitivity models. Such an approach should also prove useful in evaluating parametric sensitivity in exposure calculations and risk analysis which are based on multimedia compartmental screening models.

## REFERENCES

Bird, R.B., M.E. Stewart and E.M. Lightfoot, "Transport Phenomena", p. 363, Wiley, New York (1960).

Brkto, W.J. and R.L. Kabel, J. Phys Oceanogr., $\underline{8}$, 543 (1978).

Brutsaert, W. "Evaporation into the Atmosphere: Theory, History and Applications", D. Reidel Publ. Co., Boston, MA (1982).

Brutsaert, W., Water Resources Res., $\underline{11}$, 4 (1975).

Cohen, Y., W. Cocchio and D. Mackay, Envir. Sci. Technol., $\underline{12}$, 553 (1978).

Cohen, Y., M.A.Sc. Thesis, Department of Chemical Engineering, University of Toronto, Toronto, Canada (1976).

Cohen, Y. Int. J. Heat Mass Transfer, $\underline{26}$, 1284 (1983).

Cohen, Y., "Modeling of Pollutant Transport and Accumulation in a Multimedia Environment", in the conference on "Geochemical and Hydrologic Processes and their Protection", Council on Environmental Quality, September 25, 1984, Washington, D.C. (also to be published in an Environmental Series by Plenum Press).

Cohen, Y. and P.A. Ryan, Envir. Sci. Technol, $\underline{9}$, 412 (1985).

Cohen, Y. and P.A. Ryan, Int. Comm. Heat Mass Transfer, $\underline{12}$, 139 (1985).

Cohen, Y., Envir. Sci. Technol., feature article, in press 1986.

Dana, M. and M. Wolf, in "Precipitation Scavenging", Champaign, IL, Oct. 14-18, 1974, Semonin, R.G. and R.W. Beadle (eds.) ERDA Symp. Ser., No. 41 (CONF-741003) (1977).

Davies, C.N., "Aerosol Science", Academic Press, New York (1966).

Deacon, E.L., Tellus, $\underline{29}$, 363 (1977).

Dipprey, D.F. and R.H. Sabersky, Int. J. Heat Mass Transfer, $\underline{6}$, 329 (1963).

Eloubaidy, A.F. and E.J. Plate, J. Hydraulic Div. ASCE, $\underline{98}$, 153 (1972).

Englemann, R.J., And W.G. Slinn (eds.), "Precipitation Scavenging", AEC Symp. Series, No. 22 (CONF-700601), Richland, WA., Jove 2-4 (1970).

Englemann, R.J., Meteorology and Atomic Energy, USAEC Report TID-24190, Env. Sci. Services Administration (1968).

Eskinazi, S., "Fluid Mechanics and Thermodynamics of Our Environment", Academic Press, New York, pp. 83-129 (1972).

Fernandez, J. de la Mora and S.K. Friedlander, Int. J. Heat Mass Transfer, $\underline{25}$, 1725 (1982).

Fitzgerald, J.W., J. Appl Met., $\underline{14}$, 1044 (1975).

Frank, P.M., "Introduction to System Sensitivity Theory", Academic Press (1978).

Gunn, R. and G.D. Kinzer, J. Meteorology, $\underline{6}$, 243 (1949).

Haase, L. and P.S. Liss, Tellus, 470 (1980).

Hales, J.M. Atm. Envir. $\underline{6}$, 635 (1972).

Hales, J.M., Atm. Envir. $\underline{12}$, 389 (1978).

Hwang, S.T. and L.J. Thibodeaux, Envir. Progress, $\underline{2}$, 81 (1983).

Jirka, G.H. and W. Brutsaert, in "Gas Transfer at Water Surfaces", Brutsaert, W. and G.H. Jirka (eds.), D. Reidel Publishing Co., Boston (1984).

Knuth, R.H., E.O. Knutson, H.W. Feely and H.L. Volchok in "Precipitation Scavenging, Dry Deposition and Resuspension", Pruppacher, H.R., R.G. Semonin and W.G. Slinn (eds.), $\underline{2}$, 1325 (1983).

Kondo, J., Boundary-Layer Met., $\underline{9}$, 91 (1975).

Kramer, M.A., J.M. Calo, H. Rabitz and R.J. Kee, "AIM: The Analytically Integrated Magnus Method for Linear and Second Order Sensitivity Coefficients", Sandia Report SAND82-8231 (1982).

Lewis, W.K. and W.G. Whitman, Ind. Eng. Chem., 16, 1215 (1924).

Liss, P.S. and P.G. Slater, Nature, 247, 181 (1974).

Lunny, P.D. and L.J. Thibodeaux, Envir. Progress, 4, 203 (1985).

Lusis, M.A., W.H. Chan, A.J.S. Tang and N.D. Johnson, "Scavenging Rates of Sulfur and Trace Metals from a Smelter Plume", in Precipitation Scavenging, Dry Deposition and Resuspension (1982), Santa Monica, CA, Nov. 29 - Dec. 3, 1982, H.R. Pruppacher, R.G. Semonin and W.G. Slinn (eds.), Volume 1, 369-382 (1983).

Lyman, W.J., W.F. Reehl and D.H. Rosenblatt, "Handbook of Chemical Property Estimation Methods: Environmental Behavior of Organic Compounds", McGraw-Hill (1982).

McKone, T.E., "The Use of Environmental Health-Risk Analysis for Managing Toxic Substances", Lawrence Livermore National Laboratory Report UCREL-92329, Livermore, California.

Mackay, D. and A.T.K. Yeun, Envir. Sci. Technol., 17, 211 (1983).

Mackay, D. and S. Paterson, Envir. Sci. Technol., 15, 106 (1981).

Mackay, D. and S. Paterson, Envir. Sci. Technol., 16, 654A (1982).

Mackay, D., S. Paterson and M. Joy, in "Fate of Chemicals in the Environment" Swann, R.L. and A. Eshenroeder (eds.), ACS Symposium Series No. 225, ACS, Washington, D.C. (1983).

Mackay, D. and W.Y. Shiu, "The Aqueous Solubility and Air-Water Exchange Characteristics of Hydrocarbons under Environmental Conditions", paper presented to Electrochem. Soc. Sym., Toronto, May 1975, and published in the Proceedings "The Chemistry and Physics of Aqueous Solutions", 93-110.

Mackay, D., W.Y. Shiu and R.P. Sutherland, Envir. Sci. Technol., 13, 333 (1979).

Mackay, D. and W.Y. Shiu, in "Gas Transfer at Water Surfaces", Brutsaert, W. and G.H. Jirka (eds.), D. Reidel Publishing Company (1984).

Mackay, D. and M.R. Paterson, in "Workshop on Pollutant Transport and Accumulation in a Multimedia Environment", January 21-24, 1986, Santa Monica, CA.

Mangarella, P.A., A.J. Chambers, R.L. Street and E.Y. Hsau, J. Phys. Oceanography, 3, 93 (1973).

Marshall, J. and W. Palmer, J. Meter., 5, 165 (1948).

Miguel, A.H. and S.K. Friedlander, Atm. Envir., 12, 2407 (1978).

Monin, A.S. and A.M. Yaglom, "Statistical Fluid Mechanisms", Vol. 1, the MIT Press, Cambridge, MA (1971).

Munnich, K.O., W.B. Clarke, K.H. Fischer, D. Flothmann, B. Kromer and W. Roether, In "Turbulent Fluxes through the Sea Surface, Wave Dynamics, and Prediction" (Farre, A. and K. Hasselman), Plenum Press, New York (1978).

O'Connor, D., J. Envir. Engr., 109, 731 (1983).

Owen, P.R. and W.R. Thompson, J. Fluid Mech., 15, 321 (1962).

Peng, T.H., W.S. Broecker, G.G. Mathiu, Y.-H. Li and A.E. Bainbridge, J. Geophys. Res., 84, 2471 (1979).

Plate, E.J. and R. Friedrich, in "Gas Transfer at Water Surfaces", Brutsaert, W. and G.H. Jirka (eds.), D. Reidel Publishing Co., (1984).

Radke, L.F., P.B. Hobbs and M.W. Eltgrath, J. Appl. Meter., 19, 715 (1980).

Radke, L.F., E.E. Hindman and P.V. Hobbs, in "Precipitation Scavenging", Champaign IL, Oct. 14-18, 1974, Semonin, R.G. and R.W. Beadle (eds.), ERDA Symp. Ser., No. 41, (CONF-741003), 425 (1977).

Rathbun, R.E., J. Hydraulic Div., ASCE, 98, HYI, 1733 (1972).

Ryan, P.A., M.S. Thesis, Department of Chemical Engineering, University of California, Los Angeles, Los Angeles, CA. (1984).

Ryan, P.A. and Y. Cohen, Chemosphere, 15, 21 (1986).

Kramer, M.A., J.M. Calo, H. Rabitz and R.J. Kee, "AIM: The Analytically Integrated Magnus Method for Linear and Second Order Sensitivity Coefficients", Sandia Report SAND82-8231 (1982).

Lewis, W.K. and W.G. Whitman, Ind. Eng. Chem., 16, 1215 (1924).

Liss, P.S. and P.G. Slater, Nature, 247, 181 (1974).

Lunny, P.D. and L.J. Thibodeaux, Envir. Progress, 4, 203 (1985).

Lusis, M.A., W.H. Chan, A.J.S. Tang and N.D. Johnson, "Scavenging Rates of Sulfur and Trace Metals from a Smelter Plume", in Precipitation Scavenging, Dry Deposition and Resuspension (1982), Santa Monica, CA, Nov. 29 - Dec. 3, 1982, H.R. Pruppacher, R.G. Semonin and W.G. Slinn (eds.), Volume 1, 369-382 (1983).

Lyman, W.J., W.F. Rechl and D.H. Rosenblatt, "Handbook of Chemical Property Estimation Methods: Environmental Behavior of Organic Compounds", McGraw-Hill (1982).

McKone, T.E., "The Use of Environmental Health-Risk Analysis for Managing Toxic Substances", Lawrence Livermore National Laboratory Report UCREL-92329, Livermore, California.

Mackay, D. and A.T.K. Yeun, Envir. Sci. Technol., 17, 211 (1983).

Mackay, D. and S. Paterson, Envir. Sci. Technol., 15, 106 (1981).

Mackay, D. and S. Paterson, Envir. Sci. Technol., 16, 654A (1982).

Mackay, D., S. Paterson and M. Joy, in "Fate of Chemicals in the Environment" Swann, R.L. and A. Eshenroeder (eds.), ACS Symposium Series No. 225, ACS, Washington, D.C. (1983).

Mackay, D. and W.Y. Shiu, "The Aqueous Solubility and Air-Water Exchange Characteristics of Hydrocarbons under Environmental Conditions", paper presented to Electrochem. Soc. Sym., Toronto, May 1975, and published in the Proceedings "The Chemistry and Physics of Aqueous Solutions", 93-110.

Mackay, D., W.Y. Shiu and R.P. Sutherland, Envir. Sci. Technol., 13, 333 (1979).

Mackay, D. and W.Y. Shiu, in "Gas Transfer at Water Surfaces", Brutsaert, W. and G.H. Jirka (eds.), D. Reidel Publishing Company (1984).

Mackay, D. and M.R. Paterson, in "Workshop on Pollutant Transport and Accumulation in a Multimedia Environment", January 21-24, 1986, Santa Monica, CA.

Mangarella, P.A., A.J. Chambers, R.L. Street and E.Y. Hsau, J. Phys. Oceanography, 3, 93 (1973).

Marshall, J. and W. Palmer, J. Meter., 5, 165 (1948).

Miguel, A.H. and S.K. Friedlander, Atm. Envir., 12, 2407 (1978).

Monin, A.S. and A.M. Yaglom, "Statistical Fluid Mechanisms", Vol. 1, the MIT Press, Cambridge, MA (1971).

Munnich, K.O., W.B. Clarke, K.H. Fischer, D. Flothmann, B. Kromer and W. Roether, In "Turbulent Fluxes through the Sea Surface, Wave Dynamics, and Prediction" (Farre, A. and K. Hasselman), Plenum Press, New York (1978).

O'Connor, D., J. Envir. Engr., 109, 731 (1983).

Owen, P.R. and W.R. Thompson, J. Fluid Mech., 15, 321 (1962).

Peng, T.H., W.S. Broecker, G.G. Mathiu, Y.-H. Li and A.E. Bainbridge, J. Geophys. Res., 84, 2471 (1979).

Plate, E.J. and R. Friedrich, in "Gas Transfer at Water Surfaces", Brutsaert, W. and G.H. Jirka (eds.), D. Reidel Publishing Co., (1984).

Radke, L.F., P.B. Hobbs and M.W. Eltgrath, J. Appl. Meter., 19, 715 (1980).

Radke, L.F., E.E. Hindman and P.V. Hobbs, in "Precipitation Scavenging", Champaign IL, Oct. 14-18, 1974, Semonin, R.G. and R.W. Beadle (eds.), ERDA Symp. Ser., No. 41, (CONF-741003), 425 (1977).

Rathbun, R.E., J. Hydraulic Div., ASCE, 98, HYI, 1733 (1972).

Ryan, P.A., M.S. Thesis, Department of Chemical Engineering, University of California, Los Angeles, Los Angeles, CA. (1984).

Ryan, P.A. and Y. Cohen, Chemosphere, 15, 21 (1986).

Scott, B.C., Atm. Environ., $\underline{10}$, 1753 (1983).

Semonin, R.G. and R.W. Beadle (eds.), "Precipitation Scavenging", ERDA Symp. Series, 41 (CONF-741003), Champaign, IL, Oct. 14-18 (1974).

Slinn, W.G.N., in "Precipitaton Scavenging", Champaign IL, Oct. 14-18, 1974, Semonin, R.G. and R.W. Beadle (eds.), ERDA Symp. Ser. No. 41 (CONF-741003), 1-60 (1977).

Smith, J.H. and D.C. Bomberger, AIChE. Symp. Series, 190, No. 75, 375 (1979).

Smith, J.H., D.C. Bomberger and D.L. Haynes, Envir. Sci. Technol., $\underline{14}$, 1332 (1980).

Street, R.L., Int. J. Heat Mass Transfer, $\underline{22}$, 885 (1979).

Swann, R.L. and A. Eshenroeder, "Fate of Chemicals in the Environment", ACS Symposium Series, No. 225, American Chemical Society, Washington, D.C. (1983).

Thibodeaux, L.J., "Chemodynamics", John Wiley and Sons, New York, NY (1979).

Tomovic, R. and M. Vukobratovic, "General Sensitivity Theory", Elsevier, New York (1970).

Topalian, J.H., S. Mitra, D.C. Montague, A. Quintanar and H.R. Pruppacher, J. Atm. Sci., $\underline{1}$, 325 (1984).

Vali, G., in "Precipitation Scavenging", Champaign, IL, Oct. 14-18, 1974, Semonin, R.G. and R.W. Beadle (eds.), ERDA Symp. Series, No. 41, (CONF-741003), 1-60 (1978).

Walcek, C., Ph.D. dissertation, Dept. Atm. Sciences, University of California, Los Angeles, Los Angeles, CA.

Walcek, C.J. and H.R. Pruppacher, J. Atm. Chem., $\underline{1}$, 269 (1984a).

Walcek, C.J. and H.R. Pruppacher, J. Atm. Chem., $\underline{1}$, 307 (1984b).

Walcek, C.J., H.R. Pruppacher, J.H. Topalian and S.K. Mitra, J. Atm. Chem., $\underline{1}$, 291 (1984).

Wesely, M.L. and B.B. Hicks, J. Air Pol. Control Assoc., $\underline{27}$, 1110 (1977).

Wiersma, G.V., "Kinetic and Exposure Commitment Analysis of Lead Behavior in the Bioshpere Reserve", MARC Report 15, Monitoring and Assessment Research Center, Chelsea College, University of London, London, England (1979).

Williams, R.M., Atm. Env., $\underline{16}$, 1933 (1982).

Wu, J., J. Phys. Oceanogr., $\underline{9}$, 1064 (1979).

Yu, J., J. Phys. Oceanogr., $\underline{10}$, 727 (1980).

Yaglom, A.M. and B.A. Kader, J. Fluid Mech. $\underline{62}$, 601 (1974).

Yasuda, N., Science Reports Tohoku Univ., (Sendai Japan), Ser. 5, Geophys., $\underline{22}$, 87 (1975).

Yung, Y.L., M.B. McElroy and S.C. Wofsy, Geophys. Res. Lett., $\underline{2}$, 347 (1975).

TRANSPORT OF CHEMICAL CONTAMINANTS IN THE MARINE ENVIRONMENT

ORIGINATING FROM OFFSHORE DRILLING BOTTOM DEPOSITS - A VIGNETTE MODEL

L.J. Thibodeaux and D.D. Reible     C.S. Fang
Louisiana State University          University of Southwestern LA
Baton Rouge, LA                     Lafayette, LA

INTRODUCTION

Drilling fluids are required in rotary drilling for oil and gas exploration and development. Eventually drilling fluids, also called mud, and cuttings must be disposed from the drilling vessel. Although these drilling discharges can be barged ashore or to other sites at sea for disposal, cost and operational considerations favor onsite disposal, by either overboard discharge or shunting through a pipe to some depth. A vignette model is developed to track the movement and depletion of selected pollutants in this multimedia marine environment. The seabed media involved are the solid/sediment, the porewater and the benthic boundary layer.

The principal bulk constituents of drilling fluids are water, barite (barium sulfate), clay minerals, chrome lignosulfonate, lignite and sodium hydroxide. All of these constituents are nontoxic to marine organisms at the dilutions reached shortly after discharge. There is limited information on the compositions and quantities of additives in the fluids discharged on the outer continental shelf. Several common drilling-fluid additives, including biocides and diesel fuel (No. 2 fuel oil), are much more toxic to marine organisms than the bulk constituents.

The rate of bulk discharges range from 500 to 2,000 bbl/h. Over the life of an exploratory well, some 5,000 to 30,000 bbl of fluid are discharged (Costlow, et.al., 1983). Total additives vary from 520 to 2118 tons per well while drill cuttings vary from 823 to 1285 tons per well. These data are for exploratory wells (Menzie, 1982). Less fluid and cuttings are usually discharged from development wells. The following description of the phenomena observed during drilling fluid discharges is obtained from the Offshore Operators Committee (OOC) mud discharge model (Brandsma and Sauer, 1983).

Upon discharges, the material goes through three distinct phases: convective descent (jet), dynamic collapse, and passive diffusion. In the jet phase, the plume, influenced by gravity, descends through the water column, entraining ambient fluid while bending in the direction of the ambient current. The collapse phase begins when the plume

encounters the level of neutral buoyancy or the ocean floor, where
descent is retarded and horizontal spreading dominates. It is during
the convective descent and dynamic collapse phase that large quantities
of mud and cutting particles are deposited on the ocean floor near the
well site. The upper plume spreads into a thin layer, the transport
and spreading of this plume is determined by the ambient current and
turbulence than by any dynamic character of its own. Here passive
diffusion begins.

The OOC model predicts the fate of drilling discharges with major
emphasis on the passive diffusion phase and concentrations in the water
column. In the case of drilling fluids, most of the discharged
materials (>90%) sinks to the bottom near the well site, with the
distance from the discharge point (within 500m for most OCS areas)
depend on depth of water, lateral transport, particle size and density
of material. How long the settled material remains at the well site
depends on environmental factors that govern sediment resuspension,
transport, and dispersion. The redistribution of settled drilling
solids depends on the shear between the bottom forms and the flowing
seawater. Drilling solids may build up on extended periods at certains
times of the year, but one major storm event may be sufficient to move
the entire layer the solids have formed. In low-energy areas the layer
may remain for a year or longer.

The important factor in determing the fate of pollutants is the
resuspensive transport of sediments that tends to dilute and disperse
particle contaminants. Transport depends on the hydrodynamic regime of
an environment. No accumulation of contaminants in bottom sediments
were observed where strong currents resuspended and dispersed the
discharged material. On the other hand pollutants persist in the more
quiescent benthic environments of the Gulf of Mexico and the shelf break
of the Middle Atlantic Bight. Barium was found incorporated to at least
15 cm depth near a production platform more than 5 years after drilling
had ceased (Costlow, et.at.,1983).

The NRC Panel concludes that effects of discharges are confined
mainly to the benthic environment. It also notes that a significant
shortcoming in understanding the long-term effects of drilling
discharges concerns the fate of particulate components once they reach
the seabed. Uncertainties regarding effects exist for low energy
depositional environments, which experience large inputs of drilling
discharges over long periods of time.

Diffuse transport models have been applied to problems of chemical
diagenesis and a number of analytical solutions of model equations under
certain boundary conditions have been reported (Lerman 1979). These are
steady-state diffusion models. By contrast, transient-state models are
used seldom, not only because of the more intricate mathematics, but
also because steady-state models have in many instances been shown to
give satisfactory fit to the field data (Maris and Bender, 1982) so that
the need for transient-state models is relatively small.

Some simple transient-state models are presented by Duursma and
Smies(1982) that address aspects of chemical behavior in the benthic
boundary layer. The models include a sorption-distribution term in the
diagenetic process. However, none of these can be applied to the
desorption and transport of chemical species in a finite sediment layer.
Steady-state models are used along with a defined mean chemical
residence time in order to force a transient flavor to the analysis.
Calculations, using $^{134}C_s$ as the absorbed species with an effective

diffusion coefficient of 1E −8 cm$^2$/s and a convective pore water velocity of 1 cm/y, yield a residence time of 7.9 to 10.6 years in a 1-cm layer of sediment.

The "vignette" model presented in this manuscript is concerned with fate aspects of pollutants deposited on the seabed as a consequence of mud and cuttings discharges. It addresses some of the major short-comings noted in the NRC report (Costlow et.al., 1983). The emphasis is on chemical desorption and multimedia transport in the sediment zone and the adjoining benthic boundary layer.

The multimedia model is termed vignette because, as a brief incident or scene in a play, movie or literary sketch is so termed, it is a short but descriptive scientific sketch. It portrays a small but important portion of the drilling fluid discharge fate. Without attempting to be global and accounting for all known phenomenon, a vignette model can nevertheless realistically quantify important fate and transport aspects of chemical in the seabed sediment. The major assets of a vignette model are: simple mathematics, narrow scope, few input variables, verified physicochemical phenomenon, ease of use and ease of interpretating results.

## Model Development

Chemical leaching undoubtly occurs as the particles move from the exit of the platform discharge chute to the seabottom. However in the case of oil based muds Poley and Wilkinson (1983) report on experiments which suggest that during the journey downward the cuttings lose little "free oil". The movement of chemical species from these waste occurs predominantly as it lies on the bottom. This present model is concerned with the natural leaching (or dissolution) processes as the cuttings reside in a layer on the sea bottom. For the purpose of the model it is assumed that all discharges from the platform have ceased and a fixed quantity of cuttings reside in-place on the bottom. The thickness of this layer is h (cm) as indicated in Figure 1.

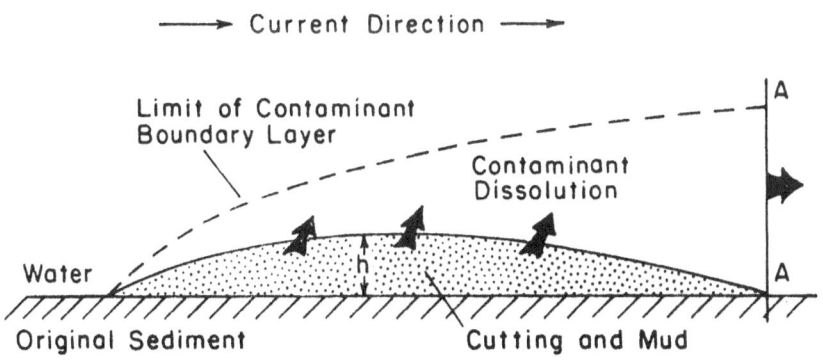

Figure 1. Cutting and Mud Pile

The cuttings form a layer that reside above the original sediment. For the purposes of this model no further mixing of the cuttings and sediment is assumed to occur after the deposition process. Once the sediment is in place the chemical dissolution processes commence and soluble components leave the cutting layer and move into the overlying seawater. The process is initiated as soluble constituents, which are sorbed onto the cutting particles located at the sediment-water interface, desorb and diffuse through the aqueous benthic sublayer. This dissolution mechanism is fairly rapid and the cutting particles in the upper sediment layer soon become depleted of soluble constituents. Later the deeper soluble constituents must diffuse through more layers of sediment to reach the sediment water interface. A leached-out zone develops as indicated in Figure 2. The length of this zone increases with time as the dissolution/diffusion process continues.

## The Initiation Period of the Leaching Process

During the initiation period of this natural leaching process the flux rate of the soluble chemical species to the overlying water is controlled by water-side transport processes. Initially the dissolution plane concides with the sediment-water interface. During this period a significant "leached-out" zone has not had time to develop. The leaching rate from the cutting particles residing on the sediment-water interface is:

$$\eta_A = {}^3k_{A2} \, (\rho_{A2}^{\,*} - \rho_{A2}) \tag{1}$$

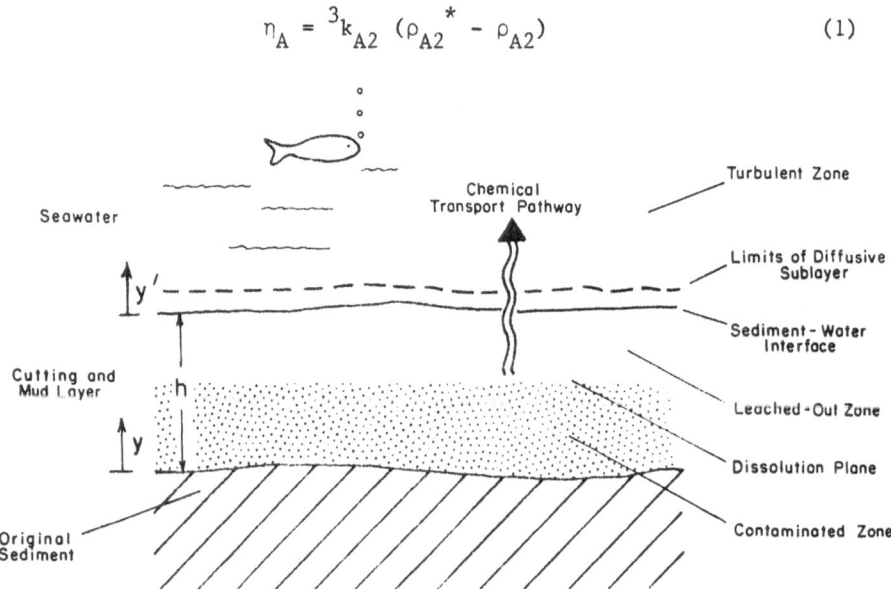

Figure 2. Dissolution Processes and Pathways in Sediment and Seawater

where $\eta_A$ is the chemical rate, $g/cm^2 \cdot s$,

$\quad {}^3k_{A2}$ is the water-side mass-transport coefficient, cm/s,

$\quad \rho_{A2}^{\,*}$ is the chemical concentration in the surface sediment porewater, $g/cm^3$ and

$\quad \rho_{A2}$ is the chemical concentration in the seawater at a point well removed from the sediment surface, $g/cm^3$.

During the initiation process the leaching rate is highest.

The desorption process is assumed to be rapid and not rate controlling. This process is one in which the soluble constituents originally sorbed onto the cutting particles, desorb into the adjoining porewater. Equilibrium partitioning of the chemical between particle and adjoining porewater is therefore assumed. This condition is related by the partition coefficient:

$$K_{A32} = \rho_{A2}*/\omega_A \qquad (2)$$

where $\omega_A$ is the concentration of the chemical in the particle or solid phase, g/g, and $K_{A32}$ is the distribution coefficient, g/cm$^3$.

The leaching rate for chemicals sorbed at the sediment-water interface is extremely rapid because the process is enhanced by turbulence in the benthic boundary layer. The rate given by Eq. 1 cannot occur for a every long period. Once the upper-most layer of cuttings is free of solubles, further leaching must occur from within the sediment layer. If the origin of these chemicals is at a distance $\Lambda$, cm below the interface, the two resistance theory applies. Boudreau and Guinasso (1982) used this additive resistance theory to quantify soluble chemical transport through the sublayer and adjoining porewater in the marine environment:

$$1/^3K_{A2} = 1/^3k_{A2} + \Lambda/D_{A3} \qquad (3)$$

where $^3K_{A2}$ is the overall mass-transport coefficient, cm/s, and $D_{A3}$ is the effective diffusion coefficient for the chemical in the sediment layer, cm$^2$/s. The first term on the right of Eq. 3 is the water-side resistance and the second is the sediment-side resistance.

When the respective resistance terms in Eq. 3 are equal then both processes are clearly significant. It is assumed here that the bed-side processes dominate when the second term accounts for at least 95% of the overall resistance. When this occurs the depth of the desorption plane is given by:

$$\Lambda = 19 \ D_{A3}/^3K_{A2} \qquad (4)$$

$\Lambda$ is in cm. For an effective diffusivity of 4E $-7$cm$^2$/s and mass transfer coefficient of 1E $-4$ cm/s, Eq. 4 yields $\Lambda$ <1mm. Therefore the plane of dissolution need retreat only a short distance before the sediment-side resistance dominates. As will be demonstrated in the results section the time of leaching to this small distance is also short in comparison to the leaching time of the entire cutting and mud layer.

## Dissolution and Diffusion Processes Within the Bed

This segment of the model, that will be applied to the bed-side transport processes, is an adaptation of an existing laboratory verified algorithm. The original was concerned with the movement of volatile chemicals through the soil to the air/soil interface. Upon leaving the interface plane the volatiles enter the overlying air mass. A gradientless model for the evaporation and vapor phase diffusion of liquids spilled or otherwise placed into the soil was proposed (Thibodeaux, 1979). The model as developed accounts for the primary mechanisms that influence the equilibrium and transport phenomena, yielding flux rate and life-time algorithms. Successful application of the model has been demonstrated for the desorption and vapor phase diffusion through the air/soil interface of two pesticides incorporated into a laboratory soil column (Thibodeaux and Hwang, 1982). Carvanos (1985) and Buff (1984) have successfully tested the model with several other volatile organics in soil and sand systems. Since the dissolution and liquid phase diffusion process of interest in this manuscript involves analogous processes, it will be applied to the movement of soluble constituents from the cutting and mud mound into the overlying sea water.

The chemical transport zones near the interface are illustrated in Figure 2. This illustration is at a time well into leaching process. The chemical contaminant, which is sorbed onto the cutting and mud particles, desorb and go into solution with the porewater at the desorption plane identified in Figure 2. Once in the porewater the contaminant molecule moves through the leached-out zone by diffusion. The rate of movement in this zone is given by the one-dimensional integrated form of Fick's first law:

$$\eta_a = \frac{D_{A3}}{h-y} (\rho_{A2}^* - \rho_{A2i}) \tag{5}$$

where $\rho_{A2i}$ is the concentration at the sediment water interface, $g/cm^3$

　　h　is the depth of the cutting and mud layer, cm,

　　y　is the distance from the original sediment interface, cm.

As y decreases from h to 0 the flux rate of the soluble constituent decreases. This occurs as time proceeds and the leached-out zone increases in length at the expense of the contaminated zone. Performing a change of variable from distance (y) to time (t) yields a time dependent flux rate:

$$\eta_a = [\frac{D_{A3}}{2t} (\rho_{A3}(\rho_{A2}^* - \rho_{A2}))]^{1/2} \tag{6}$$

where t is time, s and $\rho_{A3}$ is the mass of A per unit bulk volume of sediment (i.e., cuttings, mud, water), $gA/cm3$; $\rho_{A3} = \rho_{A2}^* [\epsilon + (1 - \epsilon) \rho_3 K_{A32}]$.

In reality $\rho_{A2i}$ is also a function of time, but this has been ignored in the development of Eq. 6. The time dependence of $\rho_{A2i}$ is an insignificant factor in the flux computation when the sediment-side resistance controls.

The dissolution and liquid phase diffusion model contains an algorithm for estimating the chemical leaching time for the cutting and mud pile, This equation is

$$\tau = h^2 [\epsilon+(1-\epsilon)K_{A32} \rho_3]\rho_{A2}/2D_{A3}(\rho_{A2} - \rho_{A2i}) \tag{7}$$

where

　　$\rho_3$ is the average particle density of the cutting and mud layer, $g/cm^3$,

　　$\epsilon$ is bed porosity, $cm^3/cm^3$ and

　　$\tau$ is the leaching time, s.

This equation is an adaptation of an analogous result for the vaporization time of a liquid from soil pores. For a bed with h = 1mm, $\epsilon$ = .5, $K_{A32}$ = 1 $cm^3/g$, $\rho_3$ = 2.8 $g/cm^3$ and $D_{A3}$ = 4E-7$cm^2/s$ the dissolution time is approximately 5 hours. For a h= 60 cm. the time is approximately 200 years. This comparison of times further illustrates that the initiation time is extremely brief and that the sediment-side resistance dominate the leaching process.

## Transport Processes Within the Benthic Boundary Layer

As the contaminant emerges from the sediment-water interface it enters the diffusive sublayer which resides immediately above the surface. The contaminant then moves into the turbulent zone of the benthic boundary layer. These concepts give a simple two-layer structure to the benthic boundary layer. The respective rate equations for each layer are

$$n_A = \frac{D_{A2}}{\delta_A} (\rho^*_{A2i} - \rho_{A2}|\delta_A) \tag{8}$$

for the sublayer and

$$n_a = \frac{\kappa_1 \nu_*}{\ln(y'/\delta_A)} (\rho^*_{A2}|\delta_A - \rho_{A2}) \tag{9}$$

for the turbulent zone,

where $D_{A2}$ is the molecular diffusivity of A in water, $cm^2/s$,

$\delta_A$ is the sublayer thickness, cm,

$\rho_{A2}|\delta_A$ is the concentration at the edge of the sublayer, $gA/cm^3$

$\kappa_1$ is von Karman's constant (0.4)

$\nu*$ is the friction velocity, cm/s

y' is the distance above the sublayer, cm, and $\rho_{A2}$ is the concentration at y', $gA/cm^3$.

The diffusivity and sublayer thickness are related to the mass-transfer coefficient by $K_{A2} = D_{A2}/\delta_A$. The flux, $n_A$, in Eq. 8 and 9 is controlled by Eq. 6 since the transport process of the contaminant within the cutting and mud pile dominates.

## Concentration and Velocity Profiles in the Benthic Boundary Layer

As illustrated in Fig. 1 the quantity of the contaminant that leaves the cutting and mud pile enters the benthic boundary layer and exits the pile vicinity by moving through the plane A-A. Turbulence within the boundary layer dominates the chemical transport phenomenon and rapidly mixes entering the contaminant. In this zone of constant shear stress the turbulent diffusivity, $D^{(t)}_{A2}$, $(cm^2/s)$, can be estimated by

$$D^{(t)}_{A2} = 1.5\kappa_1 \nu_* y' \tag{10}$$

A similar eddy diffusivity parameter has been suggested by Costlow, et.al.(1982) for the bottom boundary layer. The 1.5 factor accounts for the observed increase in chemical diffusivity over the viscous eddy diffusivity. The mean square distance of the contaminated plume is $\sigma^2 (cm^2)$ and is related to the turbulent diffusivity and time, t (s), by

$$\sigma^2 = 2D^{(t)}_{A2} \tag{11}$$

The appropriate time scale is that necessary for the bottom current to traverse the length or width of the pile. The contaminant boundary layer thickness, $y'(cm)$, can be approximated by $2\sigma$. This multiple of $\sigma$ accounts for 97.5% of the contaminant in the boundary layer as it departs the pile vicinity.

The turbulent boundary layer velocity profile is

$$v_x = \frac{v_*}{\kappa_1} \ln (y'/y'_o) \tag{12}$$

where $v_x$ is the velocity at $y^1$, cm/s and $y_o^1$ is the bottom roughness height, cm the average velocity in the boundary layer between $Y_o^1$ and $Y_b^1$ is defined as:

$$\langle v_x \rangle = \frac{1}{y'_b - y'} \int_{y_o^1}^{y^1} v \, dy \tag{13}$$

Substituting Eq. 13 for $v_x$ and integrating yields:

$$\langle v_x \rangle = \frac{v_*}{\kappa 1} \quad \frac{y'b}{y'_b - y'_o} \left[ \ln(y'_b/y'_o) - 1 \right] + \frac{y'_o}{y'_b - y'_o} \tag{14}$$

Referring to Fig. 1 it is apparent that the quantity of the contaminant leaving the cutting and mud pile by diffusive processes enters the overlying benthic boundary layer and then moves by the convective water flow through section A-A. Equating the chemical movement rates entering the boundary layer from the cuttings and mud to the rate departing the pile vicinity is:

$$w_A \equiv \eta_A w1 = \langle v_x \rangle w \, y'_b \, \langle \rho_{A2} \rangle \tag{15}$$

Extracting $\langle \rho_{A2} \rangle$ from Eq. 15 yields

$$\langle \rho_{A2} \rangle = \eta_A 1/\langle v_x \rangle \, y'_b \tag{16}$$

where    $w_A$ is the mass rate of A leaving the pile, gA/s,

   w is the average width of the pile, cm,

   1 is the average length of the pile cm and

   $\langle \rho_{A2} \rangle$ is the average concentration of the contaminant in the boundary layer "stream" departing the pile vicinity, gA/cm$^3$.

The vignette model is complete at this point. Useful results include simple alogorithms for predicting the contaminant leaching time, Eq. 7, the flux rate from the cutting and mud pile, Eq. 6, and the average concentration in the boundary layer "stream", Eq. 16. After discussing limitations of the model the remaining sections of the manuscript will be devoted to selecting proper transport and equilibrium parameters for the marine environment and demonstrating the model with example calculations based on chemicals in the mud and cutting sediment layer.

Limitations and Caveats of the Vignette Model

The model is limited in that it considers only the transport of the soluble consituents from the cutting and mud deposits. Particulate transport is an important process in the overall loss of contaminants from sediment. This has been alluded to only when addressing the spread of solids on the seabed by periodic storm events. The more difficult problem of modeling the scour and resuspension of particles to the water column was ignored. A truly realistic model of the fate of this material must have the capability of yielding estimates of chemical concentrations in water due to both the soluble and particulate fractions. The approach taken by Pavlou, et at. (1982) in estimating sediment entrainment may be extended to account for the concentration of the particulate fraction.

The vignette model is not a true gradient model that results when applying Fick's second law. The vignette model contains a time varying, linearized distance gradient which simplifies the mathmatics greatly. During the early stages of the leaching process the vignette model rate prediction is virtually indentical to those of the Fickian slab model, however in the later stages of the leaching process the algorithms of the vignette model are only rough approximations.

The vignette model overestimates the leaching rate and under-estimates the leaching time. At the time the vignette model predicts leaching to be completed (i.e., 100%) the Fickian slab model predicts only approximately 80% leached. Neither model accounts for the movement of contaminates downward futher into the sediment. This process of downward movement undoublty occurs in nature. It therefore appears that both models will underestimate the leaching time by a significant factor. For these reasons the times, rates and concentrations appearing in Tables 2 through 6, for leached values of 80 % and greater, are likely very crude approximations of the real values.

The vignette model does not account for chemical or biochemical transformations that may degrade contaminats within the sediment zone or boundary layer. It is likely that some aerobic or anaerobic processes may be occuring. Bioturbation processes are also excluded from the model. Bottom sediment mixing is handled in a very simplistic fashion as is presented in the next section.

Although the model is based on processes known to occur in the marine environment (Boudreaux and Guinasso, 1982; Duursma and Smies, 1982; Lerman, 1982; Maris Bender, 1982) there are insufficient data for rigorous validation exercises. Laboratory simulations of the basic fate and transport processes involving waste drilling solids in the sediment and benthic boundary layer region are apparently lacking. A limited amount of field data is available that allows some crude inferences abut general model validity. These are briefly discussed in the results section with respect to specific chemical species.

PARAMETER ESTIMATES

The model is very general in its final form. It can be applied to a variety of chemicals and aquatic sediment environments. The chemicals can be organic and metal species. The aquatic setting must be flowing but it can be fresh water or marine waters. Parameter estimates in this section are chosen to simulate chemical transport from oil drilling cutting and mud waste in bottom deposits of the Gulf of Mexico.

Chemical Parameters

Two parameters are required that reflect the specific chemical molecule or ion properties. The molecular diffusivity of the chemical species in seawater at the temperature and pressure of sediment-water interface is required. The other chemical parameter is the equilibrium partition coefficient between the cutting/mud solids and the adjoining porewater. Table 1 gives the $D_{A2}$ and $K_{A32}$ values used in this simulation. The molecular diffusivity of benzene in water, equal to 1.02E-5 cm$^2$/s, was chosen as a basis. The diffusivities of the other organic species were estimated using Graham's law. Values of diffusivities of metal ions were obtained from Lerman (1979). Various sources were used to obtain estimate the sediment-water partition coefficients. Experimental values were used when available. When organic carbon-water (i.e., Koc) values were available a 2% organic carbon content in sediment was assumed. When experimental values of the octanol-water (i.e., Kow) values were available Karickhoff's (1981) relation was used to estimate Koc.

Table 1  Diffusivities and Partition Coefficients

| Chemical | Diffusivity $D_{A2}$ (cm$^2$/s) | Partition Coefficient $K_{A32}$ (g/cm$^3$) | Source of $K_{A32}$ |
|---|---|---|---|
| Benzene | 1.02E-5 | 1.94 | (Chiou, 1979) |
| Naphthalene | 7.96E-6 | 1070 | (JRB, 1984) |
| Heptane | 9.0E-6 | 276 | (Lyman, et al., 1982) |
| Phenanthrene | 6.75E-6 | 348 | (JRB, 1984) |
| Chromium | 6E-6 | 300000 | (Delos, et al., 1984) |
| Lead | 7.59E-6 | 200000 | (Delos, et al., 1984) |
| Generic | 1E-5 | 0 | - |

## Marine Parameters

Marine parameters are: the background concentration of the chemical in water, $\rho_{A2}$, water depth, H, temperature, $T_2$, and viscosity, $\nu_2$. In all the calculations a zero background concentration and 25° was used. Water depths of 10m to 200m were assumed. For this range of depths a viscosity of 0.011 cm$^2$/s (at 25°) was chosen. Other marine parameters that characterize turbulent transport in the benthic boundary layer are friction velocity, $\nu_*$, roughness height, $y_o$, and mass-transfer coefficient, $k_{A2}^3$. The studies performed by Adams, et al. (1982) on the Louisiana Continental Shelf were used for $\nu_*$ and $y_o$. Values of $\nu_* = 0.6$ cm/s and $y_o = 1.3$ cm were used in all calculations. The formulas for the mass transfer coefficient at the sea floor compiled by Boudreau and Guinasso (1982) were reviewed and that attributed to Sideman and Pinczewski was chosen. The equation is $k_{A2} = D_{A2} \nu_*/\nu_2$ and is applicable for Schmidt Numbers of approximately 1000.

## Cutting and Mud Sediment Parameters

Barite is a mineral commonly used in drilling fluids to increase the mud density. It has a density of approximately 4.5 g/cm$^3$. Discharges to the ocean are typically 65% solids and have bulk densities approaching 1.7 g/cm$^3$. Most natural particles have a density of 2.5 g/cm$^3$. An average particle density of 3.0 g/cm$^3$ was chosen to represent the solid cutting and mud solids. Although most bottom surface sediments have a porosity of 0.6 and higher (Lerman, 1979) a value of 0.5 was chosen because of the very small solid particles in the mud. The effective diffusion coefficient in sediment is a function of porosity. Lerman (1979) recommends an equation that is related to the square of the porosity: $D_A3 = D_A2 \, \varepsilon^2$.

Estimates of chemical concentrations in sediment and porewaters were also obtained from various sources. Fang and Smith (1985) present data on the concentration of organic substances in water discharged to the ocean from a sediment cleaning operation tested in the Gulf of Mexico. The total hydrocarbon content of the sample was 3,480 g/m$^3$. The reported benzene and heptane concentrations were 158.5 and 12.0 g/m$^3$ respectively. The NRC study (Costlow, et al.; 1983) reported 20.1 ppm napthalenes and biphenyls in a sample of drilling fluid which was treated with diesel fuel. The above organic chemical concentrations were used in the model calculations. Besides barium, other metals such as chromium, iron, lead and mercury, are restricted to the near-rig region and show moderate to no concentration elevation at distances from the rig (Costlow, et al.; 1983). A 3.33 part per million concentration of chromium and lead on solids was chosen for calculation purposes.

Cutting and mud solids deposited upon the seabed may be moved about by bottom currents. Periodic storm events can effectively transport sediments away from the rig site. Two geometric configurations were chosen for the bottom sediment. In the "low-energy" case the waste deposit was assumed to be 30 cm. in depth and occupy an area of 100 m. x 100 m. In the "high energy" case the waste deposit was assumed to be 1 cm. in depth and occupy an area 3 km x 3 km. Chemical concentrations for the "high energy" bottom case were reduced by a factor of 30 to maintain the same mass of chemical as the "low energy" case.

RESULTS AND DISCUSSION

Selected equations were coded in BASIC and calculations of the vignette model were performed on personal computers. Simulations were performed to investigate some general aspects of the transport processes in the sediment region and to obtain specific behavioral information on the seven chemicals listed in Table 1. The following list identifies the important equations and summarizes the quantitative information they deliver:

> Eqn. 1; maximum flux rate from the sediment,
> Eqn. 2; maximum concentration in water,
> Eqn. 4; sediment depth when pore diffusion controls,
> Eqn. 7; time lapse until pore diffusion controls, and total leaching time,
> Eqn. 6; flux rate during the pore diffusion period,
> Eqn. 10,11 boundary layer height at the edge of pile,
> Eqn. 14; average water velocity in the boundary layer,
> Eqn. 15; leaching rate from the pile, and
> Eqn. 16; concentration in the boundary layer.

The following contains some numerical results and a discussion of the results.

General Aspects of the Transport Process

A generic chemical was chosen in order to explore the purely transport aspects of the leaching. This required that the substance have a partition coefficient of zero. For an initial pore water concentration of 100g/m$^3$, and low energy bottom current conditions, the initial leaching rate is 102 g/min. This rate is controlled by the low resistance that exist on the water side. The time lapse until pore diffusion (i.e., sediment processes) controls is 13 minutes. At this time the leaching rate is 5.2 g/min. The depth to the diffusion plane when this occurs is 0.87 mm. The total leaching time, that is, the time lapse until all the chemical has departed the sediment layer, is 2.85 years. It appears that the in-sediment processes must be well understood and properly quantified if realistic model predictions are to be made. This same general behavior was observed in calculations with the other chemicals modeled.

Turbulent processes in the water above the cutting and mud pile contribute significantly to the chemical concentration in the benthic boundary layer. For conditions in which $v_* = .6$ cm/s and $y_o = 1.3$ cm the layer reaches a height of 6.2 meters at the far edge of the pile and has an average velocity of 5 cm/s. The height is independent of $v_*$ but is function of the roughness height, $y_o$, and the length of the pile, 1. A similar dependence on these variables was observed in using the von Karman integral method with the one-seventh power velocity profile. It appears that realistic predictions of water concentrations in the pile vicinity are related to a proper quantification of turbulent transport parameters in the boundary layer immediately above the pile.

Specific Leaching Characteristics of Chemicals

Calculated results of various leaching related parameters for benzene appear in Table 2. This simulation is for the "low energy" bottom case and relates to cuttings and mud solids which remain underneath or very near the rig. Bottom currents are low and deposited solids are assumed not to move beyond a 100 m x 100 m area. The first two line entries in the table are for the condition when the water-side resistance controls (line 1) and when the sediment-side resistance controls (line 2). The sediment-side controls in all other entries in the table.

Initially the benzene leaching rate is rapid, but the rate falls by an order of magnitude in less than an hour and two more orders of magnitude in seventy days. It is not suprising that benzene is diffi- cult to detect in the benthic boundary layer above waste solid piles. The results indicate that quantities of benzene will persist within the pile for nineteen years at which time the concentration in water is 44 ppt.

Table 2 Benzene Leaching from a  30 cm. Pile

| Percent Leached | Time (y) | Rate (g/h) | Boundary Layer Concentration $(g/m^3)$ |
|---|---|---|---|
| 0 | 0 | 533(g/min) | 2.87E-1 |
| .29 | 1.4(h) | 27.3(g/min) | 1.47E-2 |
| 10 | 70(d) | 48.9 | 4.37E-4 |
| 20 | .764 | 24.4 | 2.19E-4 |
| 30 | 1.72 | 16.3 | 1.46E-4 |
| 40 | 3.05 | 12.2 | 1.09E-4 |
| 50 | 4.76 | 9.78 | 8.75E-5 |
| 60 | 6.88 | 8.15 | 7.29E-5 |
| 70 | 9.35 | 6.98 | 6.25E-5 |
| 80 | 12.2 | 6.11 | 5.47E-5 |
| 90 | 15.4 | 5.43 | 4.86E-3 |
| 100 | 19.1 | 4.89 | 4.38E-5 |

In the case where bottom currents are occasionally high the discharged cutting and mud solids are spread onto a larger area of the sea bottom and incorporated into the upper layer of natural sediment. To simulate these conditions it was assumed that the pile depth was one centimeter and the area was 3km x 3km. The same mass of benzene, 1.63E6 g, was deposited on-bottom in each energy level case. Table 3 gives the important leaching related parameters for the "high energy" case. The time and rate columns are in days and gram per second respectively, to scale this rapid leaching process. The estimated total leaching time is less than eight days. Concentration in the water are however in the same general range as during the early period of the "low energy" case. This occurs because as the leaching rate increases so does the volume of water in the boundary layer. The "high energy" case boundary layer height was 33.9 m as compared to 6.2 m for the "low energy" case. For benzene it appears that spreading the waste solids over a large bottom area aids in very rapid leaching of the contaminants from the sediment.

Table 3.   Benzene Leaching from a 1 cm. Pile

| Percent Leached | Time (d) | Rate (g/s) | Boundary Layer Conc. $(g/m^3)$ |
|---|---|---|---|
| 0 | 0 | 266 | 5.22E-2 |
| 8.71 | 1.41(h) | 13.7 | 2.69E-3 |
| 10 | 1.86(h) | 11.9 | 2.34E-3 |
| 50 | 5.80 | 2.43 | 4.76E-4 |
| 100 | 7.74 | 1.22 | 2.38E-4 |

Benzene does not sorb strongly onto solid particles. It has a partition coefficient of approximately 2 $g/cm^3$. Heptane has a partition coefficient which is approximately three hundred times larger. Table 4 contains the leaching related parameters for heptane. In the low energy environment heptane persist for approximately three thousand years. In general heptane is more persistant than benzene. Concentration levels are similar to those of benzene because of the relatively high heptane concentration on the waste solids (i.e., 3300 ppm).

Table 4.   Heptane Leaching from Piles

| Percent Leached | Time (y) | Rate (g/h) | Concentration $(g/m^3)$ |
|---|---|---|---|
| Low energy; h = 30 cm, l = w = 100m | | | |
| 0 | 0 | 2120. | 1.9E-2 |
| .29 | 8.1(d) | 109. | 9.74E-4 |
| 10 | 26.3 | 3.24 | 2.9E-5 |
| 50 | 656 | .648 | 5.8E-6 |
| 100 | 2630 | .324 | 2.9E-6 |
| High energy; h = 1 cm, l = w = 3km | | | |
| 0 | 0 | 17.6(g/s) | 3.46E-3 |
| 8.71 | 8.1(d) | 3240 | 1.77E-4 |
| 10 | 10.7(d) | 2840 | 1.55E-4 |
| 50 | .73 | 578 | 3.15E-5 |
| 100 | 2.92 | 270 | 1.58E-5 |

Naphthalene and phenanthrene are very strongly sorbed onto solid particles. Respectively, the partition coefficients are 1070 and 348 $g/cm^3$. Table 5 contains the naphthalene leaching parameters. This chemical is very persistant in the sediment, however the water concentrations are also very low. Table 6 contains the phenanthrene leaching parameters for the "low energy" sea bottom case only. It behaves somewhat similar to heptane in leaching time. The water concentrations are low because the concentration on the cutting and mud solids is only 20 ppm (wt).

Table 5.  Naphthalene Leaching from Piles

| Percent Leached | Time (y) | Rate (g/y) | Concentration $(g/m^3)$ |
|---|---|---|---|
| Low energy; h = 30 cm,  l = w = 100 m | | | |
| 0 | 0 | 69.8(g/d) | 2.61E-5 |
| .29 | 35(d) | 3.58(g/d) | 1.34E-6 |
| 10 | 115. | 39.0 | 3.99E-8 |
| 50 | 2,880. | 7.81 | 7.98E-9 |
| 100 | 11,500. | 3.90 | 3.99E-9 |
| High energy; h = 1 cm, l = w = 3 km | | | |
| 0 | 0 | 87.5(g/h) | 4.78E-6 |
| 8.71 | 35(d) | 4.50(g/h) | 2.46E-7 |
| 10 | 1.28 | 3.92(g/h) | 2.14E-7 |
| 50 | 3.2 | .80(g/h) | 4.35E-8 |
| 100 | 12.8 | .40(g/h) | 2.18E-8 |

Table 6.  Phenanthrene Leaching from Seabed Piles[*]

| Percent Leached | Time (y) | Rate (g/h) | Concentration $(g/m^3)$ |
|---|---|---|---|
| 0 | 0 | 7.61 | 6.83E-5 |
| .29 | 13.6(d) | .391 | 3.51E-6 |
| 10 | 44.1 | .278(g/d) | 1.04E-7 |
| 50 | 1100. | 20.4(g/y) | 2.08E-8 |
| 100 | 4410. | 10.2(g/y) | 1.04E-8 |

* Low energy case

Two metal species were chosen for leaching calculations.  Due to the extremely high partition coefficients of chromium and lead the leaching behavior was investigated for the "high energy" case only.  The two metals are extremely persistent in the sea bottom sediment.  The calculated total leaching time for lead was 2510 years and 4760 years for chromium.  For an assumed 33.3 ppm(wt) concentration in the sediment solids the maximum concentrations in water were calculated to be $1.22E-7(g/m^3)$ for lead and $6.39E-8(g/m^3)$ for chromium.  It is not surprising that metals are being detected in the sediment underneath and in the general vicinity of drilling rigs.

CONCLUSION

A vignette model is a short descriptive mathematical sketch emphasizing limited but important phenomena on pollutant transport and accumulation in a natural multimedia environment. This type model was developed to investigate the transport of soluble chemical contaminants from a seabed sediment to the adjoining benthic boundary layer. The model, developed to predict chemical movement rates, life-times and concentration levels, was applied to the problem of oil well drilling waste deposited onto the bottom of the Gulf of Mexico. The behavior of chemicals in this waste, such as benzene, heptane, naphthalene, phenanthrene chromium and lead, was studied. Calculation scenario were created to simulate both low and high seabed water current energy conditions.

In general it was found that the natural leaching process was dominated by processes within the bottom sediment and not in the adjoining benthic boundary layer once the material was in-place. Rates were controlled by chemical desorption from solid particles and pore diffusion from the sediment. Chemical concentration levels in the water column were dominated by water turbulence levels above the sediment. Realistic predictions of concentration and life time can be made provided the processes both within the sediment and adjoining boundary layer are theoretically sound and robust in order to reflect specific site conditions. A limited of field data is available to perform crude checks of the model predictions. Laboratory data from controlled simulation experiments with cutting and mud waste in sediment is apparently nonexistant however, data on analogous processes lend a high degree of overall validity to the vignette model.

Field data on specific chemicals, waste characteristics and marine parameters in the Gulf of Mexico was chosen in order to perform calculations with the model. The following is some of the model predictions. Benzene, which is not strongly sorbed onto sediment solid particles leaches relatively rapid to yield concentration of 4E-5 to 0.3 $g/m^3$. The leaching half-life of benzene is 5.8 days and 4.76 years, respectively for the high and low energy seabed scenario cases. Heptane is pore persistant than benzene. Its half-life is 0.73 to 660 years. Concentration levels in water vary from 2E-2 to 2E-5 $g/m^3$. Naphthalene and phenanthrene are very strongly sorbed and display half-lifes of 2900 and 1100 years for low energy seabed case. Concentrations in water were in the range of 1E-5 and lower. Due to the extremely high sorption coefficients of chromium and lead leaching half-lifes were calculated to be 1200 and 730 years respectively for the high energy seabed case.

REFERENCES

Adams, C.E., Jr., J.T. Wells, and J.M. Coleman, Contributions in
        Marine Science (1982), Vol. 25: 133-148
Boudreaux, B.P. and N.L. Guinasso, Jr., Chapter 6 in "The Dynamic
        Environment of the Ocean Floor", Edited by K.A. Fanning and
        F.T. Manheim, Lexington Books, Lexington MA(1982).
Brandsma, M.G. and R.C. Sauser, The OOC Model: prediction of short
        term fate of drilling fluids in the ocean. Part two: Model
        results. In: Proceedings of Minerals Management Service
        Workshop on Discharges Modeling, Santa Barbara, CA, Feb.
        7-10, 1983.
Buff, R.B., M.S. Thesis, University of Arkansas, Fayetteville, AR
        (1984).
Caravanos, J., Ph.D. Dissertation, Columbia University, School of
        Public Health, NY (1984).

Chiou, G.T., Peters and V.H. Freed, _Science_, Vol. 206, 16 Nov., 1979, p. 331-832.

Costlow, J.D., Chairman of Panel, "Drilling Discharges in Marine Environment", National Academy Press, Wash., D.C. 1983.

Delos, C.G., et al., Technical Guidance Manual for Performing Waste Load Allocations, Book II Streams and Rivers, Ch. 3 Toxic Substances, U.S. EPA, Wash., D.C. Aug. 1984, p, 62.

Duursma, E. K. and M. Smies, Sediments and Transfer at and in the Bottom Interfacial Layer, Ch. 3 in "Pollutant Transfer and Transport in the Sea", Volume II, G. Kullengberg, Editor, CRC Press, Inc. Boca Raton, Florida, 1982.

Fang, C.S. and Smith, Jr.,S.A., Cleaning of Ocean Floor Near Offshore Platform in the Gulf Coast, Paper No. 6b, Amer. Inst. Chem. Eng. Mtg., Houston, Texas, March 24-28, 1985.

J.R.B. Associates, Report on sediment-water equilibrium partitioning for Puget Sound. Seattle, Wash, 1984.

Karickhoff, S.W., D.S. Brown and T.A. Scott. _Water Res._, Vol. 13, 291 (1979).

Lerman, A., "Geochemical Processes: Water and Sediment Environments", J. Wiley and Sons, NY, 1979.

Lyman, W.J., et. al., 1982, "Handbook of Chemical Property Estimation Methods", McGraw-Hill, New York.

Maris, C.R.P. and M.L. Bender, Upwelling of Hydrothermal Solutions Through Ridge Flank Sediments Shown by Pore Water Profiles, _Science_, Vol. 216, 7 May 1982.

Menzie, C.A., The Environmental Implications of Offshore Oil and Gas Activities, Environ. Sci. Technol, Vol. 16, No. 8, 1982, p. 454A.

Pavlou, S.P., W. Horn, R.N. Dexter, D.E. Anderson, and E.A. Quinlan, _Environment International_, Vol. 7, pp. 99-117, 1982.

Poley and Wilkerson, (1983) (documentation source misplaced)

Thibodeaux, L.J., "Chemodynamics - Environmental Movement of Chemicals in Air, Water and Soil", Wiley, N.Y. (1979) p. 337-339.

Thibodeaux, L.J. and S.T. Hwang, _Environmental Progress_, Vol., No. 1., p. 42, Feb. 1982.

POLLUTANT TRANSPORT IN FLOW-THROUGH WETLAND ECOSYSTEMS

Robert H. Kadlec

The University of Michigan

Department of Chemical Engineering
Ann Arbor, MI   48109-2136

ABSTRACT

A simple, tractable mathematical model is developed which permits dynamic simulation of wetland hydrology and of interactions between wastewater and the wetland ecosystem. Spatial variations due to surface water flow are described, and material balance calculations carried out for phosphorus, nitrogen, and chloride. A hydrology model predicts overland flow. Ecosystem phenomena are represented, using a one-dimensional, spatially distributed compartmental model. Compartments include soil, surface water, interstitial soil water, and various types of live biomass, standing dead and litter. Solutions to the partial differential equations which comprise these spatial models are demonstrated using finite-difference methods. Computer simulations are compared to operating data from the Porter Ranch wastewater treatment facility at Houghton Lake, Michigan. They accurately predict solute concentrations in surface water, biomass growth patterns, and other major ecosystem features.

INTRODUCTION

This paper is a summary of wetland modeling work which spans the period 1970-1985, centered on advanced wastewater treatment by a wetland located near Houghton Lake, Michigan.

The Site

This community has a seasonally variable population, averaging approximately 5000. Wastewater from this residential community is collected and transported to two aerated lagoons, which provide initial treatment. Sludge accumulates on the bottom of these lagoons, below the aeration pipes. Effluent is then stored in a 12 hectare pond for summer disposal, resulting in depth variation from 0.5 meters (fall) to 3.0 meters (spring). The final disposal is to a 700 hectare peatland, which provides advanced treatment.

Wastewater is pumped through a 30 cm diameter underground force line to the edge of the Porter Ranch peatland. There the transfer line

surfaces and runs along a raised platform for a distance of 800 meters
to the discharge area in the wetland. There the wastewater may be split
between two branches of the discharge pipe which runs 500 meters in each
direction. The water is distributed across the width of the peatland
through small gated openings in the discharge pipe. Each of the 100
gates discharge approximately 100 cubic meters per day, under typical
conditions, and the water spreads slowly over the peatland. The
wastewater progresses eventually to the Muskegon River twelve kilometers
away. The peatland irrigation site originally supported two distinct
vegetation types. One called the sedge-willow community included
predominantly sedges (Carex spp.) and willows (Salix spp.). The second
community was leatherleaf-bog birch consisting of mostly Chamaedaphne
calyculata (L.) Moench and Betula pumila L., respectively. The
leatherleaf-bog birch community also had sedge and willow vegetation,
but only in small proportions. Standing water was present at most times
throughout most years, with lower levels in late summer. The
leatherleaf-bog birch cover type generally had less standing water than
the sedge-willow cover type. Soil in the sedge-willow community is one
to two meters of a highly decomposed sedge peat, while in the
leatherleaf-bog birch area there is two to five meters of sphagnum peat,
with a medium degree of decomposition.

Small, natural water inflows occur on the north and east margins of
the wetland. Overland flow proceeds from northeast down a 0.02%
gradient to a stream outlet (Deadhorse Dam) and beaver dam seepage
outflow (Beaver Creek), both located three kilometers from the
discharge. Wastewater adds to the surface sheet flow.

The treated wastewater arriving at the peatland is a good effluent
which contains virtually no heavy metals or refractory chemicals. This
is due to the absence of agriculture and industry in the recreational
community. Phosphorus and nitrogen are present at 3-5 ppm, mostly as
orthophosphate and ammonium. BOD is about 15 ppm, and solids are about
20 ppm. Typical levels of chloride are 100 ppm, pH 8, and conductivity
700 μmho/cm. The character of the water is dramatically altered in its
passage through the wetland. Phosphorus, nitrogen, pH, BOD and
conductivity are reduced to background levels. Chloride passes through
the ecosystem without significant retention.

## Wetland Simulation

Many investigators have developed detailed compartmental models,
often with great complexity, but very few spatially distributed models
have emerged (Mitsch, 1983). The mobility of waterborne components has
been neglected in most wetland models. In the case of wastewater
irrigation, such hydraulically driven transport is very important. From
the mathematical standpoint, this has the effect of converting the
wetland model from a set of ordinary differential equations to one which
also contains partial differential equations.

Hydrological models of wetlands have been reviewed by Mitsch, et
al., (1982). They describe three basic types: 'ecosystem' models and
'regional' models which simply account for water inventory by material
balance, and 'hydrodynamic transport' models which describe stream flow
and storm runoff. None of these approaches address thin sheet
hydrology.

Prior modeling efforts which are applicable to wastewater/ecosystem
interactions include the spatial wetland model of Parker (1974). The
computer simulation routine, REBUS (Routine for Executing Biological

Unit Simulations), utilizes a compartmental representation of the ecosystem, and combines this with a topological structure which divides the wetland into spatial blocks. Each block is assumed to be homogeneous throughout (e.g., uniform water level, solute concentrations, biomass, etc.) and connected to other blocks by flow streams. Detailed and quite complex models for vascular plants, standing dead material, litter, and soil have been developed for use with the REBUS simulator (Dixon, 1974).

Subsequently, six years of experience with wastewater irrigation at the Porter Ranch site (Houghton Lake, MI) identified a number of limitations in REBUS. First, hydrological predictions were poor. Data had not been available to adequately test the model. Secondly, the compartmental models did not predict the observed trends in plant biomass, despite the considerable complexity of the submodels. Finally, some important wetland entities were not incorporated into the REBUS model. These included interstitial soil water, which is now considered to be of substantial importance.

## DETAILED BALANCE EQUATIONS

The components of the peatland are the water sheet, the soil, and several categories of biomass, as depicted in Figure 1. Each of these exhibits unsteady spatial variability in three dimensions, both in total material and individual constituent species. Transfers between compartments are limited in number, but occur at greatly different speeds. Advective transport is the most rapid, while soil accretion represents the slowest rate process. Length scales are also quite different. Horizontal distances are normally measured in kilometers or fractions thereof, but vertical scales are much smaller. Plants inhabit

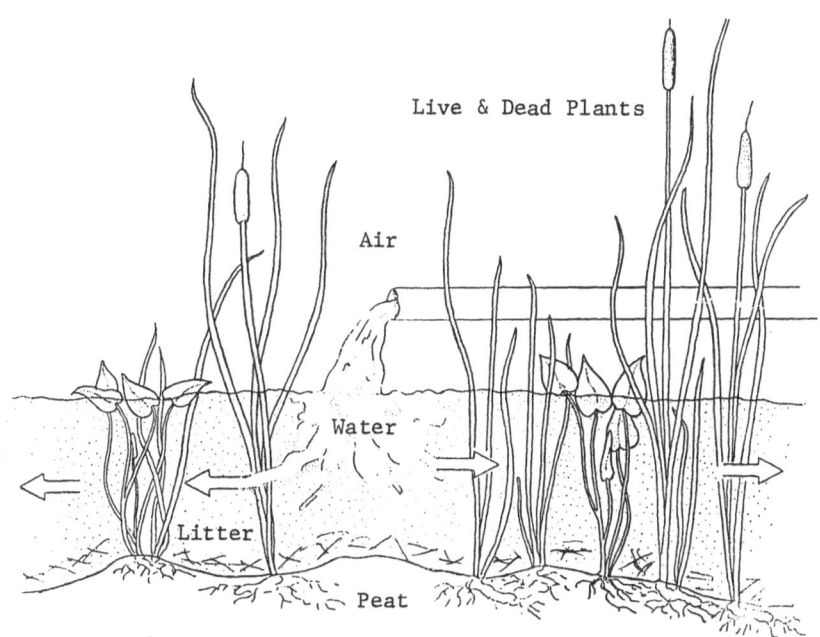

Figure 1. Peatland ecosystem compartments. From Hammer and Kadlec, 1981.

vertical distances measured in meters, while the active soil zone may comprise only the top few centimeters. Wetlands, by definition, have water depths which do not normally exceed one or two meters, and are frequently less than one meter. These disparities may be exploited to some extent, to develop simplifications in the mathematical models. But they also lead to computational difficulties due to the resulting stiffness of the describing equations.

Each major compartment will next be examined in turn, with simplifications for the general wetland situation, and further for the specific site for which data fitting and validation have been attempted.

## The Water Sheet

Total Flow. The movement of water in the ecosystem is governed by mass conservation in the form of the equation of continuity:

$$- \vec{\nabla} \cdot (\phi \rho \vec{v}) = \frac{\partial}{\partial t} (\phi_s \rho) \tag{1}$$

where a distinction has been made between the volume fraction available for flow ($\phi$) and the total void volume fraction ($\phi_s$). Figure 2 illustrates the porous structure of a wetland. Total saturation is presumed to occur within the water sheet. Concentrations of dissolved and suspended materials are herein presumed to be low--in the parts per million range. Thus, no concentration effects alter the behavior of essentially pure water. Given the present state of knowledge of wetland structure, and the relative thinness of the water sheet, it is profitable to use vertical averaging, via depth integration:

$$\int_o^h - \vec{\nabla} \cdot (\phi \rho \vec{v}) dz = \int_o^h \frac{\partial}{\partial t} (\phi_s \rho) dz \tag{2}$$

which reduces to:

$$- \frac{\partial}{\partial x} (\phi \rho v_x h) - \frac{\partial}{\partial y} (\phi \rho v_y h) - [\phi v_z \rho]_o^h = \frac{\partial}{\partial t} (\phi_s \rho h) \tag{3}$$

where $\phi$, $\phi_s$, $v_x$ and $v_y$ now represent vertical averages. The natural mass influx at the upper water surface is the excess of precipitation over evapotranspiration, and the mass flux downward into the soil is the infiltration. Pumped additions may also be present. Density may be regarded as constant, leading to:

$$\frac{\partial}{\partial x} (\phi v_x h) + \frac{\partial}{\partial y} (\phi v_y h) + \frac{\partial}{\partial t} (\phi_s h) = p + a - i - e \tag{4}$$

For the Houghton Lake peatland, infiltration has been determined to be negligible (Haag, 1979). This flow system is also nearly one-dimensional, resulting in the site-specific relation:

$$\frac{\partial}{\partial x} (\phi v h) + \frac{\partial}{\partial t} (\phi_s v h) = p + a - e \tag{5}$$

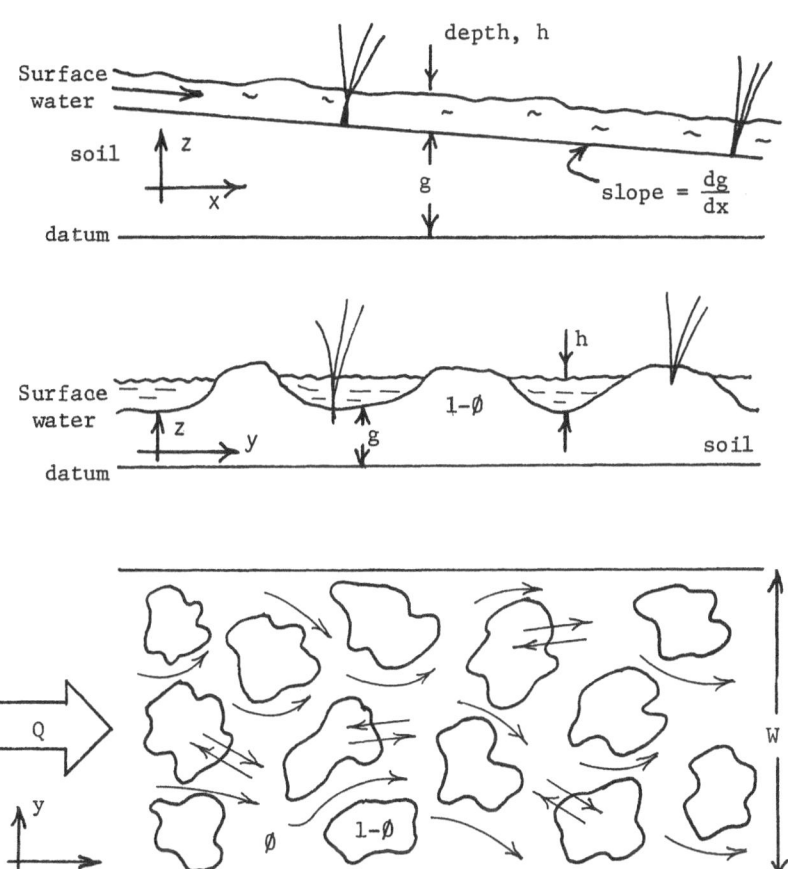

Figure 2. Flow configuration for a peatland ecosystem.
From Hammer and Kadlec, 1983.

A rate law is required for the calculation of water velocity. The boundary flows p and e are computable from meteorological data, and any pumped addition will be known. Some starting condition must be known or presumed. The upstream and downstream boundary conditions will be addressed separately, in a subsequent section.

The site-specific rate law for flow has been determined to be (Kadlec, et al., 1981)

$$\phi v = \varepsilon h^\beta \ (-\frac{ds}{dx}) \qquad (6)$$

where s = g + h = elevation of the water surface. This correlation covers both frictional and porosity effects, in that it calculates only the superficial velocity, $\phi v$. The exponent $\beta$ is approximately equal to two for this site.

<u>Dissolved Components</u>. For species i, which may undergo bulk transport, diffusion, chemical reaction, and transfer to or from another phase:

$$-\vec{\nabla} \cdot (\delta \vec{v} C_i - D_e \vec{\nabla} C_i) - R_i - \sum_j T_{ij} = \frac{\partial}{\partial t} (\delta_s C_i) \qquad (7)$$

If the species do not undergo homogeneous chemical reaction, which is the case for many nutrients, pollutants and simple inorganic ions, then $R_i = 0$. The $T_{ij}$, which are the volume-specific transfers to or from other immersed compartments, include leaching of above-ground vegetation, utilization by attached and suspended organisms, and sorption by suspended solids.

Lateral diffusion is expected to be negligible compared to bulk flow for a 'flow through' wetland. Vertical averaging is again dictated by water sheet thinness:

$$-\frac{\partial}{\partial x} (\delta v_x h C_i) - \frac{\partial}{\partial y} (\delta v_y h C_i) - [N_i]_o^h - \sum_j h T_{ij} = \frac{\partial}{\partial t} (\delta_s h C_i) \qquad (8)$$

where $\delta$, $\delta_s$, $v_x$, $v_y$, $T_{ij}$ and $C_i$ now represent vertical averages.

Surface fluxes are represented by mass transfer to the soil phase as well as by precipitation and infiltration. Evapotranspiration would not normally transport a dissolved species, and volatile constituents are not considered here. Consequently, the general wetland situation is represented by:

$$\frac{\partial}{\partial x} (\delta v_x h C_i) + \frac{\partial}{\partial y} (\delta v_y h C_i) + \frac{\partial}{\partial t} (\delta_s h C_i) =$$

$$\qquad (9)$$

$$pC_{ip} + aC_{ia} - iC_{ii} - \sum_j h T_{ij} - N_{is}$$

For the Houghton Lake peatland, a one-dimensional version is adequate, infiltration is not present, and only one significant interphase transfer is thought to be present. That transfer is the utilization of nutrients by suspended organisms of short life spans, such as algae and bacteria. Thus, for this specific site:

$$\frac{\partial}{\partial x} (\delta v h C_i) + \frac{\partial}{\partial t} (\delta_s h C_i) = pC_{ip} + aC_{ia} - U_i h - N_{is} \qquad (10)$$

Solution of these equations requires initial and boundary conditions (to be discussed later), and the surface and volume fluxes. Precipitation and pumped additions are of known amount and concentration. The transfer rate to the soil surface has been found (Kadlec and Hammer, 1982) to be:

$$N_{is} = k(C_i - C_{is}) \qquad (11)$$

$$k = \gamma (\delta v h)^\delta \qquad (12)$$

Equation (12) is a modified form of the usual Reynolds number–Schmidt number correlations for mass transfer coefficients (Hammer and Kadlec, 1983). The exponent $\delta = 1$ for this site. The volumetric utilization $U_i$ is zero for biologically inactive constituents, for example chloride ions. For nutrients, such as ammonium and orthophosphate ions, an uptake is typically present. For a stable fully–developed consumer population, a constant (Zero–order) rate might be expected, and can explain the observed phenomena (Kadlec, 1985a). For growth–phase conditions, a biomass expansion model must be employed to compute the $U_i$ (Schwegler, 1978).

Suspended Components. Particulate material will be present in the water, and may be either organic or inorganic in character. It is presumed to move with the water, with possible relative motion:

$$- \vec{\nabla} \cdot (\phi \vec{v}_s C_s) + G_s - T_s = \frac{\partial}{\partial t} (\phi_s C_s) \tag{13}$$

The possibilities of biological generation ($G_s$) and volumetric filtration, or trapping, ($T_s$) have been included. Vertical averaging, together with horizontal non–slip conditions ($v_{sx} = v_x$, $v_{sy} = v_y$) yield:

$$- \frac{\partial}{\partial x} (\phi v_x h C_s) - \frac{\partial}{\partial y} (\phi v_y h C_s) - [N_s]_0^h + G_s h - T_s h = \frac{\partial}{\partial t} (\phi_s h C_s) \tag{14}$$

where, as before, $C_s$ and other quantities represent vertical averages.

At the upper water surface, atmospheric dry fall may occur, in the form of dust, plant debris and invertebrate herbivory by–products. Precipitation can scour the atmosphere and produce a wetfall of particulates. At the soil surface, the processes of gravity–driven deposition and resuspension can occur simultaneously, together with filtration if there is an infiltration water flow. For the general wetland ecosystem:

$$\frac{\partial}{\partial x} (\phi v_x h C_s) + \frac{\partial}{\partial y} (\phi v_y h C_s) + \frac{\partial}{\partial t} (\phi_s h C_s) =$$

$$aC_{as} + pC_{ps} + f - iC_{is} - S + R + G_s h - T_s h \tag{15}$$

For the Houghton Lake peatland, $i = 0$ and there are negligible atmospheric inputs. This is a one–dimensional wetland flow, with apparently very little volumetric filtration ($T_s = 0$). Consequently:

$$\frac{\partial}{\partial x} (\phi v h C_s) + \frac{\partial}{\partial t} (\phi_s h C_s) = aC_{as} + G_s h - S + R \tag{16}$$

Solution of this equation requires initial and boundary conditions, together with the right–hand–side functions. Additions ($aC_{as}$) must be known. The gravitational sedimentation rate is, to a first approximation, the product of concentration and a settling velocity:

$$S = uC_s \tag{17}$$

Generation of suspended material is related to the nutrient status of the water as well as the vegetation type (Kadlec, 1985a). It is the parent function for the utilization functions $U_i$, in that organisms are constructed from carbon and nutrients. These subsequently die and form a source of suspended material. Thus, Michaelis-Menten rate laws may well apply, but a zero order, nutrient driven rate appears to fit the data.

Resuspension in wetlands is neither well understood, nor much studied. To a first approximation, it appears to be proportional to the water velocity (Kadlec and Hammer, 1982) and to the amount of settled material on the soil surface (Kadlec, 1985a).

$$R = \varepsilon(\phi v)B \tag{18}$$

## The Soil

Total Solids. The soil is here defined as the immobile, consolidated substrate underlying the peatland ecosystem. The upper surface is not easily defined, but does not include suspendable material, or loose plant litter. The elevation of the upper surface (g) must be defined by some ad hoc procedure, such as the bottom elevation of a standard surveyors staff when placed on the wetland surface. Even then, local variability can be very large, requiring multiple measurements and statistical interpretation.

In general, continuity requires:

$$-\vec{\nabla} \bullet (\rho_d \vec{v}_d) - D_d + D_R = \frac{\partial \rho_d}{\partial t} \tag{19}$$

This may be simplified to one vertical dimension due to the absence of lateral transport:

$$\frac{\partial}{\partial z}(\rho_d v_d) + \frac{\partial \rho_d}{\partial t} = D_R - D_d \tag{20}$$

The death rate of roots ($D_R$) is included, and is a function of vertical distance. This is a coupling of the biomass models. The soil decomposition rate ($D_d$) is a decreasing function of depth, with very little soil loss occurring at depths below 10 cm, due to the lack of the requisite oxygen, which is in turn due to the water blanket.

Adaptation to the Houghton Lake site involves only the experimental observation that the density is nearly constant, leading to:

$$\frac{\partial v_d}{\partial z} = \frac{D_R}{\rho_d} - \frac{D_d}{\rho_d} \tag{21}$$

Solution of this equation requires an initial condition, and a moving upper boundary condition. The accumulation flux of litter and sediment decay products, less the decomposition-driven subsidence rate, yields the net upward (or downward) movement of the ground level, g. Since this is in the range of millimeters per year (radioisotope methods), the effect on the water coordinate system is slight except for long times.

The required decomposition function, $D_d$, has been inferred from litter decomposition and soil surface measurements at this site. Only coarse detail is possible, yielding a 10:1 rate reduction over two ten-centimeter depth increments.

Structural Components of the Solids. The effect of exchange and decomposition processes upon the chemical composition of soil solids is even less understood then the gross accretion (or subsidence) of soil. Until further knowledge becomes available, the only recourse is to impose measured or presumed concentration profiles on the soil column. Significant variability does exist: several-fold concentration changes occur over a few centimeters.

Interstitial Water. Equation (1) applies, with appropriate alteration of the porosities. There are still two void fractions: one available for flow, and one occupied by an impermeable gel phase. Darcy's law is presumed to apply (Hemond and Fifield, 1982). However, the peat at the Houghton Lake site is several orders of magnitude less permeable than the biomass it supports, and subsurface flow is thus negligible.

Dissolved and Adsorbed Species. In the most general case, there will be tranfers between the interstitial water and the sorption surfaces on the peat particles, requiring separate balance equations for each phase. However, interstitial water movement is often so slow that local sorption equilibrium exists. Then a single component balance may be coupled with the equilibrium statement:

$$-\vec{\nabla} \cdot (\vec{v}\phi C_i - D_e\, C_i) - R_i - P_i = \frac{\partial}{\partial t}\, (\phi_s C_i + (1-\phi_s)\rho_d C_{Ai}) \qquad (22)$$

$$C_{Ai} = f(C_i) \qquad (23)$$

where $D_e$ is the effective diffusion coefficient within the porous matrix. Reactions are not likely for many chemical species of interest; but can be important for some, as for the consumption of nitrate by denitrifying bacteria (Engler and Patrick, 1974). The most important transfer for many dissolved species is the uptake by plants via the root system $(P_i)$. Lateral diffusion is of little importance due to the extreme differences in vertical and horizontal distance scales.

In addition, the specific site for this study has negligible underground flow. Therefore:

$$D_e\, \frac{\partial^2 C_i}{\partial z^2} = P_i + R_i + \frac{\partial}{\partial t}\, (\phi_s C_i - (1-\phi_s)\rho_d f(C_i)) \qquad (24)$$

The diffusion coefficient is reduced by the porous matrix, due to void fraction and tortuosity. The effective diffusivity $(D_e)$ is thus 2-10 times smaller than the molecular diffusivity (Patrick and Reddy, 1976; Kadlec, 1984a). The sorption isotherms are known for some constituents at this site: chloride, phosphate, ammonium and copper (Kadlec and Rathbun, 1984; Hammer and Kadlec, 1980). These are representable by power law (Freundlich) models.

Biomass Compartments

The live and dead above ground biomass, coupled with the below-ground plant parts, comprise the vegetative biotic components of

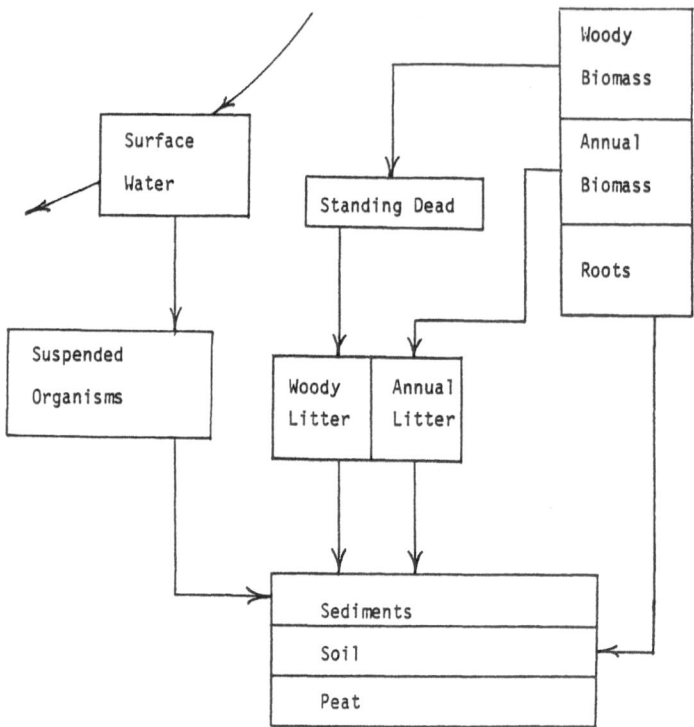

Figure 3.  Simplified compartmental model of the wetland.  Adapted from Kadlec and Hammer, 1985a.

the ecosystem. Certain subsets of this material are conveniently separable and identifiable in the field. One such compartmentalization is shown in Figure 3, which has been found useful at the Houghton Lake site. Live emergent and submerged plants are separated into woody biomass and annual biomass. Wood has a considerably longer life span than leaves, grasses, sedges and other annually senescent plants. The mobile aquatic plants, algae, fungi and bacteria are lumped in a suspended organism category.

These three compartments undergo death processes, but at significantly different time scales. Woody plants live for years, the annuals for months and smaller organisms for only days in many cases. The woody dead material remains standing for significant time periods, which can be several years. Annual biomass is transferred directly to a litter pool in some cases (leaves, for instance), but may remain standing for up to a year (cattail leaves, for example). Standing dead also eventually transfers to a litter pool. The litter compartment must be divided on the basis of the decay rates of its constituents. Woody litter decays at much slower rates than leaf litter (Chamie. 1975). Suspended organisms die and contribute to the suspended solids compartment.

The decomposition of litter contributes directly to soil building, and sedimentation of fine organic detritus contributes also. Roots die, and thus contribute directly to soil accumulation. There is no convenient technique to measure root litter separately from the peat matrix; thus soil is defined to contain dead roots at all stages of decomposition.

Very detailed models can be constructed to describe the growth, death and accumulation of these biomass groupings. Dixon (1974) constructed a four compartment model containing 47 equations, requiring 80 parameters and several time function specifications. In view of the overall complexity of the ecosystem equations, including those in the preceding sections, it is necessary to use a simpler approach, so that the simulation is of feasible size.

A balance equation may be written for each stationary compartment:

Woody
$$\frac{\partial B}{\partial t} = G_B - D_B \tag{25}$$

Annual
$$\frac{\partial A}{\partial t} = G_A - D_A \tag{26}$$

Roots
$$\frac{\partial R}{\partial t} = G_R - D_R \tag{27}$$

Standing Dead
$$\frac{\partial S}{\partial t} = D_B - D_S \tag{28}$$

Annual Litter
$$\frac{\partial L}{\partial t} = D_A - D_L \tag{29}$$

Woody Litter
$$\frac{\partial W}{\partial t} = D_S - D_W \tag{30}$$

and, the suspended organisms are subject to accounting according to:

$$-\vec{\nabla} \cdot (\phi \vec{v}_o C_o) + G_o - D_o = \frac{\partial (\phi_s C_o)}{\partial t} \tag{31}$$

For the one-dimensional site in question:

$$\frac{\partial}{\partial x} (\phi v_o C_o) + \frac{\partial (\phi_s C_o)}{\partial t} = G_o - D_o \tag{32}$$

Integration in the vertical direction yields the averaged equation:

$$\frac{\partial}{\partial x} (\phi v_o C_o h) + \frac{\partial (\phi_s C_o h)}{\partial t} = G_o h - G_s h \tag{33}$$

where it has been presumed that the death of suspended organisms results in the generation of suspended organic solids.

The parameters and rate laws needed here are many less than in more detailed models. First order dependencies of all G and D functions, with nutrient-mediated rate constants, appear to work reasonably well (Hammer, 1984). Mediation of growth was presumed to follow a simple set of rules:

1. Plants grow at the maximum rate permitted by the nutrient supply in the water in the root zone, but not exceeding an upper limit determined by the maximum carbon fixation rate.
2. The growth rate is determined by the limiting nutrient (the one in shortest supply).
3. Nutrients can be extracted only at concentrations above a threshold value in the interstitial water. This value may be zero, indicating that the plant cannot break down the soil to extract structural nutrients.

The effect of these rules is similar to a Michaelis-Menten rate law. However, there are far fewer parameters, and over-use of nutrients is prevented. There must, in addition, be seasonal modification of growth: the rules above apply only during the growing season, and growth is zero at other times.

Seasonal restrictions also apply to the death rates and fall rates for litter. Leaf senescence occurs over the fall period. Standing dead enters the litter compartment at all times, but preferentially due to snow pressure. The rates of such processes are difficult to measure, and must be indirectly obtained from biomass measurements on the compartments involved (Kadlec and Hammer, 1985a).

Constituents of the Biomass. The chemical composition of plants changes in response to the soil-water chemistry in the root zone. Changes in foliar nutrient content can vary by $\pm30\%$ in wetland plants under different nutrient regimes. Likewise, trace elements are taken up to differing degrees by different plants, and in amounts dependent on water content. Lack of knowledge of these processes limits the ability to calculate responses; hence the only avenue open is to impose measured or presumed compositions on the biomass compartments.

INITIAL CONDITIONS AND BOUNDARY CONDITIONS

The calculation of the wetland ecosystem behavior requires the solution of equations (5), (10), (16), (21), (24), (25)-(30), and (33); together with composition specifications for the solids compartments. The initial state, boundary conditions and driving functions must all be specified as well.

Initial Conditions

In simulation of small-scale processes, characterized by short time-constants, the choice of initial conditions is not critical because their effect soon disappears in response to system driving forces. Often, the response of a system is measured with reference to an initial steady-state. In ecosystem simulation, two difficulties occur that require special attention: initial values do not 'wash out' quickly, and the initial state is nearly periodic on an annual basis.

The chief culprits in delaying wash-out are woody biomass and litter, because of the long characteristic times involved in their growth and death. Dixon (1974) found that time periods of 5-10 years were required for his detailed model to reach a stationary state. The cost of this evolutionary production of a stationary starting condition is prohibitive for complex, spatially-varying systems. It is therefore necessary to establish a consistent stationary-state starting condition, from which a starting point state may be selected. Gupta (1977) and Hammer (1984) provide computational procedures for direct calculation of periodic stationary states. Gupta solved the entire set of coupled ecosystem equations, while Hammer separated out biotic compartments and adjusted behavior to match nutrient supplies.

These baseline calculations have intrinsic value, apart from initial state specification. Attempts to balance ecosystem functions elucidate the rather narrow ranges for certain parameters. For example, soil accretion cannot exceed the nutrients available for its construction. Litter decay must be fast enough to avoid buildups in excess of observations. Friction law coefficients must be such as to provide reasonable water depths. These and other constraints serve to guide parameter estimation.

## Boundary Conditions

Because the biomass compartments do not require boundary conditions, due to immobility, attention is entirely on water flow and chemical composition at system boundaries. Further, any inflow must, in theory, be completely specified for simulation purposes. However, this may not always be convenient, or even possible, as in the case of overland flow into the wetland from surrounding uplands. In that case, driving forces may become parameters to be estimated from simulation and indirect data.

Boundary conditions for water movement at outflow points generally fall into one of five categories, illustrated for the one-dimensional case:

1.  Impermeable shore: $\frac{\partial h}{\partial x} = 0$ (34)

2.  Weir control: $Q = a_w (h-h_w)^{b_w}$ (35)

3.  Controlled flow: $Q = Q_{BC}$ (36)

4.  Controlled depth: $h = h_{BC}$ (37)

5.  Pool and Weir: $Q = A_p \left( \frac{\partial h}{\partial x} + e - p \right) + a_w (h-h_w)^{b_w}$ (38)

Concentrations of water-borne material at these outflow points are computed from the model in the course of simulation.

## TIME AND DISTANCE AVERAGING

The several partial differential equations which comprise the model contain so much detail that overall ecosystem response is obscured. As an aid to exploration of overall system response, it is useful to simplify the model considerably. As an illustration, consider the response of the Houghton Lake wetland to an input of nutrients and water. The global observations are two: there is an expanding area of increased biomass, and all excess nutrients are stripped from the water (see Figure 4).

The steps in further simplification are: (1) lump all biomass compartments and lump all soil compartments, (2) deal with a single

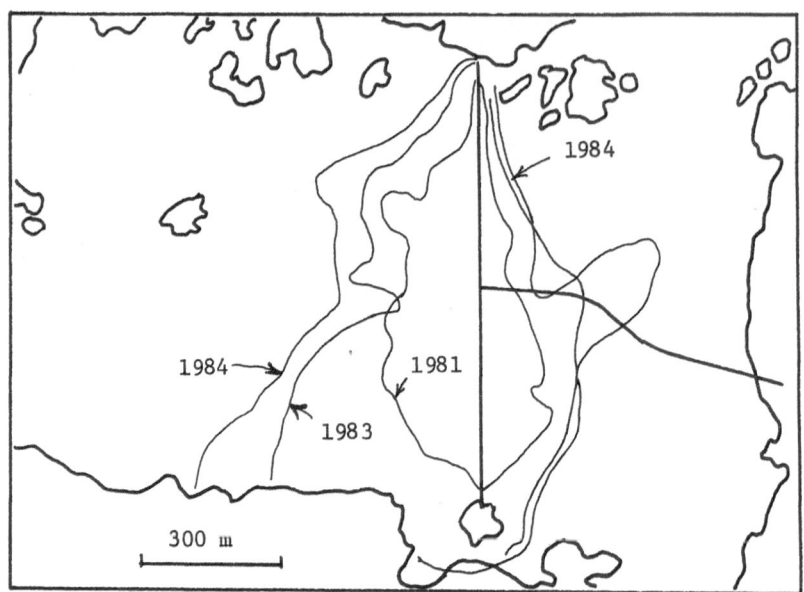

Figure 4. Zones of enhanced vegetative productivity, f1981–1984.
Boundaries of the enhanced zones are determined from oblique
aerial color infrared photography contrasted to similar
control photography (pre-irrigation) (1976).

limiting nutrient, (3) average over distance, and (4) average depth,
flow, concentration, and biomass rates over each year.

Lumping all biomass leads to:

$$\frac{\partial J}{\partial t} = G - \alpha J \tag{39}$$

where

$$J = B + A + R + S + L + W \tag{40}$$

Here, $\alpha$ is the decay rate constant and G is the average annual growth
rate. In a starting stationary state:

$$0 = G^* - \alpha J^* \tag{41}$$

Lumping all soil compartments leads to

$$\frac{\partial S}{\partial t} = b(\alpha J) \tag{42}$$

where b represents the fraction of dead biomass which becomes soil (the
balance decays, returning nutrients to the ecosystem). In an initial
stationary biomass state soil does accrete:

$$\frac{\partial S^*}{\partial t} = b(\alpha J^*) \tag{43}$$

It is assumed that water-borne organisms grow, die and decay within one year; thus having no effect on the annual averages.

The water balance equation is no longer necessary, since annual average depths and flows are to be used. The balance for the limiting nutrient in the water sheet becomes:

$$\phi vh \frac{\partial C}{\partial x} + \phi_s h \frac{\partial C}{\partial t} = pC_p + aC_a - C_b \frac{\partial J}{\partial t} - N_s \tag{44}$$

The flux of limiting nutrient to the soil consists of soil building and increases in the interstitial-sorbed amount:

$$N_S = (C_S + C_{AS}) \frac{\partial S}{\partial t} + d_S \frac{\partial C_{AS}}{\partial t} \tag{45}$$

where $d_s$ represents the depth of the sorption zone in the soil. A constant sorption zone depth has been presumed, which is partially justified by both data, and increasing diffusion resistance. In the stationary reference state, $a=0$, $C_{AS}$ is constant, and there are no changes in annual averages. Consequently, soil building results from unused precipitation and flow inputs:

$$(C_S^* + C_{AS}^*) \frac{\partial S^*}{\partial t} = pC_p - [(\phi hvC)^*]_o^L \tag{46}$$

where integration with respect to x has been done, and where L is the affected length of the system. For no distance variability in the water sheet, the bracketed terms drop out.

Next, reference all quantities to the original stationary state:

$$J' = J - J^* \tag{47}$$

$$S' = S - S^* \tag{48}$$

$$C' = C - C^* \tag{49}$$

In terms of these departure variables, equations (39), (42), and (44) become, respectively:

$$\frac{\partial J'}{\partial t} = G' - aJ' \tag{50}$$

$$\frac{\partial S'}{\partial t} = baJ' \tag{51}$$

$$\phi hv \frac{\partial C'}{\partial x} + \phi_s h \frac{\partial C'}{\partial t} = aC_a - C_b \frac{\partial J'}{\partial t} - [(C_S + C_{AS}) \frac{\partial S}{\partial t}'] - d_s \frac{\partial C_{AS}'}{\partial t} \tag{52}$$

The surface water concentration and adsorbed concentration may be related by a locally linearized equilibrium:

$$C_{AS}' = KC' \tag{53}$$

Equation (61) states that added nutrients (LHS) must be used to expand the affected area, build an increased biomass pool, or be immobilized in and on the soil. It can equally well by generalized to two dimensions, which for the one-dimensional case reduces to A = WL.

For constant parameters, equation (61) assumes the form

$$a_1 \int_0^\theta e^{-(\theta-\phi)} \frac{df}{d\phi} \, d\phi + a_2 \frac{df}{d\theta} + f = \rho = 1 \tag{62}$$

with $f(0) = f_0 = 0$ for a startup process. Here, $\theta = at$, $\phi = a\tau$, $f = L/L_{max}$ and $a_1$ and $a_2$ are constants. This linear integro-differential equation may be solved analytically to yield:

$$f(\theta) = \frac{\rho}{a_2} \left[ \frac{1}{r_1 r_2} - \frac{A_1}{r_1} e^{r_1\theta} + \frac{A_2}{r_2} e^{r_2\theta} \right] + f_0 \left[ - B_1 e^{r_1\theta} + B_2 e^{r_2\theta} \right] \tag{63}$$

where $A_1$, $A_2$, $B_1$, $B_2$, $r_1$, $r_2$ are constant groupings.
Now suppose that $\rho = 1$, but a sequence of annual additions $\{\rho_j\}$. Select the maximum annual addition as the generator of $A_{max}$, so that $\rho_j = 1$. At the end of year one

$$f(a) = \frac{\rho_1}{a_2} \left[ \frac{1}{r_1 r_2} - \frac{A_1}{r_1} e^{r_1 a} + \frac{A_2}{r_2} e^{r_2 a} \right] + f_0 \left[ - B_1 e^{r_1 a} + B_2 e^{r_2 a} \right] \tag{64}$$

$$= C_1 \rho_1 + C_2 f_0$$

This procedure may be repeated for each succeeding year, leading to:

$$f(na) = C_1 \left( \rho_n + C_2 \rho_{n-1} + \ldots + C_2^{n-1} \rho_1 \right) \tag{65}$$

This procedure has been applied to the Houghton Lake data: all parameters were derived from field measurements, and the area expansion was independently measured from aerial infrared photography. Figure 5 compares predictions and results.


THE FULL EQUATION SET

The lumping and averaging of the preceding section provides some global results of interest, but cannot answer more detailed questions involving fast processes or distributed phenomena. Accordingly, the entire set was programmed for the one dimensional case. The nutrients were restricted to nitrogen and phosphorus; and chloride was included as an inert tracer. The total number of material balances is 35, as shown in Table 1.

Solution Strategy

The ecosystem biota react slowly to nutrient status alterations, compared to the speed of surface water processes. Further, moderate changes in plant size, suspended organism density or the amount of standing dead have negligible impact on the rate or direction of water

80

with this substitution, integration of (52) gives:

$$
\frac{Q}{W} C_a - (\phi h v) \int_0^L \frac{\partial C'}{\partial x} = \phi_s h \int_0^L \frac{\partial C'}{\partial t} dx + C_b \int_0^L \frac{\partial J'}{\partial t}
$$

$$
+ \int_0^L [(C_S + C_{AS}) \frac{\partial S}{\partial t}]' dx + K d_s \int_0^L \frac{\partial C'}{\partial t} dX \tag{54}
$$

where Q is the volumetric addition rate and W is the width of the wetland. Define the areal average concentration as:

$$
\hat{C'} = \frac{1}{L} \int_0^L C' \, dX \tag{55}
$$

Then for $C' =$ constant, and $C'(0) = C'(L)$:

$$
\frac{Q}{W} C_a = (\phi_s h + K d_s) \hat{C'} \frac{dL}{dt} + C_b \int_0^L \frac{\partial J'}{\partial t} dx + \int_0^L [(C_S + C_{AS}) \frac{\partial S}{\partial t}]' dx \tag{56}
$$

Upon integration of (50), $J'$ is known:

$$
J' = \begin{cases} J'_1 (1 - \exp(-\alpha(t-\tau))), & t > \tau \\ 0, & t < \tau = \text{arrival time} \end{cases} \tag{57}
$$

where

$$
J'_1 = \frac{G'}{\alpha} \tag{58}
$$

A change of variable may be made by acknowledging that internal positions correspond to unique frontal arrival times, and hence we may write

$$
dx = \frac{dx}{d\tau} d\tau \tag{59}
$$

and
$$
L = L(t) \tag{60}
$$

Then

$$
\frac{Q}{W} C_a = (\phi_s h + K d_s) \hat{C'} \frac{dL}{dt} + (C_b - b(C_S + C_{AS})) \alpha J'_1 \int_0^t e^{-\alpha(t-\tau)} \frac{dL}{d\tau} d\tau
$$

$$
+ b\alpha[J_0 (C_S + C_{AS})' + J'_1 (C_S + C_{AS})] L \tag{61}
$$

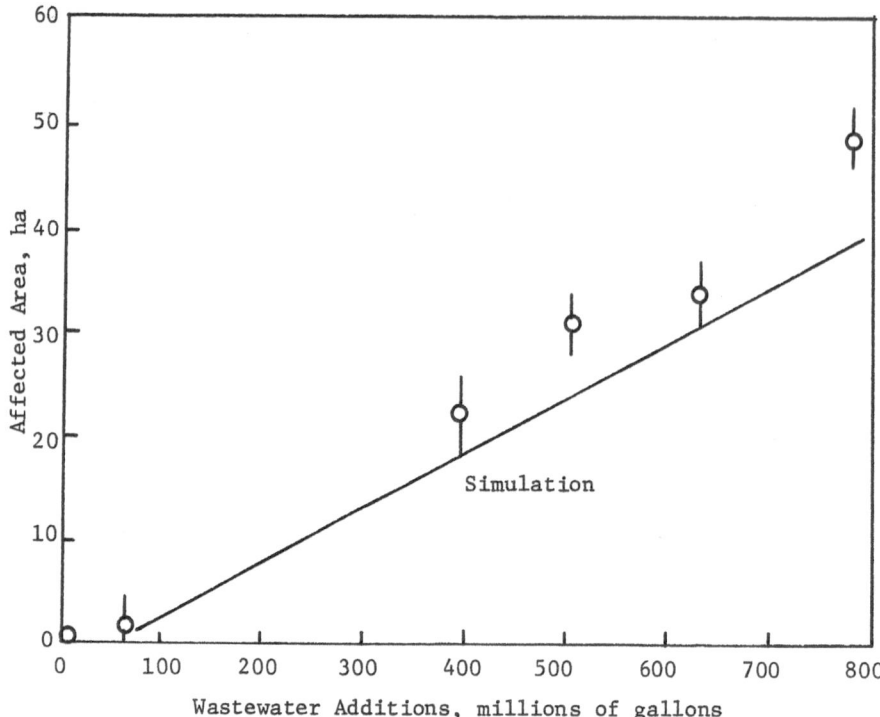

Figure 5. Affected area expansion with time. Color IR photo interpretation, Houghton Lake.

flow. Thus, the surface water flow equation was decoupled from the rest of the model, and solved first. Should parameters change, annual (or more frequent) adjustments may be made. Likewise, the effects of the moving soil surface are slight during one or more years, simply because only millimeters of accretion or depletion are likely. A fine time and space grid was used for hydrology—typically 0.05 days and 5 meters. An entire years' depth and velocity results, on a daily basis, were then stored for use in other computations.

The water-borne constituents exhibit exchanges which require the same time and distance scales. Rain events, and variations in the pumping schedule affect nutrient concentrations on an hourly basis. It is still advantageous to decouple hydrology, since different nutrient scenarios may be imposed on the same hydrology. The same time step, but a coarser (50 m) grid is acceptable for interstitial species balances,

Table 1. Species and balance equation enumeration.

| Compartment | Species Present | Number of Balances |
|---|---|---|
| Surface water | Total, N, P, $Cl^-$, SS | 5 |
| Sediment bed | Total, N, P | 3 |
| Soil | Total, N, P | 3 |
| Interstitial | N, P, $Cl^-$ | 3 |
| Biomass (7) | Total, N, P | 21 |
| | | 35 |

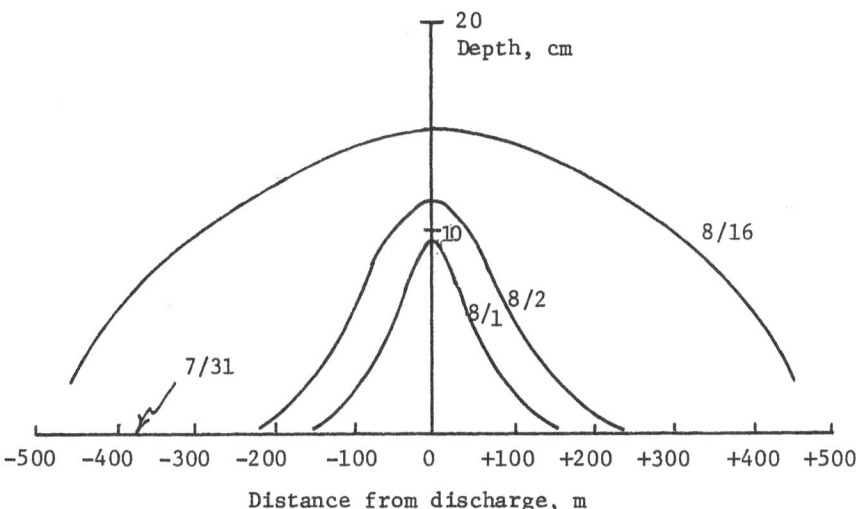

Figure 6. Water wave development on pump startup, 1978. Simulation is for a flat peatland surface.

and both time and distance steps may be large for the immobile biomass compartments (one day, 50 m). Details of the algorithms may be found in Hammer (1984). In summary, short and medium time scales were handled with grid size adjustment, while very long term phenomena were handled by program interruption for parameter adjustment.

### Results

The basic structure of the model equations is hyperbolic, and consequently chromatographic or wave behavior is anticipated. Several examples are presented here to provide understanding of wetland phenomena. First, the water sheet displays complex behavior. The pump start-up wave shown in Figure 6 is anticipated, but multiple flow directions can occur, and are not so easily guessed. These are associated with pump starting and stopping sequences. Further results of hydrological simulation and field data are given in Kadlec, et al., (1981) and Hammer and Kadlec, (1985).

Dissolved species should move at speeds somewhat slower than the water, due to interaction with immobile wetland compartments. The weakest interaction is for chloride, which is not biologically active. Simulation predicts this swift permeation of the wetland, and field data support this. On the other hand, nutrients interact strongly with soil, sediments and plants. Figure 7 gives a seven year progression of the ammonium front, and includes a sample of simulation output. Further data and simulation results are presented in Kadlec and Hammer, 1984 and 1985b.

Suspended organisms also exhibit wave phenomena, as shown in Figure 8, in which chlorophyll-a is taken to be a measure of algal biomass. This in turn results in a vertical sediment flux, measured and simulated in Figure 9. A further discussion of these phenomena is given in Kadlec, 1985a. The immobile plants and litter show the slowest response as indicated by Figure 10, taken from Kadlec and Hammer, 1985b. Here the progression is on a year-to-year basis.

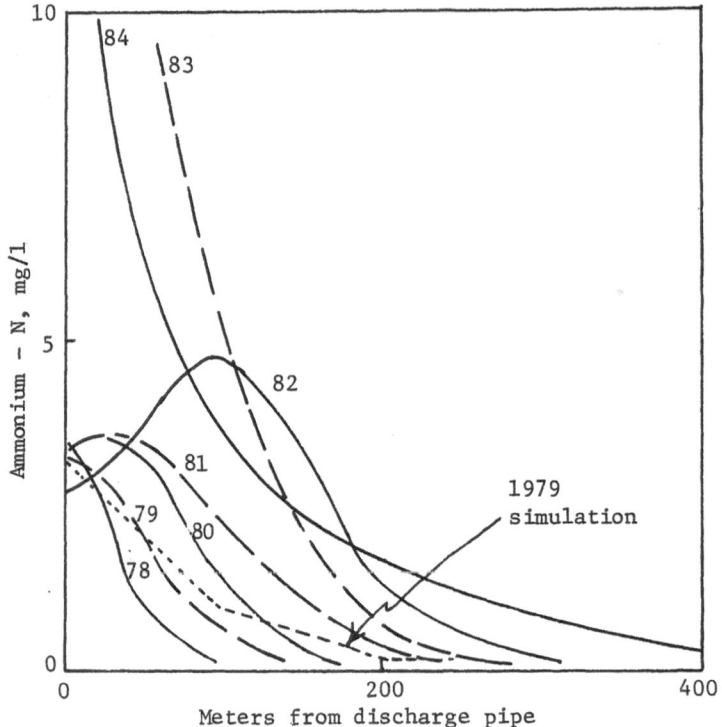

Figure 7. Progression of the ammonium front, Houghton Lake.

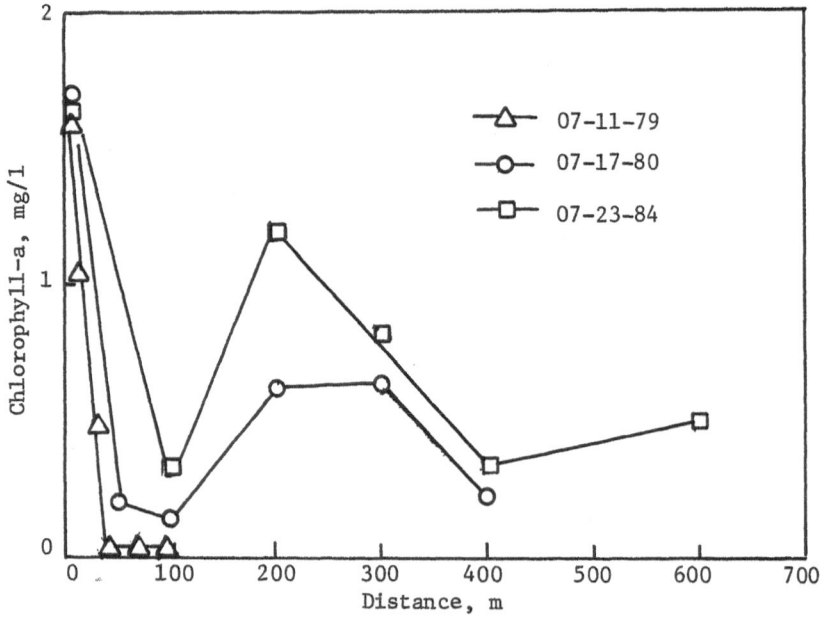

Figure 8. Annual variation in chlorophyll-a. Distance profiles in July, Houghton Lake wetland.

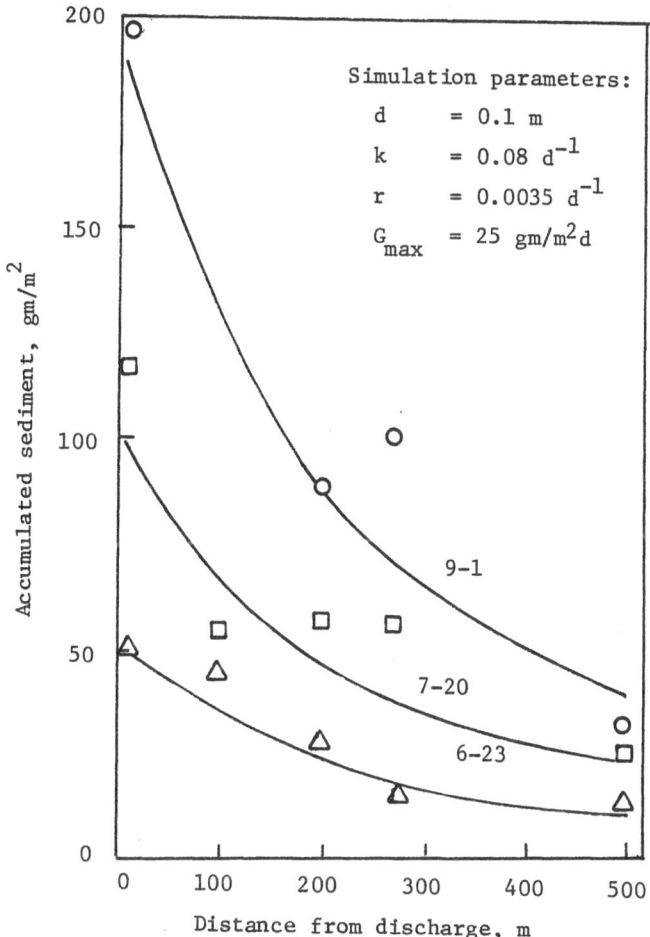

Figure 9. Accumulated sediment, leatherleaf cover type, versus distance. Lines are computed. Houghton Lake wetland from Kadlec, 1985a.

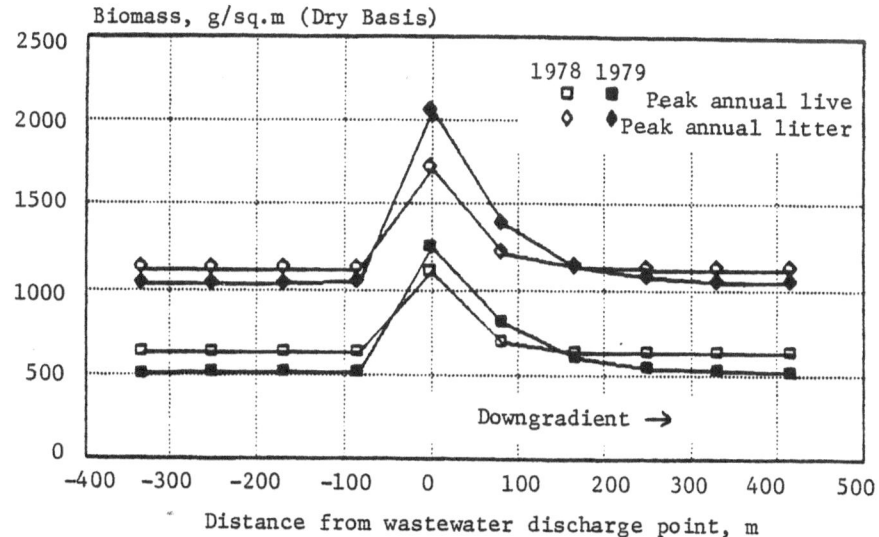

Figure 10. Spatial variation of the peak annual live and peak annual litter biomass pools. Simulation of the Porter Ranch treatment facility (Houghton Lake, MI).

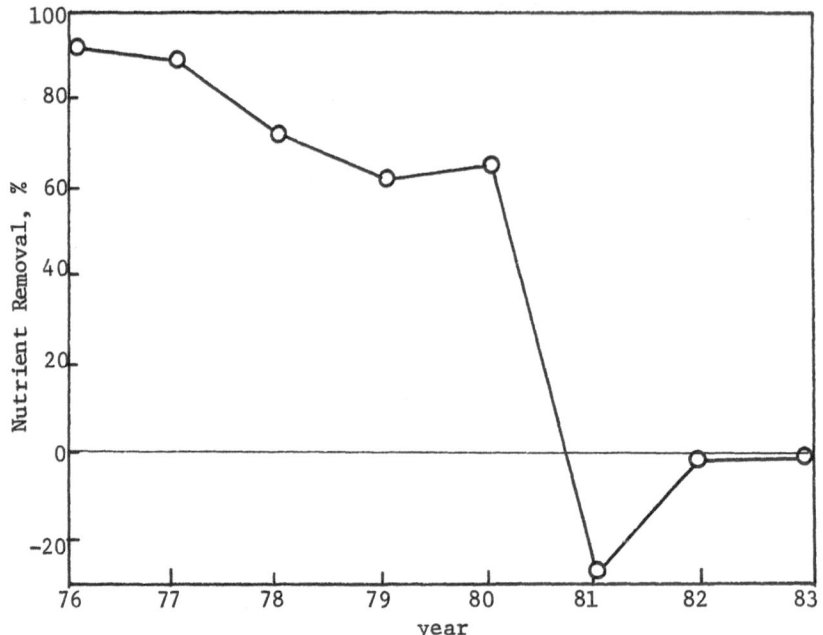

Figure 11.   Phosphorus front breakthrough at the Bellaire wetland.

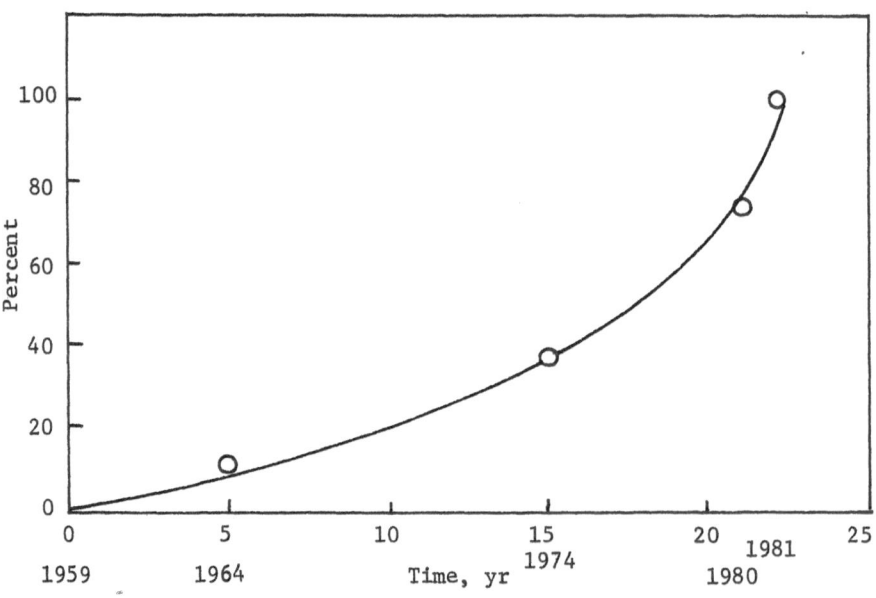

Figure 12.   Cattail replacement of original vegetation, Kincheloe wetland.

These and many more results can be obtained via full simulation. The expense is not small, since CPU times of many minutes per year of real time are required on an Amdahl 5860 computer. However there are many interesting phenomena which are correlated with this model. The preceding results were all for the Porter Ranch site, where solute concentration profiles were fully contained within the wetland (chloride excepted). After a sufficient period, in some cases, nutrient breakthrough can be expected, as shown in Figure 11 for a different wetland irrigation system.

## BENEFITS AND DIFFICULTIES

There is more incentive for modeling and simulation in the ecosystem context than in other more mechanical, non-biological situations. The time scales of the wetland processes examined here range up to years, making experimentation difficult for several obvious reasons. Experimentation is also perilous, in that negative results are viewed as catastrophic failure by certain segments of the public, and by environmental regulators.

Simulation, imperfect as it may be, permits the exploration of key variables, thus facilitating design. Management strategies may be explored without unpleasant consequences in the field. For example, the termination of nutrient additions to wetlands may produce a timed release of stored chemicals. It is clearly quicker and less costly to explore such questions with a simulator than with field experiments.

A second, equally important benefit of simulation is the organization and interpolation of sparse and irregular field data sets. For most ecosystem experiments, an enormous amount of detail is of interest--the 35 compartmental variables of Table 1, over space and time, in the present case. Computations based on mass conservation permit the reordering of data in a consistent way.

There are significant difficulties in wetland simulation, over and above those found in more deterministic, non-biological situations. In many cases, there is no history of investigation into the form of rate laws for use in balance equations. In this wetland simulator, examples are plant growth, sediment re-entrainment, and flow through vegetation. Thus simulation is best supported by a large amount of research into the associated transport phenomena. Even if this can be accomplished, there remains a major task of parameter estimation to fit generalized transport laws to a specific site. This frequently must be done in an indirect manner, by fitting the response of the entire simulator to data by adjusting individual parameters.

The wetland ecosystem also produces effects which will probably never by modelable. In terms of wetland vegetation, there is a long term change in species composition as well as biomass. This is clearly apparent for short times and short distances at the Houghton Lake site, and has been observed in virtually all disturbed wetlands. At the Kincheloe wetland, subjected to wastewater for nearly 30 years, a complete replacement of the original vegetation with cattail has occurred. The course of the replacement can be followed from historical aerial photography, and the results are shown in Figure 12. Such alterations in species have large effects on simulation parameters. Nutrient uptake is altered, flow resistance changes, and sediment transport is altered. Because of the long time scale, such changes have been omitted in this wetland simulator.

There are also significant interactions with animals. Their trails channelize flow (deer), their dams alter boundary conditions (beaver), and their consumption of vegetation changes flow porosity (muskrats). Waterfowl can influence nutrient status: Horicon marsh in Wisconsin experiences $10^7$ goose-use-days each fall, at 38 grams of feces per goose. Motile invertebrates (Daphnia) respond to light, currents and food, and do not follow deterministic rules.

Numerical problems also occur. There is a huge data input requirement. For just the hydrology simulator in this work, one year's simulation requires 1537 separate numbers as input. Costs are significant, both for internal computer storage and for execution. Accuracy problems occur due to the very large number of time steps required to span even one year of simulation. This forces special forms of the finite difference equations. Failure to do so can result in enormous errors: one meter of water can be lost in one year due strictly to error accumulation.

Stability is also troublesome in ways which are not common in other simulation situations. In the hydrology routine, multiple flow directions occur simultaneously, necessitating special cell-marker procedures to stabilize the algorithm. The dry-out phenomenon potentially causes extremely small time constants for the incremental water puddles, necessitating other special procedures. Spurious waves in concentration can result from improper grid sizing. And, even when all is working properly, the simulator generates an overwhelming volume of output, which contains the detail of interest.

One final difficulty is that so many complicated mathematical and computer techniques have been brought to bear, that ecologists and regulators may tend to be skeptical of the results.

## ACKNOWLEDGEMENTS

The efforts of Drs. Ken Dixon, Peter Parker, Prem Gupta and David Hammer form the basis for this paper. Financial support has included NSF, USEPA, the Rockefeller Foundation, and FMC Corporation.

## NOTATION

| | |
|---|---|
| a | additional rate; constant |
| A | area; constant |
| b | constant |
| B | wood biomass, sediment in bed, constant |
| C | concentration; constant |
| $d_s$ | sorption zone depth |
| D | death or decomposition rate |
| $D_e$ | effective diffusicity |
| $e$ | evaporation rate |
| f | dryfall rate, dimensionless area |
| g | ground elevation |
| G | growth rate |
| h | depth |
| i | infiltration rate |
| J | biomass |
| k | mass transfer coefficient |
| K | distribution coefficient |
| L | litter biomass, length |
| N | component flux |

| | |
|---|---|
| p | precipitation rate |
| P | plant uptake rate |
| Q | volumetric flow rate |
| r | constant |
| R | reaction or resuspension rate; root biomass |
| S | sedimentation rate; standing dead |
| t | time |
| T | transfer or trapping rate |
| u | settling velocity |
| U | utilization rate |
| v | velocity |
| W | width |
| X | horizontal coordinate |
| Y | horizontal coordinate |
| Z | vertical coordinate |

## Greek

| | |
|---|---|
| $\alpha$ | constant |
| $\beta$ | constant |
| $\gamma$ | constant |
| $\delta$ | constant |
| $\varepsilon$ | constant |
| $\theta$ | time |
| $\rho$ | density; ratio |
| $\tau$ | arrival time |
| $\phi$ | porosity |

## Subscripts

| | |
|---|---|
| a | addition |
| A | annual biomass; adsorbed |
| b | biomass |
| B | wood biomass |
| BC | boundary condition |
| d | soil |
| i | component; infiltration |
| j | compartment |
| L | litter |
| o | suspended organism |
| p | pool |
| R | roots |
| S | solids; standing dead |
| W | wood litter; weir |
| x,y,z | directions |

## Superscripts

| | |
|---|---|
| $\wedge$ | average |
| $'$ | deviation |
| $*$ | stationary state |

## REFERENCES

Bartlett, M. S., Brown, L. C., Hanes, N. B., and Nickerson, N. H., 1979, Denitrification in freshwater wetland soil, _J. Environ. Qual._, 8(4):460.

Chamie, J. P. M., 1975, 'The Effects of Simulated Sewage Effluent upon Decomposition, Nutrient Status and Litterfall in a Central

Michigan Peatland,' Ph.D. Thesis, The University of Michigan, Ann Arbor, MI.

Engler, R. M., and Patrick, W. H., Jr., 1974, Nitrate removal from floodwater overlying flooded soils and sediments, J. Environ. Qual., 3:409.

Dixon, K. R., 1974, 'A Model for Predicting the Effects of Sewage Effluent on a Wetland Ecosystem,' Ph.D. Thesis, The University of Michigan, Ann Arbor, MI.

Gupta, P. K., 1977, 'Dynamic Optimization Applied to Systems with Periodic Disturbances,' Ph.D. Thesis, The University of Michigan, Ann Arbor, MI.

Haag, R. D., Jr., 1979, 'The Hydrogeology of the Houghton (Lake) Wetland,' M.S. Thesis, The University of Michigan, Ann Arbor, MI.

Hammer, D. E., and Kadlec, R. H., 'Orthophosphate Adsorption on Peat.' 6th International Peat Congress, Duluth, MN. 1980.

Hammer, D. E., and Kadlec, R. H., 1981, 'Wetland Utilization for Management of Community Wastewater: Concepts and Operations in Michigan,' The Universtiy of Michigan, Institute of Science and Technology, Ann Arbor, MI.

Hammer, D. E., and Kadlec, R. H., 1983, Design principles for wetland treatment systems, Report to USEPA, Kerr Laboratory.

Hammer, D. E., and Kadlec, R. H., 1985, A model for wetland surface water dynamics,' in: 'Water Resources Research,' accepted.

Hammer, D. E., 1984,'An Engineering Model of Wetland/ Wastewater Interactions,' Ph.D. Thesis. The University of Michigan, Ann Arbor, MI.

Hemond, H. F., and Fifield, J. L., 'Subsurface Flow in Salt Marsh Peat: A Model and Field Study,' Limnol. Oceanogr., 27(1):126, 1982.

Kadlec, R. H., Hammer, D. E., Nam, I.-S., and Wilkes, J. O., 1981, The hydrology of overland flow in wetlands, Chem. Eng. Comm., 9:331.

Kadlec, R. H., and Hammer, D. E., 1982, Pollutant transport in wetlands, Environ. Prog., 1(3):206.

Kadlec, R. H., 1984a, Freezing-induced vertical solute movement in peat, Seventh International Peat Congress Proceedings, Dublin, Ireland.

Kadlec, R. H., and Hammer, D. E., 1984, 'Wastewater rennovation in wetlands: six years at Houghton Lake,' in: 'Proceedings of the Water Reuse Symposium III', San Diego, CA.

Kadlec, R. H., and Rathbun, M. A., 1984, Copper sorption on peat, in: 'Proceedings of the International Symposium on Peat Utilization,' S. A. Spigarelli and C. A. Funhsman, Eds., Bemidji State University, MN.

Kadlec, R. H., 1985b, 'Aging phenomena in wastewater wetlands,' in: 'Ecological Considerations in Wetland Treatment of Municipal Wastewaters,' E. R. Kaynor, P. J. Godfrey and J. Benforado, Eds. Van Nostrand.

Kadlec, R. H., and Hammer, D. E., 1985b, Modeling nutrient behavior in wetlands, Ecological Modeling, accepted.

Kadlec, R. H., 1985a, Sediment processes in wetlands, Water Research, in review.

Kadlec, R. H., and Hammer, D. E., 1985a, Simplified computation of wetland vegetation cycles, in: 'Coastal Wetlands,' F. M. D'Itri and H. Prince, Eds., Lewis, Chelsea.

Mitsch, W. J., 1983, 'Ecological models for management of freshwater wetlands,' in: 'Application of Ecological Modeling in Environmental Management, Part B,' S. E. Jorgensen and W. J. Mitsch, Eds. Elsevier Science Publishers B. V., Amsterdam, The Netherlands.

Mitsch, W. J., Taylor, J. R., and Madden, C., 1982, Models of North
    American freshwater wetlands, <u>Int. J. Scol. Environ. Sci.</u>,
    51:109.
Parker, P. E., 1974, 'A Dynamic Ecosystem Simulator,' Ph.D. Thesis,
    The University of Michigan, Ann Arbor, MI.
Patrick, W. H., Jr. and Reddy, K. R., 1976, Nitrification-denitrification
    reaction in flooded soils and water bottoms: dependence on
    oxygen supply and ammonium diffusion, <u>J. Environ. Qual.</u>,
    5(6):469.
Schwegler, B. R., 1978, 'Effects of Sewage Effluent on Algal Dynamics
    of a Northern Michigan Wetland,' M.S. Thesis, The University
    of Michigan, Ann Arbor, MI.

# UNIFIED TRANSPORT MODEL FOR ORGANICS

M. R. Patterson

Computing and Telecommunications Division
Oak Ridge National Laboratory
Oak Ridge, TN 37831

The Unified Transport Model UTM-TOX is a computer simulation model funded by the USEPA at the Oak Ridge National Laboratory to develop transport capabilities for new chemicals and pesticides that are organic in nature. This model combines atmospheric transport with hydrologic transport and couples sediment transport, plant growth models, and biodegradation within one unified package. It is an outgrowth of an original study [1] sponsored by NSF which was concerned with the transport of trace heavy metal ions. As a part of that development of the Unified Transport Model, the atmospheric model and the hydrologic transport model were coupled in order to study the concentrations of trace heavy metal ions from deposition on the vegetation through the soil profile into the stream channel. The current model builds heavily on this initial model. The present additions to the model are, in fact, logical extensions of those capabilities that already existed within the model.

A typical source that has been considered by the UTM-TOX is shown in Fig. 1. In this figure we see a coal fired steam plant which generates 800 megawatts for the TVA system sited at Bull Run. This plant has a very tall stack, on the order of 900 feet, and it emits various types of particles and gases. This source would be handled in the model as a point source. In the next figure, Fig. 2, we see an idealization of one of the scenarios that we considered for the USEPA. These sources consist of a point source emission to the atmosphere from an industrial plant, direct release to the stream channel, and some domestic releases. These sources can be handled within the model, as well as several others.

In the next figure, Figure 3, we see a very simple representation of the Unified Transport Model. Within the umbrella of the Unified Transport Model fall the Atmospheric Transport Model (ATM) and the Hydrologic Transport Model (WHTM). We will see as this paper progresses a continuing buildup of the model in terms of both complexity and scope. However, at the lowest level of resolution, Fig. 3 shows how the model functions: it is a hydrologic model coupled to an atmospheric model; the atmospheric model provides a source of deposition to the hydrologic transport model and the latter provides washout data to the atmospheric model.

In Figure 4 we see a breakdown in somewhat more detail of the coupling of the atmospheric transport model with the hydrologic model [2]. The Atmospheric transport model is coupled into one of the three major

Fig. 1.  Aerial view of Bull Run Steam Plant, Tennessee.

Fig. 2. Idealization of releases to air and water.

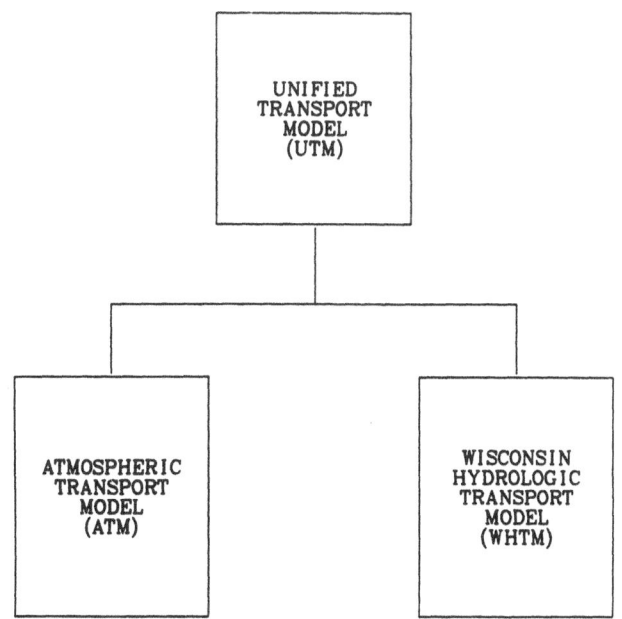

Fig. 3. Basic representation of UTM-TOX.

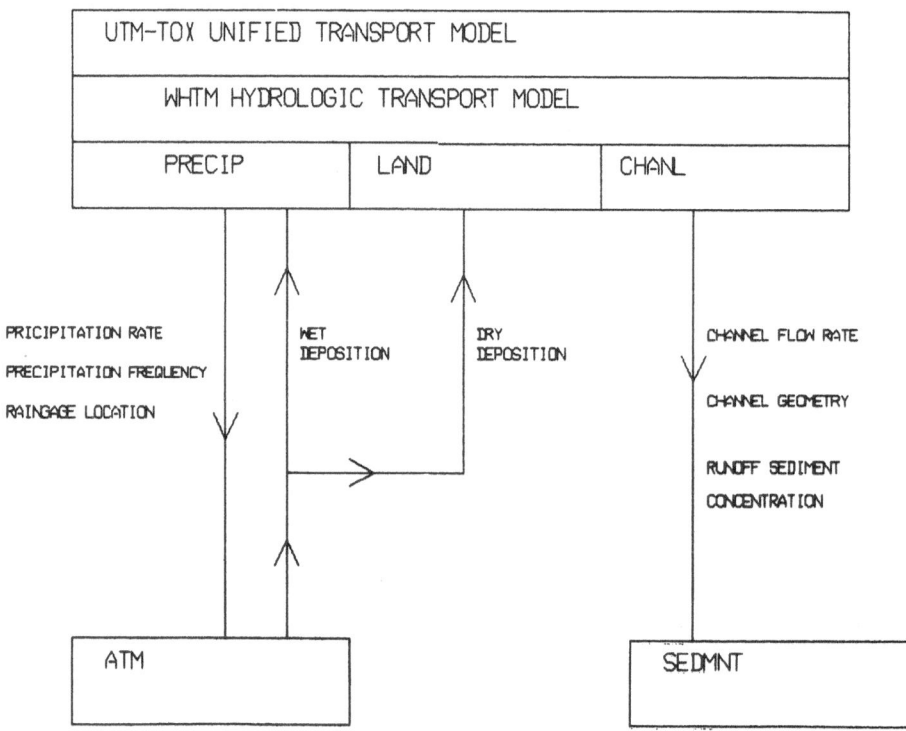

Fig. 4.  Coupling of the Atmospheric Transport Model (ATM) to the
Hydrological Transport Model (WHTM)

links of the hydrologic model WHTM, the PRECIP link, which handles precipitation and wet deposition data in the hydrologic transport model. The atmospheric transport model also furnishes dry deposition to the LAND link of the hydrologic transport model. Both wet and dry deposition then progress through the runoff from the soil to the stream channel and wind up in the portion of the model called CHANL, which routes the flow from the landscape through the stream channel to the outfall of the watershed.

Figure 5 shows the familiar subcomponents of the Unified Transport Model broken out again in a slightly different organization. In this version of the Unified Transport Model, higher level models such as TEHM [3], SCEHM [4], AGTEHM [5], and PROSPER [6] are coupled to the Unified Transport Model either at the beginning or to a routine such as LAND. This routine then provides all the data necessary to higher level models in order for it to function and to return to the Unified Transport Model those quantities that it calculates. This diagram can be further generalized as we will see in later figures.

The Atmospheric Transport Model

We have coupled together an Atmospheric Transport Model [7,8] that currently has a resolution time of one month with the WHTM which operates on shorter time scales that are shown in Fig. 6. The PRECIP link operates on a time scale of one hour. The recorded precipitation is generally taken from hourly recording gauges, and finer time resolution is not generally available. The LAND link, in turn, operates on a time scale of 15 minutes in order to distribute the one hour input of precipitation over four smaller time periods, and its output to the stream channel is calculated once each half hour. The stream CHANL, in turn, computes on a time scale from 3 to 30 minutes based on the variation in the flow and how large flow rate is at the moment. During periods of low or steady flow the computational interval increases up to 30 minutes.

Figure 7 shows the type of deposition pattern that can be calculated with the Atmospheric Transport Model. Only three isopleths are shown. These concentrations are for a lead smelter in southeastern Missouri at the Crooked Creek Watershed. The outline of the watershed is shown as a heavy black line, and the isopleths are shown in a lighter black line which is labeled by the deposition rate in mg/m sq./mo. These deposition values show how much lead was being deposited in the vicinity of this watershed, using an assumed deposition velocity of 10 cm/sec. The windrose that was used for these calculations is the closest one available, St. Louis, Missouri, about 120 miles away. The isopleths shown here can be given in more detailed fashion by obtaining more grid points before drawing the contour lines that characterize given isopleths. Note in this figure the rapid decrease of the deposition rates: in going from approximately one-half kilometer to one kilometer, the deposition rates decrease by more than a factor of two and correspondingly in going from one kilometer to two kilometers the deposition again decreases by a factor of two. The impact, then, of this deposition can be expected to be relatively local on the scale shown in the figure, that is, approximately two to four kilometers.

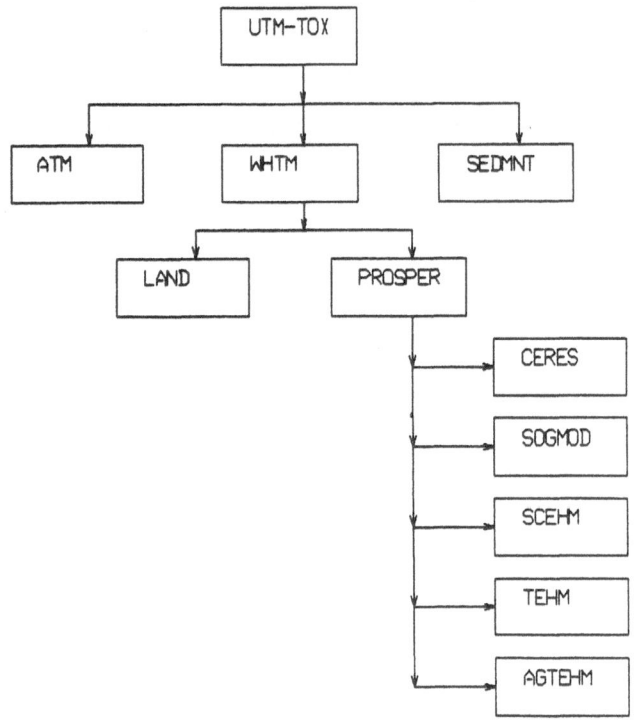

Fig. 5.   Coupling of higher level models to UTM-TOX through
          PROSPER.

## Description of the Hydrology Transport Model WHTM

        Now consider the Wisconsin Hydrologic Transport Model (WHTM) [9].
The accompanying Fig. 8 shows the well known aspects of the hydrologic
cycle that are treated within this model.  This model is primarily a
variant of the Stanford Watershed Model [10].  The hydrologic cycle shown
in Fig. 8 shows that this continuous process of transpiration and
evaporation is driven by precipitation and leads to interflow beneath the
surface of soil, maintenance of a groundwater table, and establishment of
flow in the stream channel.  These processes are quite complex within
themselves and the model provides a parametric summary that is completely
deterministic and useful for purposes of simulating runoff and transport.
The next illustration, Fig. 9, shows how one might zero in on a closer
view of this same landscape.  Again the same processes of precipitation
and transpiration are occurring, along with the uptake of a pollutant
within the vegetation.  These processes also include runoff, erosion,
solubilization of sorbed material from the litter, transport from the
litter, and infiltration into a unsaturated region of the soil where the
root zone occurs.

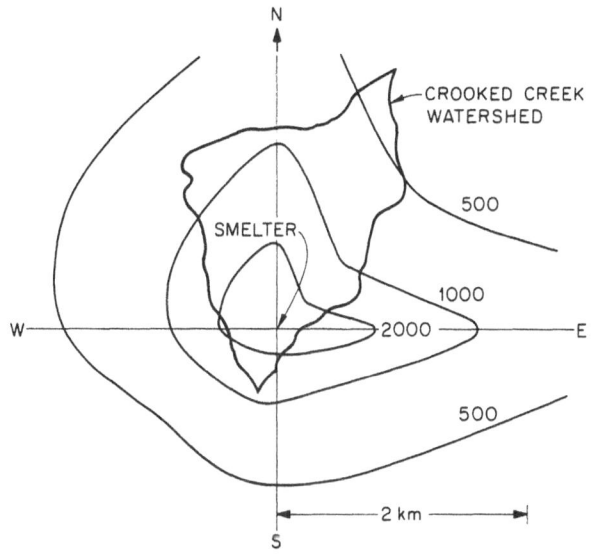

Simulated Deposition Rates (mg/m$^2$/mo) for Lead Near
the Crooked Creek Watershed, Using a Deposition Velocity
of 10 cm/sec.

Fig. 6. Computational intervals in UTM-TOX.

COMPUTATIONAL INTERVALS
IN THE UNIFIED TRANSPORT MODEL


ATMOSPHERIC TRANSPORT MODEL ............ 1 MONTH

WISCONSIN HYDROLOGIC TRANSPORT MODEL
      PRECIP...................... 1 HOUR
      LAND...................... 15 MINUTES
      CHANL............ 3 TO 30 MINUTES

Fig. 7. Deposition pattern calculated with ATM for the Crooked
Creek Watershed in Missouri.

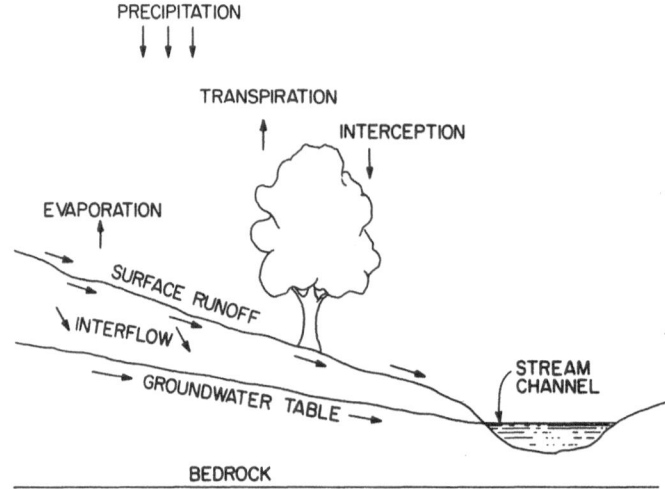

Hydrologic Processes in WHTM.

Fig. 8. Hydrologic processes in WHTM, following N. H. Crawford
and R. K. Linsley.

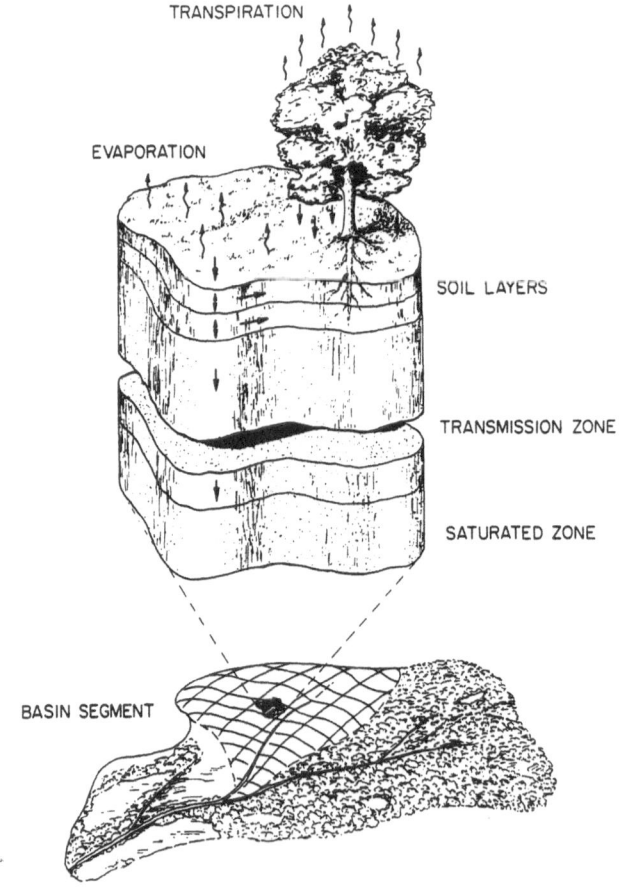

Fig. 9. Closer view of hydrologic processes in a land
area segment.

Our attention now will be focused on the macroscopic view given by the Stanford Watershed Model of the hydrologic cycle shown in Fig. 10, as realized within the model. This schematic simulates many of those processes shown in Fig. 8 and in Fig. 9 in that the incident precipitation drives the remaining parts of the hydrologic cycle. Within the model, these processes are broken down into interception storage, evapotranspiration from various compartments including upper zone storage, lower zone storage, active ground zone water storage, and inactive ground water storage.

Processes that are ongoing during the period of time that the model simulates include evapotranspiration, evaporation from interception storage, overland flow directly to the stream channel, and flow from impervious areas directly to the stream channel. Interflow, which is made up from water that has penetrated beneath the surface, also finds its way to the stream channel eventually, and flow occurs from all of the zones named, except from the inactive ground water storage, to the stream channel. Inactive ground water storage is the zone where water that is not moving on the time scale of the WHTM. The activity in that compartment might well be simulated by a ground water model [11]. In ground water models the migration of moisture occurs on a scale on the order of 5 meters per year, which is too slow to affect the runoff and the storm hydrograph. Thus groundwater movement is not of direct interest to us at the moment in accessing the runoff of toxic material. The use of groundwater models for longer terms assessment becomes, of course, very critical.

Based on the structure shown in Fig. 10, the transport of a toxic material is shown in Fig. 11. This conceptualization of the transport processes shows that for each compartment that occurs within the hydrology portion of the model there is a corresponding compartment for toxic material. In Fig. 11, we see that the wetfall and dryfall deposition are combined to form an input at the top of this diagram. As the precipitation drives the soluble portion of this toxicant through the ecosystem, some of the chemical falls directly in the stream channel by direct deposition. The rest of the chemical experiences vegetal interception and at that point can either undergo ion exchange at the soil surface or infiltration in the soil. Material erosion or solute runoff can transport a portion of the material to the stream channel again. Finally, the solute can infiltrate the soil. Part of the infiltrated material can be carried to the stream channel and part of it will penetrate to the inactive storage compartment. An exact diagram of what happens to deposition within the model is more complex than this summary figure shows. However, Fig. 11 shows the major features of the transport of the chemical within the Wisconsin Hydrologic Transport Model.

### Application of WHTM to the Walker Branch Watershed

What are the results that can be obtained from the application of the hydrologic transport model WHTM? In Fig. 12, we show both experimental and simulated values for the hydrograph. This figure shows that there is reasonably good agreement between observed values, which are shown with a solid line, and the simulated values, which are shown with a dashed line, for a 4 year period of record. In order to make this plot, the first water year, 1970, was used to calibrate the model in a very reasonable fashion.

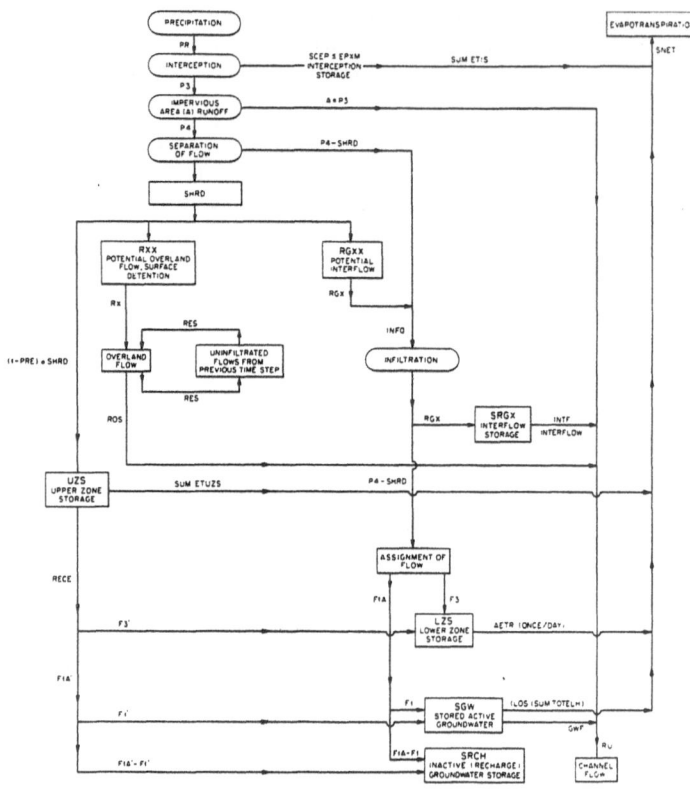

Fig. 10.  Stanford watershed model schematic of water
movement in a land area segment.

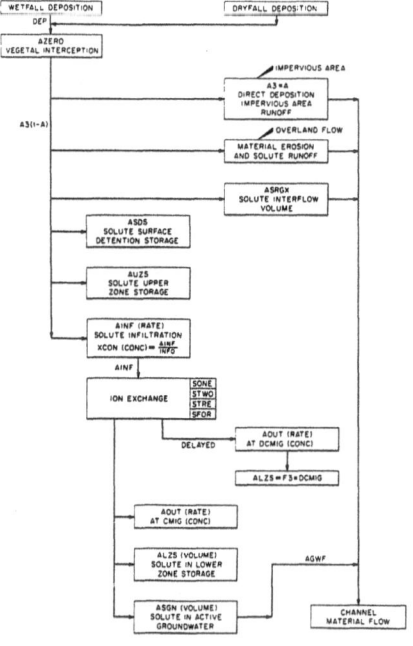

Fig. 11.  Material flow schematic in a land
area segment.

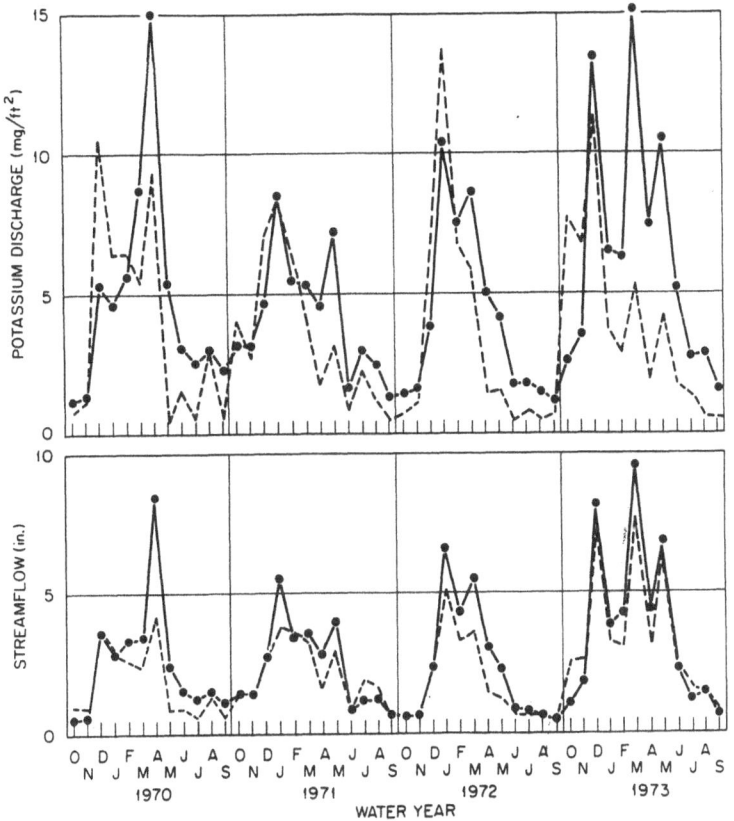

Fig. 12.  Streamflow and potassium transport for Walker Branch
          Watershed, Tennessee.  Black Dots-Experiment, Dashed
          lines-Simulation.

The transition of the scope of the UTM from consideration only of
trace heavy metals to inclusion of organic chemicals has been brought
about by the development done for the EPA [12].  The processes that have
been added during the early part of this project are shown in Fig. 13 and
Fig. 14 include, as seen there, various volatilization processes.
Volatilization occurs from the litter and from the soil surface as well as
from the soil-incorporated chemical.  Infiltrated chemical and
volatilization from the stream channel are handled by the model as well.
In addition, the degradation of organics and other chemicals has been
incorporated in the model to include hydrolysis, biodegradation in both
the soil and the stream, and finally photolysis.  These processes are
those which tend to dissipate a chemical released from a given line,
point, or area source.  In turn, volatilization serves as a secondary
source of emission for calculations in the Atmospheric Transport Model.

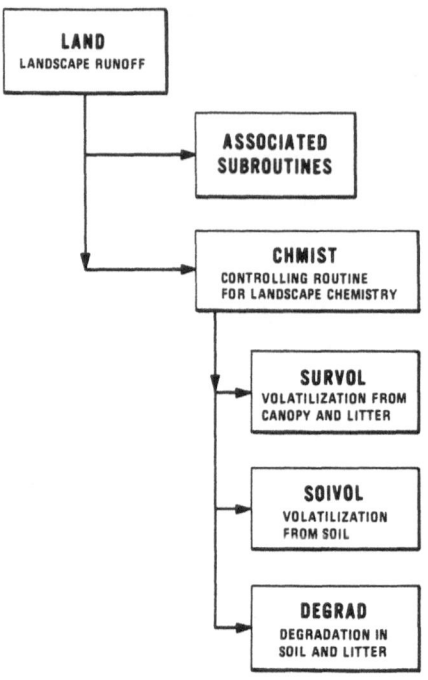

Fig. 13. Organic processes added to the LAND Routines.

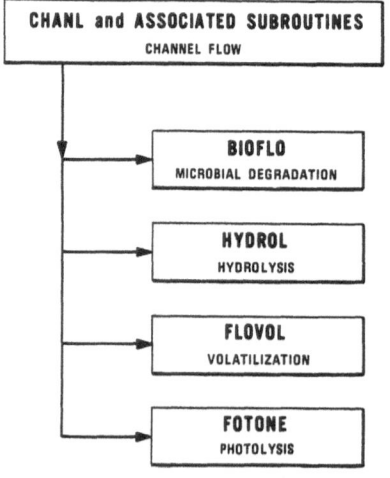

Fig. 14. Organic processes added to the CHANL Routines.

As an example of these processes within the terrestrial compartment, Fig. 15 shows the various variables that affect the processes in the terrestrial compartment. For example, during degradation the chemical in the soil experiences microbial activity, temperature dependent reactions, moisture, and adsorption by the soil in the process of undergoing degradation. Similiarly we see that rainfall, solubility, and soil adsorption affect leaching from the soil and lead finally to transport in interflow to the ground water. The processes are underlined in Figure 15, and the various factors that affect each process are shown within the arrow that points from the center to the process. Volatilization is, of course, an extremely important process for organic chemicals, leading directly to the atmosphere, and it is affected by the volatility, moisture, temperature, plant cover, and-again-soil adsorption. The surface runoff is affected as shown by rainfall, snow melt, solubility, and also by soil adsorption. Common to all these processes that are shown is the fact that soil adsorption is one of the more important processes that moderates the subsequent pathways taken by the pollutant.

The aquatic compartments are shown in Fig. 16, and the chemical processes again are underlined. Those variables that affect the processes are shown within the arrows that connect the process to the center, labeled "dissolved pollutant chemical". In photolysis, one notes that light intensity, molar adsorptivity, quantum efficiency, and sensitizers all affect the process of photolysis within the aquatic compartment. Biodegradation is affected by the microbial numbers, the nutrients available, the temperature, and biosorption. The sorbed chemical, which may well be carried by the sediment, is affected by the chemical properties of the chemical, the sediment organic carbon content, and the sediment particle size. These properties moderate sorption and desorption which then lead during runoff to hydrolysis. The amount is affected by the pH and chemical properties of the stream. Volatilization has already been discussed in terms of its effect on the terrestrial compartment, and it has an equally strong effect within the aquatic compartment. Volatilization is here also affected by the temperature of the water, by the wind and the current velocity, and by the volatility of the compound. The input of toxicant that leads to all these processes comes from runoff and from that atmospheric washout that deposits the chemical directly into the stream channel.

### Description of the Sediment Transport Model SEDMNT

The effect of sediment on the aquatic transport and sorption or desorption of the chemical has been noted in an earlier paragraph. The present sediment transport submodel is called SEDMNT [13,14]. The next figure, Fig. 17, shows the various compartments considered by the sediment transport submodels SEDMNT within each reach of a stream channel. As we see in this figure, there is a suspended load layer, a bed layer, and a resident bed layer. The fluxes among these various layers are calculated based on Stokes' law fall velocities, the shear stresses involved within each layer, particle sizes of the sediment, and the load carrying capability of the stream channel. The thicknesses of the different layers can change during the course of a run, except that the resident bed layer generally has a certain defined depth.

Matter sorbed to sediment can be transported by this submodel in the stream channel. The sediment transport model is also fed sediment material by the erosion submodel that exists within the LAND module of the Unified Transport Model. The material eroded from the landscape surface brings with it sorbed contaminant, which can in turn exchange with the sediment particles in the stream channel.

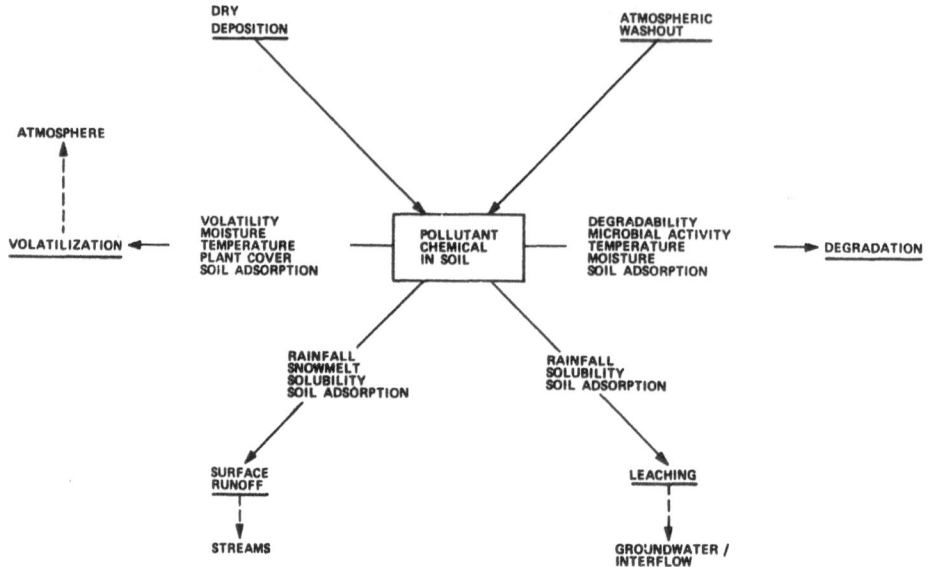

TERRESTRIAL COMPARTMENT

Fig. 15. Major physical and chemical interactions in the Terrestrial Compartment.

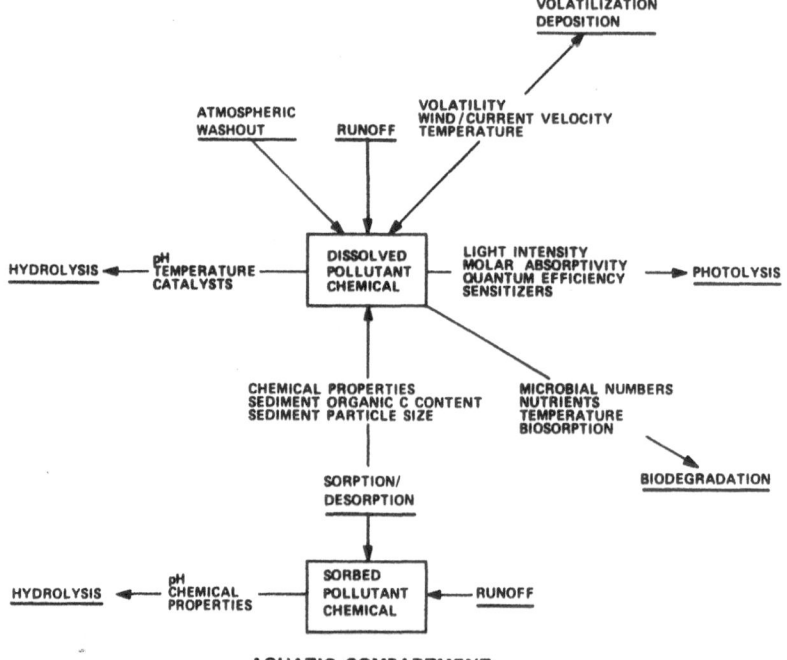

AQUATIC COMPARTMENT

Fig. 16. Major physical and chemical interactions in the Aquatic Compartment.

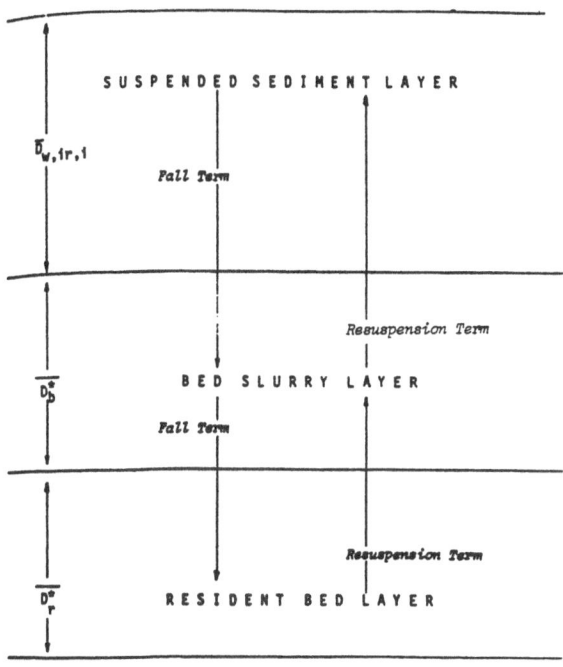

Fig. 17.  Three sediment layers and the fluxes between them.

## Variations on the UTM-TOX Model

At this point all three basic modules of the Unified Transport Model have been described.  These modules are the atmospheric transport model ATM, the hydrologic transport model WHTM, and the sediment transport model SEDMNT.  The most time consuming calculations for a simulation of one year of time with the model are incurred if the sediment transport model is used.  Sediment transport is a detailed calculation which couples together several different types of conservation and stress equations.  These equations require a considerable amount of computer time.  The results of this detailed calculation, however, are quite valuable to interpretation of the transport and fate of a contaminant, organic or inorganic.  Also, the model can be run without calling the sediment model.

In the next figure, Fig. 18, one sees the many processes that occur within a typical terrestrial compartment.  These processes are shown on a scale that is intermediate to the scale shown previously, in that it is neither an extremely large watershed nor is it an extremely small plant.  The scale here is appropriate to that for tree.  The tree experiences evapotranspiration and uptake through the roots as shown in Fig. 18.  Various types of decomposition occur within the litter layer, and food chain models can be applied within that layer.  The flow of material from the canopy down can be modeled by appropriate models coupled into the Unified Transport Model.  The scale of the Unified Transport Model, by coupling in various models with finer or coarser resolutions, can be applied to different scenarios.

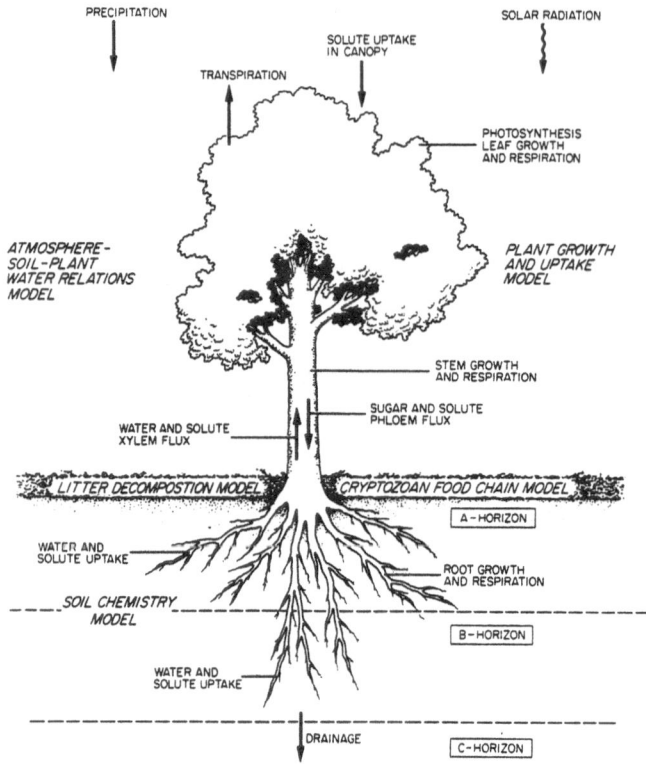

Fig. 18.  Plant growth processes in a tree and the
soil horizon.

As an example, in Fig. 19, we see another realization of the Unified
Transport Model.  Here we have the atmospheric transport model, available
to either the UTM-TOX module or the UTM-SPL (Soil, Plant, Litter) module.
Some of the UTM-SPL models have been mentioned earlier and some are under
development.  The plant growth model is called CERES [15] and replaces the
LAND module functions in many of the aspects that are concerned with soil
mechanics and transport through this matrix.  In addition, one can replace
the larger scaled LAND routine, couple other routines to it, or define
alternate soil moisture routines such as SOGMOD [11] or AGTEHM [5] (see
Fig. 5).  Further, soil chemistry models such as SCEHM [4] can be coupled
to the TEHM model for soil moisture and plant growth.

Looking at each of these models in greater detail would take too
long.  However, if one picks the SCEHM [4] model to study in more detail,
Fig. 20 shows the representation of the different profiles of the soil and
of the different chemical layers with which the contaminant reacts.  There
are several different moisture layers possible, and these can be chosen by
the user to correspond to the A-horizon, B-horizon and the C-horizon.
Within each one of these layers there are sublayers that are called "soil
chemistry layers" which can have different chemical reactivities.  For
example, the A-horizon can be broken up into two or more different soil
chemistry layers, each of which can have a different exchange coefficient.

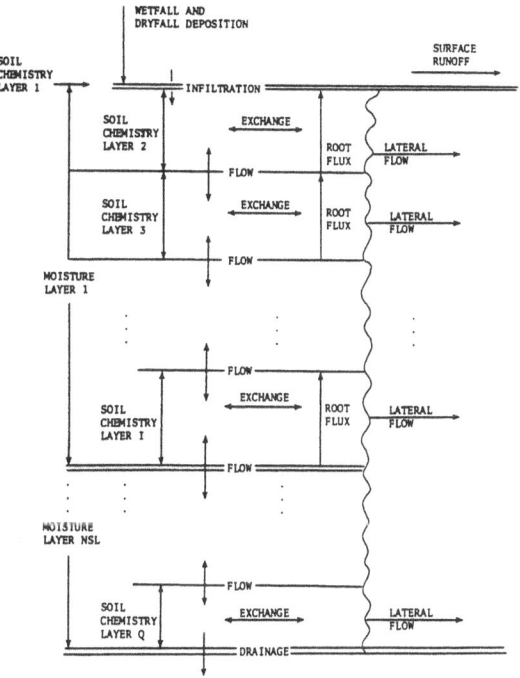

Fig. 20.  Soil chemistry layers and fluxes within
the SCEHM model.

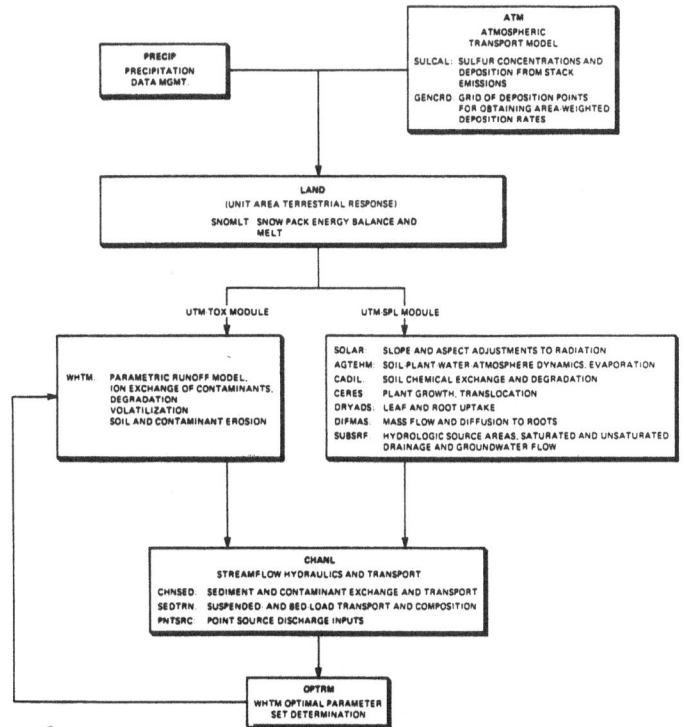

Fig. 19.  Current research with UTM–TOX couples to LAND
through the UTM–SPL (Soil, Plant, Litter)
Module.

This flexibility allows one to model the higher organic matter content in the upper part of the A-horizon differently from the lower parts of the A-horizon. Then a complete soil profile description can allow a change of exchange coefficients in traversing the A-horizon to the B-horizon, etc. The initial material is fed in at the top through wetfall and dryfall deposition and through infiltration to lower soil layers. Typically, these upper soil layers are fed from the litter, unless they lie in a part of the soil horizon that is bare and uncovered by vegetation.

## Further Crooked Creek Watershed Applications

A lead transport budget [16,17] for Crooked Creek Watershed in Southeastern Missouri is shown in Table 1, as calculated with UTM-TOX. This table shows for a one year period simulated values for the various transport mechanisms itemized there. Starting with the month of October, as shown in the left hand column of Table 1, the second successive column presents the atmospheric deposition measured in milligrams per square meter per month. These are results calculated as described previously with the Atmospheric Transport Model. The third column shows the amount of lead transported in surface waters, the surface hydrologic transport, again in the same units. The fourth column shows the amount of lead adsorbed by the soil at the site, again in milligrams per square meter per month. From these three numbers, which describe how the deposited material is distributed, one can calculate a net accumulation within the landscape which is shown in the fifth column from the left. The accumulation tends to be very large in many of the months. Finally, the next column shows how much water runs off in the process of transport. From that column, given the surface hydrologic transport of lead, one can calculate the lead concentration within the stream.

The atmospheric deposition is very large for most of these months. It dominates the amount transported away by surface runoff by more than an order of magnitude. Atmospheric deposition is of the same order of magnitude as that sorbed by the soil, but it is larger in general. Finally, the surface runoff of water is largest in the winter months and in the spring when there is a large amount of precipitation. The lead concentration calculated in the last column can be seen after the first three months to increase from about 0.3 parts per million by mass to values as large as 4.0 parts per million in the month of June. During the whole summer these values maintain levels that are larger than 2.0 parts per million, a high value for the concentration of lead in any stream [18].

## Data Needs for the UTM-TOX

The Unified Transport Model is a large transport model that consists of more than 10,000 card images. All of the different capabilities of the program do not have to be exercised at any one run of the model. However, the backbone of the model still consists of approximately 8,000 cards, without having the SEDMNT model incorporated. Execution times for the UTM are relatively small: a complete run for one watershed consisting of two segments and three reaches can be made in a CPU time of less than one minute on the IBM system 370/3033, without the sediment transport model. With the sediment transport model incorporated, this run time is increased to approximately three and a half minutes.

The data requirements for the Unified Transpot Model, considering the scope of its application and of the results that can be obtained, are reasonable. Approximately 510 data cards must be provided for a run which simulates a reasonably sized watershed. Figure 21 shows the distribution

Table 1. Lead Transport Budget for Crooked Creek Watershed

| Month | Atmospheric Deposition (mg/m$^2$/mo.) | Surface Hydrologic Transport (mg/m$^2$/mo.) | Adsorbed by Soil (mg/m$^2$/mo.) | Net Accumulation ($10^6$mg/m$^2$/mo.) | H$_2$O Runoff (ppm by mass) | Pb Concentration |
|-------|------|-----|------|------|-------|-----|
| 10 | 497 | 18 | 87 | 392 | 2.0 | * |
| 11 | 477 | 17 | 154 | 306 | 2.5 | * |
| 12 | 479 | 23 | 537 | -81 | 70.6 | * |
| 1 | 432 | 13 | 64 | 355 | 17.5 | 0.3 |
| 2 | 432 | 19 | 147 | 266 | 42.1 | 0.45 |
| 3 | 433 | 20 | 124 | 239 | 37.1 | 0.54 |
| 4 | 769 | 62 | 620 | 87 | 126.3 | 0.49 |
| 5 | 767 | 25 | 142 | 600 | 25.7 | 0.97 |
| 6 | 767 | 21 | 72 | 674 | 5.1 | 4.1 |
| 7 | 293 | 18 | 267 | 8 | 6.9 | 2.6 |
| 8 | 293 | 15 | 234 | 44 | 5.8 | 2.6 |
| 9 | 294 | 23 | 339 | -68 | 7.8 | 2.9 |
| TOTAL | 5933 | 274 | 2787 | 2872 | 349.4 | |

*Model results are not as reliable during the initial start-up period.

of the use of these cards within the Unified Transport Model. The top part of the figure shows how many of the cards are read by the Atmospheric Transport Model, the PRECIP branch of the UTM, and by the successive branches. As one can see, the largest amount of cards is read within the LAND portion of the UTM. In the bottom part of this figure, we see the number of data items that are read by each of these successive links of the UTM. The number of data items read by the atmospheric transport model ATM is relatively small as compared to those read by the PRECIP link of the model. The number of data items read by PRECIP is larger than those read by LAND, which is the reverse of the situation shown at the top of the figure, in which more cards are read by LAND. Similarly, the number of data points read by the CHANL section of the code is much less than by either the PRECIP or LAND links. Finally, the number cards or the number of data points read for the chemical routines is much less than is required for the environmental parameters that are used in the rest of the program.

Various ancillary programs are available to aid users of the Unified Transport Model in carrying out preparation of this data. For example, some programs have been written to generate grid points [19] for calculation of atmospheric concentrations and depositions on regular patterns that are either polar or rectangular. In order to calculate values needed within the hydrologic model for slope and drainage area, a program called CATCH [20] takes digital topography data and, for the region covered by the data, subdivides it into watersheds, identifies the watershed divisor line, calculates the areas of the parts of the watershed into which it has been divided, calculates the average overland flow distance to the stream channel, and calculates the average slope of the watershed. These numbers are all needed in various parts of the hydrologic model WHTM; otherwise, they must be estimated using planimeters and topography maps. Reasonably good values can be obtained without application of CATCH, but more exact values are obtained with greater ease in use of the program described here.

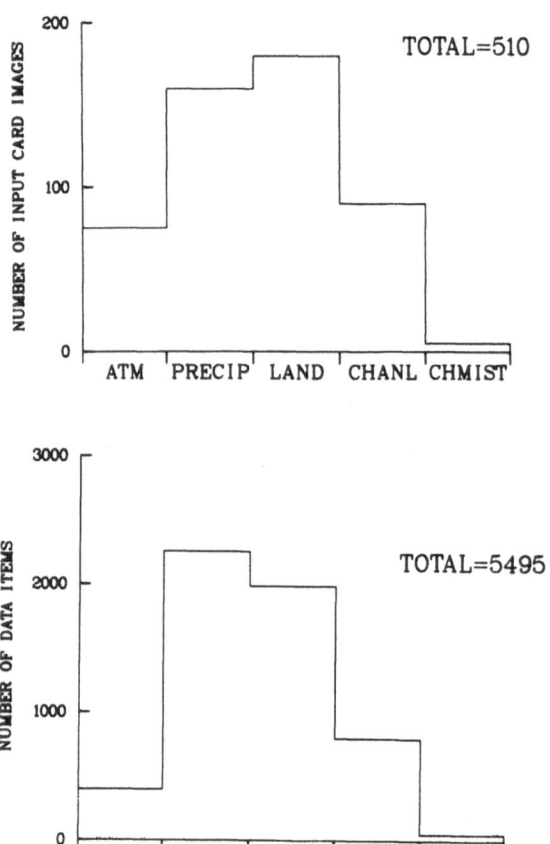

Fig. 21.  Data input requirements for the UTM-TOX.

SUMMARY

In summary, many of the capabilities of the UTM-TOX have been described in the paper. Other parts of the Unified Transport Model are still undergoing development. This model has been applied to several different cases involving research watersheds such as the Walker Branch watershed here at ORNL and a contaminated site near a lead smelter in Missouri. In all of the cases to which the model has been applied, it has yielded information that was useful to the researchers studying the watershed. It has allowed greater insight into the processes which mediate the transport and the fate of those pollutants that the model is designed to handle. It has given for the first time a reasonable means of assessing a complete budget for material that is being emitted into the atmosphere from a given source and impacts a watershed. It provides a vehicle for calculation of the concentration of a toxicant in many compartments of the ecosystem. The model further provides for the numerical coupling of that concentration and duration information directly to other models which simulate plant growth and plant uptake.

Different realizations of the model have been applied to quite different situations. The AGTEHM [5] version of the Unified Transport Model has been applied to the study of irrigation in various arid climates, while the TEHM [3] model, which is a precursor of the AGTEHM, has been applied to study the flow of material through macropores. In rapid flow through macropores the toxicant is distributed quickly from the surface of the soil to deep within the soil profile without undergoing the process of percolation through all layers of the soil profile. The TEHM model has also been applied to study source area concepts in hydrology which mediate the rapid rise of the hydrograph during an intense rain storm. By virtue of that study the user is able to couple to this type of transport, which occurs during intense storms with great carrying capacity.

The application of the UTM-TOX to aquatic systems has only begun. For studies in which the flow can be assumed to be uniformly mixed within different reaches of the channel system, the standard UTM can be applied and provides time histories of concentrations which can be used for exposure calculations. For other more detailed calculations, the finite element and finite difference codes can be applied.

The funding provided by the EPA for the UTM-TOX project has spurred the development of many organic processes that were needed within the Unified Transport Model. Within this current development of the model, some of these processes are treated in a first approximation, that is, a useful representation of the process has been built into the model. It at first represents the time-averaged result of a process such as photolysis, instead of having within the model a detailed hour-by-hour photosynthetic model. The needs of the user in application of the Unified Transport Model must be considered. For most users it is felt that a coarser resolution for such a process will be adequate. However, in application of the model to certain responses of the ecosystem, one may need a much finer time resolution for detailed simulation of, for example, photosynthesis that occurs within the stream channel. The model can accommodate such resolution, and these studies will be natural outgrowths of an ongoing effort in further development of the Unified Transport Model-TOX.

# REFERENCES

1. C. F. Baes, C. L. Begovich, W. M. Culkowski, K. R. Dixon, D. E. Fields, J. T. Holdeman, D. D. Huff, D. R. Jackson, N. M. Larson, R. J. Luxmoore, J. K. Munro, M. R. Patterson, R. J. Raridon, M. Reeves, O. C. Stein, J. L. Stolzy, and T. C. Tucker, "The Unified Transport Model," in R. I. Van Hook and W. D. Shults (eds.), Ecology and Analysis of Trace Contaminants Progress Report October 1974 - December 1975, ORNL/NSF/EATC-22 (1976), pp. 13-62.

2. M. R. Patterson, T. J. Sworski, A. L. Sjoreen, M. G. Browman, C. C. Coutant, D. M. Hetrick, B. D. Murphy, R. J. Raridon, A User's Manual for UTM-TOX, the Unified Transport Model, ORNL-6064 (1984).

3. D. D. Huff, R. J. Luxmoore, J. B. Mankin, and C. L. Begovich, TEHM: A Terrestrial Ecosystem Hydrology Model, Oak Ridge National Laboratory Report ORNL/NSF/EATC-27 (1977).

4. C. L. Begovich and D. R. Jackson, Documentation and Application of SCHEM, Oak Ridge National Laboratory Report ORNL-NSF-EATC-16 (1975).

5. D. M. Hetrick, J. T. Holdeman, and R. J. Luxmore, AGTEHM: Documentation of Modifications to the Terrestrial Ecosystem Hydrology Model (TEHM for Agricultural Applications, ORNL/TM-7856 (1982).

6. R. A. Goldstein, J. B. Mankin, and R. J. Luxmoore, Documentation of PROSPER: A Model of Atmosphere-Soil-Plant Water Flow, EDFB-IBP-73-9 (1974).

7. W. M. Culkowski and M. R. Patterson, A Comprehensive Atmospheric Transport and Diffusion Model, ORNL/NSF/EATC-17 (1976).

8. R. J. Raridon, B. D. Murphy, W. M. Culkowski, and M. R. Patterson, The Atmospheric Transport Model for Toxic Substances (ATM-TOX), ORNL/CSD-94 (1984).

9. M. R. Patterson, J. K. Munro, D. E. Fields, R. D. Ellison, A. A. Brooks, and D. D. Huff, A User's Manual for the Fortran IV Version of the Wisconsin Hydrologic Transport Model, ORNL-NSF-EATC-7 (1974).

10. N. H. Crawford and R. K. Linsley, Digital Simulation in Hydrology: Stanford Watershed Model IV, Department of Civil Engineering, Stanford University Technical Report Number 39 (1966).

11. J. O. Duguid and M. Reeves, Material Transport through Porous Media: A Finite-Element Galerkin Model, ORNL-4928 (1976).

12. M. G. Browman, M. R. Patterson, T. J. Sworski, Physicochemical Processes in the Environment: Background Information for the ORNL Unified Transport Model for Toxicants (UTM-TOX), ORNL-5854 (1983).

13. D. M. Hetrick, M. R. Patterson, and A. L. Sjoreen, SEDMNT: A Sediment Transport Submodel Based on Hydrodynamic Principls for the Unified Transport Model, ORNL/TM-7831 (1982).

14. D. E. Fields, Simulation of Sediment and Trace Contaminant Transport with Sediment/Contaminant Interaction, ORNL/NSF/EATC-19 (1976).

15. K. R. Dixon, R. J. Luxmoore, and C. L. Begovich, CERES - A Model of Forest Stand Biomass Dynamics for Predicting Trace Contaminant, Nutrient, and Water Effects, ORNL/NSF/EATC-25 (1976).

16. M. R. Patterson, C. F. Baes, Jr., C. L. Begovich, W. M. Culkowski, K. R. Dixon, D. E. Fields, D. D. Huff, N. M. Larson, R. J. Luxmoore, J. K. Munro, R. J. Raridon, M. Reeves, and T. C. Tucker, "Development and Application of the Unified Transport Model", in W. Fulkerson, W. D. Shults, and R. I. Van Hood (eds.), Ecology and Analysis of Trace Contaminants Progress Report October 1973-September 1974, ORNL-NSF-EATC-11 (1974).

17. M. R. Patterson, J. K. Munro, and R. J. Luxmoore, Simulation of Lead Transport on the Crooked Creek Watershed, 9th Annual Conference on Trace Substances in Environmental Health, University of Missouri, Columbia, June 9-12, 1975.

18. M. R. Patterson, C. L. Begovich, and D. R. Jackson, <u>Environmental Transport Modeling of Pollutants in Water and Soil</u>, NBS CONF 76-0512-1 (1976).

19. Private communication, B. D. Murphy, Generation of ATM Grid Points, ORNL (1985).

20. T. C. Tucker and D. E. Fields, <u>Digital Topography, Calculation of Watershed Area and Slope</u>, ORNL/ANSF/EATC-20 (1976).

THE FUGACITY APPROACH TO MULTIMEDIA

ENVIRONMENTAL MODELING

Donald Mackay and Sally Paterson


Department of Chemical Engineering
and Applied Chemistry
University of Toronto
Toronto, Ontario, M5S 1A4 Canada

## INTRODUCTION

It is now evident that for environmental controls and regulations to be effective they must include consideration of all media; air, water, soils, sediments, groundwater and the numerous particulate and biotic phases which may be present in dispersed form within these phases. The traditional approach of focusing on one medium in detail, studying it in one laboratory, and regulating it from one office is now clearly inadequate. There must be a comprehensive, total system approach to the environment. This is not to suggest that specialization is inappropriate. On the contrary it is an essential feature of modern environmental science that a scientist devotes an entire working life to studying the detailed science of sediments or air particulates. Rather, it is suggested that in addition to these areas of depth there must be a complementary "overseeing", "synthesising" activity in which the entire forest of environmental activities is viewed, quantified and regulated.

We thus argue that there is a need to develop models or quantitative expressions of chemical fate in multimedia environments which give the "big picture" of chemical behavior. What appears to be emerging is a suite of models of varying degree of accuracy, scope and complexity, and with various purposes. In many respects the models are analogous to maps of various scales with varying degrees of suppression or expression of detail. A map or plan of the entire US serves one purpose, that of a city another, and that of a single house yet another. The maps or models are not in competition, they have different objectives.

The largest scale maps and models are often the most difficult to compile because their scope is vast, and difficult decisions must be made about what detail to exclude and include. There has been considerable progress in recent years in developing these models, largely as a result of the innovative approach of Baughman and Lassiter (1978) who suggested using evaluative models as a means of suppressing detail. Most notable is the EXAMS model (Burns et al., 1981) which has been successfully used in air-water-sediment modeling.

In this paper we review the status of one family of environmental models, the fugacity models, which have been developed in our group in the last ten years. We discuss some advantages and disadvantages of this approach and speculate on future developments. Before describing the models it is useful to note some basic features of the approach.

BASIC FEATURES OF THE FUGACITY APPROACH

Mathematical Equivalence

Fugacity is merely a surrogate for concentration. Fugacity expressions are ultimately algebraically identical to concentration expressions, thus a calculation done in one system can be done equally well in the other. It is analogous to calculating the product of several large numbers. It can be done directly by multiplication, or it can be done by adding logarithms. The net result is the same but the nature of the task may render one method preferable over the other. We suggest that situations exist in which fugacity is the preferable vehicle for calculation, and in which its use facilitates interpetation.

Fugacity as a Potential Quantity

Fugacity is an old, well established concept devised by G. N. Lewis in 1901. It has units of pressure and expresses the escaping tendency of a dissolved chemical from a phase. It is linearly related to concentration in most cases of high dilution, and as such it is much more convenient than chemical potential which is logarithmically related to concentration. Many find it helpful to view fugacity as an analog to temperature as used in heat transfer calculations. Interestingly in heat transfer technology, heat concentrations are rarely used and two-phase heat capacity partition coefficients are never used, but these concepts are used routinely in environmental mass transfer calculations. The advantage of using temperature over heat concentration is that it expresses not only how much heat is present but also the relative potential for heat movement directly, i.e. "hot spots" are rapidly seen as such. Viewing the multimedia environment through the lens of fugacity also shows such hot spots, and more important, which phases are close to equilibrium. Those who first observed that in water containing only $10^{-5}$ mg/L of DDT there lived fish containing 1 mg/L of DDT were amazed at this "biomagnification" and invoked explanations of the "uptake rate exceeds clearance rate" type. In reality, the fish and water fugacities are often equal and these rates are merely a consequence of the thermodynamic equilibrium status. These rates are not the fundamental determinants of the biomagnification phenomenon.

In summary, we find that it is helpful to view the environment in terms of fugacity as well as concentration.

Equilibria

At the heart of fugacity calculations is the linear relationship between fugacity (f) and concentration (C), namely –

$$C = Zf$$

The Z term or fugacity capacity is an expression of the capacity of a phase for the chemical in the same sense that heat capacity characterizes the phase capacity for heat. It can be viewed as a modified phase solubility of the chemical. Methods of estimating Z values have been described elsewhere (Mackay and co-workers; 1979, 1981, 1982, 1983a, 1983b, 1985).

A point worthy of note is that when two phases have equal fugacities and are in equilibrium, the ratio of concentrations $C_1/C_2$ becomes equal to $Z_1/Z_2$ and is of course $K_{12}$ the partition coefficient. A Z value can thus be viewed as "half" a partition coefficient.

One benefit of using Z values instead of K values is that there are usually fewer of them and the confusion resulting from the two coefficients $K_{12}$ and $K_{21}$ is entirely eliminated.

## Reactions and Transfer Coefficients

The conventional method of expressing first order reaction rates is by an expression of the type VCk where V is volume, C is concentration, and k a rate constant. In fugacity terms the rate is Df where D is a transformation parameter equivalent to VZk. There is no real advantage to using fugacity in such systems.

For interphase transfer by diffusion, the conventional method is to use an expression of the type

$$\text{Flux} = k_{01}A(C_1 - C_2K_{12})$$

where $k_{01}$ is an overall mass transfer coefficient, A is area, $C_1$ and $C_2$ concentrations, and $K_{12}$ a partition coefficent. The term $C_2K_{12}$ is essentially the concentration in phase 1 which would be equilibrium with $C_2$. If there are two resistances in series, $k_{01}$ may be expressed as

$$1/k_{01} = 1/k_1 + K_{12}/k_2$$

where $k_1$, and $k_2$ are individual mass transfer coefficients. These reciprocals are resistances.

In fugacity terms the equivalent expressions are

$$\text{Flux} = D_0(f_1 - f_2)$$

$$1/D_0 = 1/D_1 + 1/D_2$$

where $D_1 = k_1Z_1A$, $D_2 = k_2Z_2A$ and $D_0 = k_{01}Z_1A$

This is a little simpler, but the principal advantage is that when there are many such expressions (as occurs in multimedia calculations) there is never any ambiguity about whether to use $C_2K_{12}$ or $C_1/K_{12}$ or $C_2K_{21}$ and there is no need to carry these partition coefficients from phase to phase through the calculation. D values automatically include the appropriate Z. This leads to algebraic simplicity and to a reduced likelihood of errors.

The non-diffusive advective or convective transport of a chemical can also be expressed in fugacity terms as Df where D is GZ and G is the convective volumetric flux such that the flux is also GC.

Some exchange process rates such as fish-water transfer are conventionally expressed as characteristic uptake times or equilibrium times $\tau$. In fugacity terms the equivalent D is $VZ\tau$, or in the case of fish $VZk_2$ where $k_2$ is the clearance rate constant and V is the fish volume.

Finally in true diffusion situations such as a soil, the conventional expression is

$$\text{Flux} = BA\Delta C/Y$$

where B is a diffusivity, $\Delta C$ a concentration difference, and Y a diffusion path length. Again, this can be expressed as $D\Delta f$ where D is $BAZ/Y$.

The principal adavantage of the fugacity approach is that by converting all reaction and transport expressions into D values it is possible to compare them directly and determine which are the most important. The rates can often be added directly leading to algebraic simplicity. It is possible to express diverse processes such as advective or convective transfer, diffusion through various series and parallel resistances (in terms of diffusivities, mass transfer coefficients or uptake times) and reactions, in identical units, add them and compare them. We illustrate these benefits later.

When using conventional concentration expressions it is possible to encounter equations containing numerous partition coefficients, mass transfer coefficients, diffusivities, path lengths, rate constants and uptake times. The algebraic complexity tends to obscure the nature of the fundamental phenomena being simulated.

MODEL LINKAGE

As confidence in models grows there will be an inevitable linking of various component models into more comprehensive descriptions of chemical behavior. Examples are linking atmospheric deposition models to water or soil volatilization; sediment-water models to include fish; human and fish uptake models to include pharmacokinetics. A strong case can be made for using consistent systems of units and expressions in order that linkage is facilitated. We find that it is useful to carry the concept of fugacity from model to model. On what basis can we compare an effluent or air concentration of PCBs with a human milk concentration? Comparison of the fugacities is however meaningful.

EXAMPLES OF FUGACITY MODELS

Having established some of the advantages of fugacity models we now present a few applications of the models to illustrate the versatility of the concept.

1.  Steady State Evaluative Models

Figure 1 shows a level III or steady state evaluative model of DDT as described by Mackay et al. (1985). The reaction, advection and transport terms are illustrated, as are the relative concentrations, amounts and fugacities. The figure contains a great deal of detail and is more conveniently summarized in Figure 2 which illustrates the environmental distribution of a PCB. The dominant behavior characteristics of the chemical become obvious. It is believed that such models are useful for chemical premanufacture notification and assessment purposes in which disparate property data are gathered in an attempt to assess potential environmental behavior. It is important to note that this is a non-equilibrium model, ie compartment fugacities differ. There is a frequent misconception that when fugacity is used, equilibrium is assumed. This is not true.

This model has been quite widely distributed in diskette form and is easily used. Copies are available from the authors.

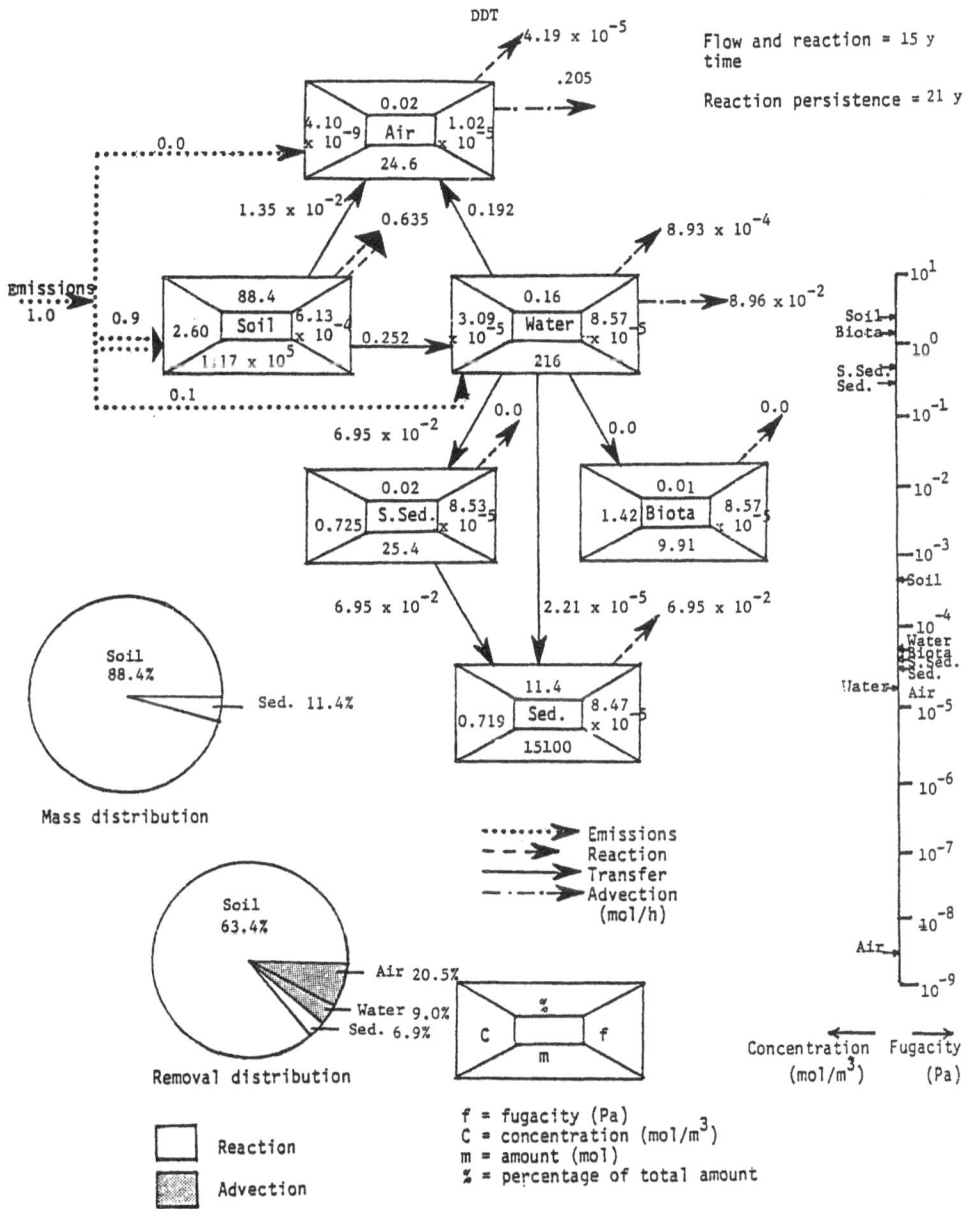

Figure 1: Level III Evaluation Model of DDT

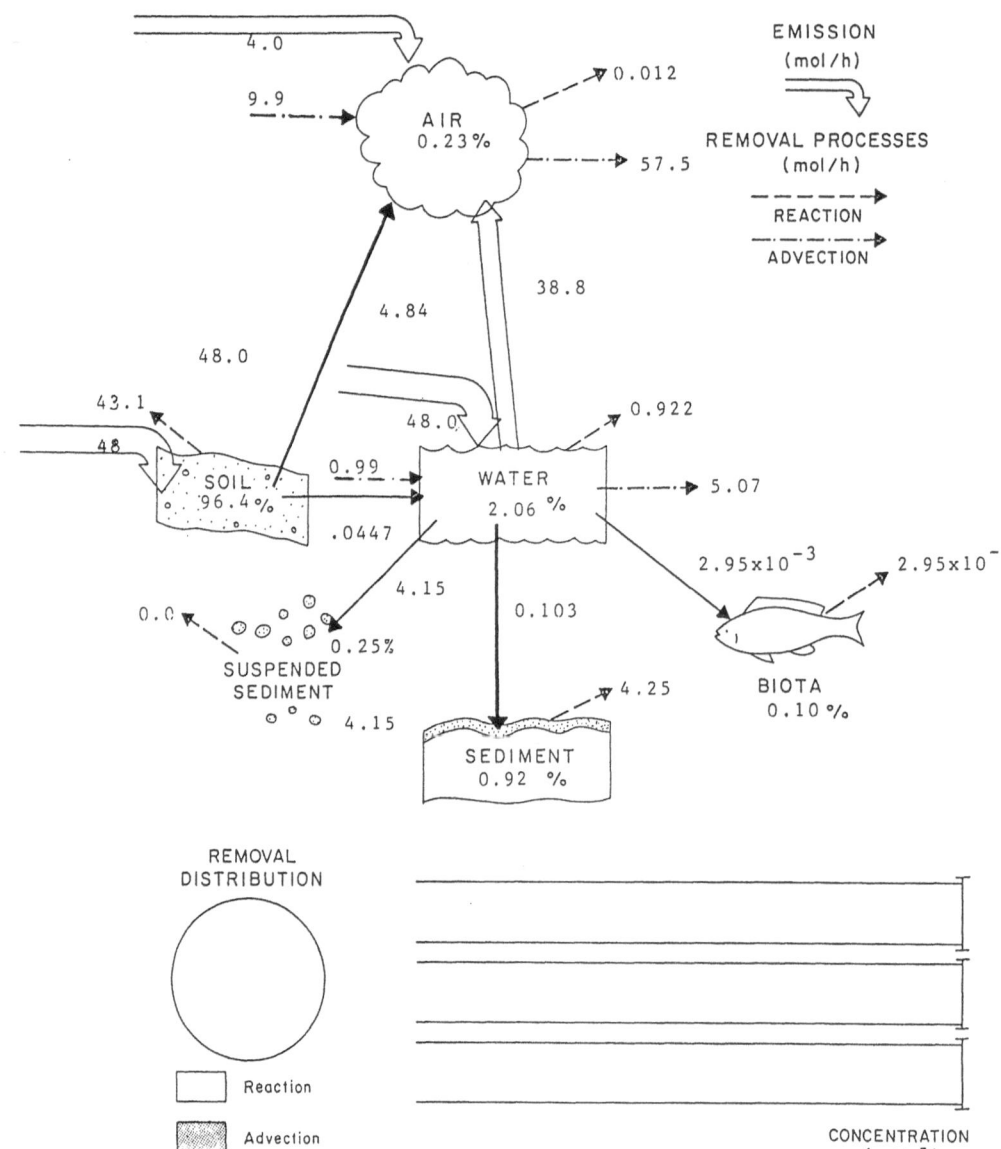

Figure 2: Environmental Distribution of a PCB

## 2. Unsteady State Evaluative Models

Figure 3 shows a level IV or unsteady state evaluative model describing system response to PCB contamination over a period of years. The relative response times are apparent. This type of model is most appropriate for application to situations in which a system is recovering from contamination by a persistent pollutant.

## 3. River Simulation

Figures 4 and 5 give an application of the Quantitative Water, Air, Soil Interaction (QWASI) Model applied to linear alkyl benzene sulphonate (LAS) in a stretch of river (Holysh et al., 1985). This example is interesting because it illustrates the inclusion of numerous reaction and transport processes in one model leading to a simple expression for water and sediment concentrations as a function of river distance. Figure 4 gives a listing of the various parameters in the model and their groupings, while Figure 5 gives the final equations and the fit to the real data.

One feature of this approach is that it may be impossible to deduce all the individual terms, but some combination of them is determined by the experimental data. In this case neither sediment-water diffusion rate or sediment deposition-resuspension rate is known but their total ($B_2$) is known. In other cases a combined reaction and volatilization rate may be estimated with the individual values being uncertain. The grouping of such process rates in terms of equivalent units facilitates data fitting and interpretation.

## 4. Fate of Soil-Applied Chemicals

A chemical present in soil is subject to volatilization, leaching, diffusion sorption and reaction. Figure 6 gives a pictorial representation of the fate of lindane in a soil using fugacity terms. Note how the various processes can be compared directly. This model is based on the excellent evaluative soil behavior model developed by Jury et al. (1983).

## 5. Air-Water Exchange

Figure 7 gives a representation of the rates of process of air-water exchange between a lake and the atmosphere including volatilization and wet and dry deposition (Mackay et al., 1985). The substance in this case is similar in properties to a PCB. The assembly of the processes in common units shows that it is possible for the chemical to cycle between air and water with intense periods of downward flux during rainfall, followed by more prolonged periods of upward flux. It is also apparent that a non-equilibrium steady state condition may exist as a result of the counteracting diffusive and non-diffusive processes.

## 6. Pharmacokinetics

For some time we have been exploring the role of fugacity as a means of elucidating the kinetics and equilibria of chemical uptake by animals. The earliest work resulted in accounts of fish bioconcentration factors and uptake and clearance rate constants (Mackay, 1982; Mackay and Hughes, 1984). More recently, we have extended this to human uptake. A fugacity version of the unsteady state (Ramsey and Andersen, 1984) styrene inhalation model has been devised (Paterson and Mackay, 1985) and a steady state version has been assembled which includes food uptake as well as inhalation. This is illustrated in Figure 8. Interestingly, the equations describing chemical fate as it circulates and partitions in body tissues are very similar in concept to those which describe behavior in the environment. A potentially

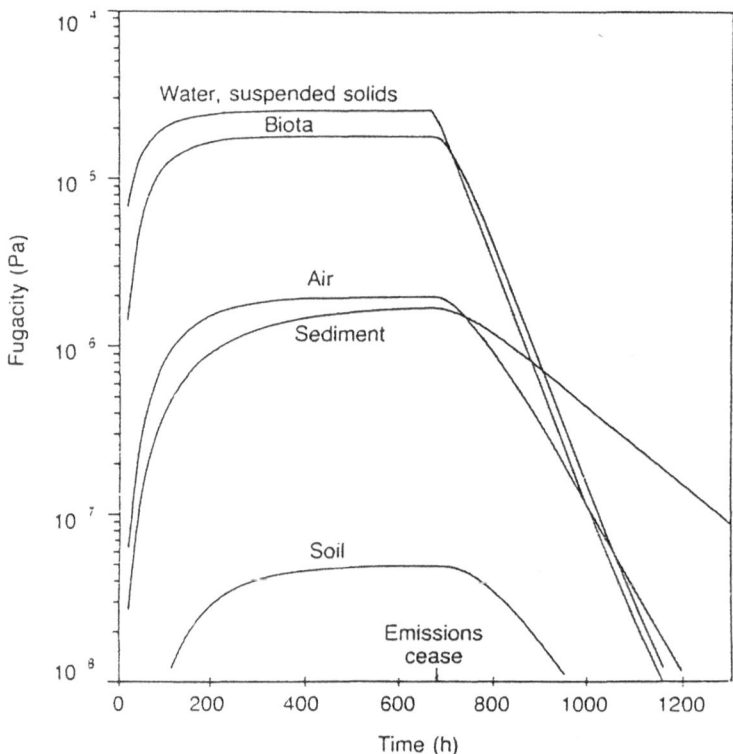

Figure 3: Level IV or unsteady-state behavior with emissions into water of 1 mol/h ceasing after 670 h

| | Kinetic Term | Dimension | Z value | B or D Value | B or D Group | Fugacity |
|---|---|---|---|---|---|---|
| Sediment Burial | $J_B = 0$ | | $Z_S = 8910$ | 0 | B1 | $f_S$ |
| Sediment Transformation | $k_S = 0.04$ | $h_S = .003$ | $Z_S = 8910$ | 1.068 | 1.068 | |
| Sediment Resuspension | $J_R = 0$ | | $Z_S = 8910$ | 0 | B2 | $f_S$ |
| Sediment→Water Diffusion | $K_T = 7.80\times10^{-3}$ | | $Z_W = 18$ | 0.140 | 0.140 | |
| Water→Sediment Diffusion | $K_T = 7.80\times10^{-3}$ | | $Z_W = 18$ | 0.140 | B3 | $f_W$ |
| Sediment Deposition | $J_D = 0$ | | $Z_P = 8910$ | 0 | 0.140 | |
| Water Transformation | $k_W = 0.021$ | $h_W = 0.47$ | $Z_W = 18$ | 0.178 | B4 | $f_W$ |
| Water→Air Volatilization | $k_V = 0$ | | $Z_W = 18$ | 0 | 0.178 | |
| Water Outflow | $G_J = 3600$ | | $Z_W = 18$ | 64800 | D5 | $f_W$ |
| Suspended Sediment Outflow | $G_Y = 0$ | | $Z_P = 8910$ | 0 | 64800 | |
| Air→Water Absorption | $K_V = 0$ | | $Z_W = 18$ | 0 | B6 | $f_A$ |
| Air Particle Deposition | $J_Q = 0$ | | $Z_Q = 0$ | 0 | | |
| Air Rain Deposition | $J_M = 0$ | | $Z_W = 18$ | 0 | 0 | |
| Water Inflow | $G_I = 3600$ | | $Z_W = 18$ | 648000 | D7 | $f_I$ |
| Suspended Sediment Inflow | $G = 0$ | | $Z = 8910$ | 0 | 64800 | |
| River Properties | Sediment Depth $h_S = 0.003$ | Water Depth $h_W = 0.47m$(distance weighted average depth) | Water Velocity $U = 1008$ m/h | Width $W = 7.6m$ | Water Flow = $G_I = G_J$ = $U \times W \times h_W$ = 3600 m³h | |

Figure 4: QWASI Calculation Form for a River

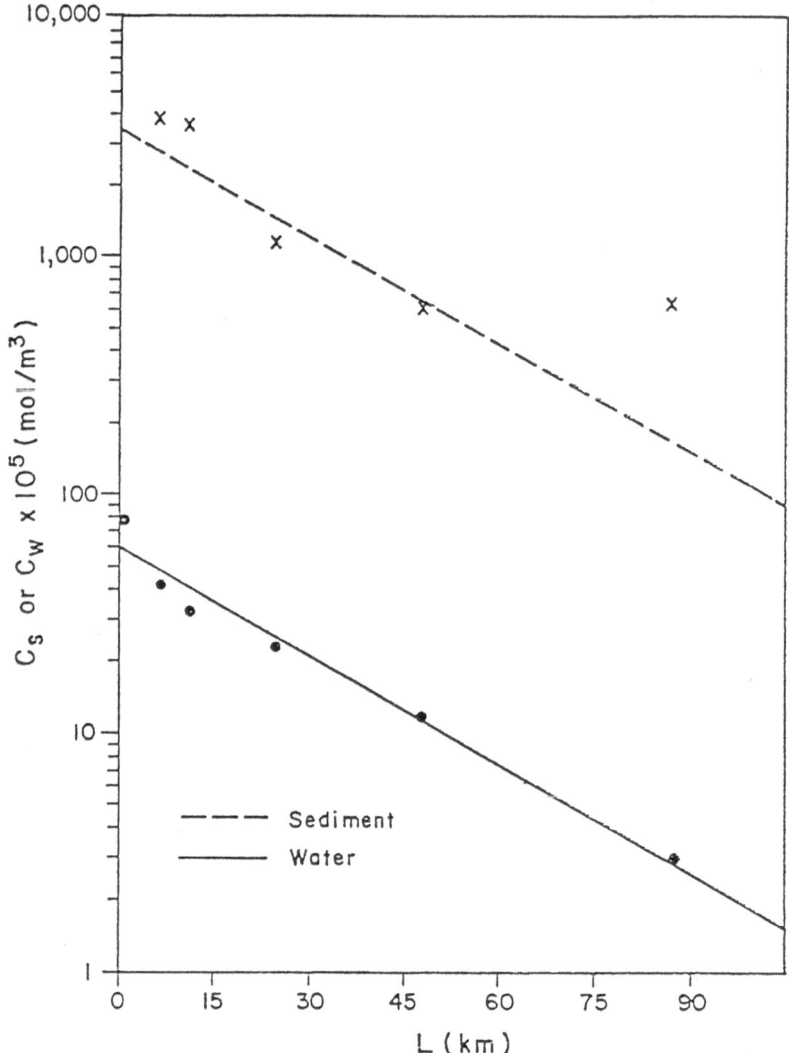

Figure 5: Plot of Water and Sediment Concentrations ($C_w$ or $C_s$) of LAS in Rapid Creek.

$$C_w = C_{wo} \exp(-K_1 L)$$

$$C_s = f_w Z_s B_2 / (B_1 + B_2)$$

$$K_1 = \frac{[B_1(B_2 + B_4) + B_2 B_4]W}{D_5(B_1 + B_2)}$$

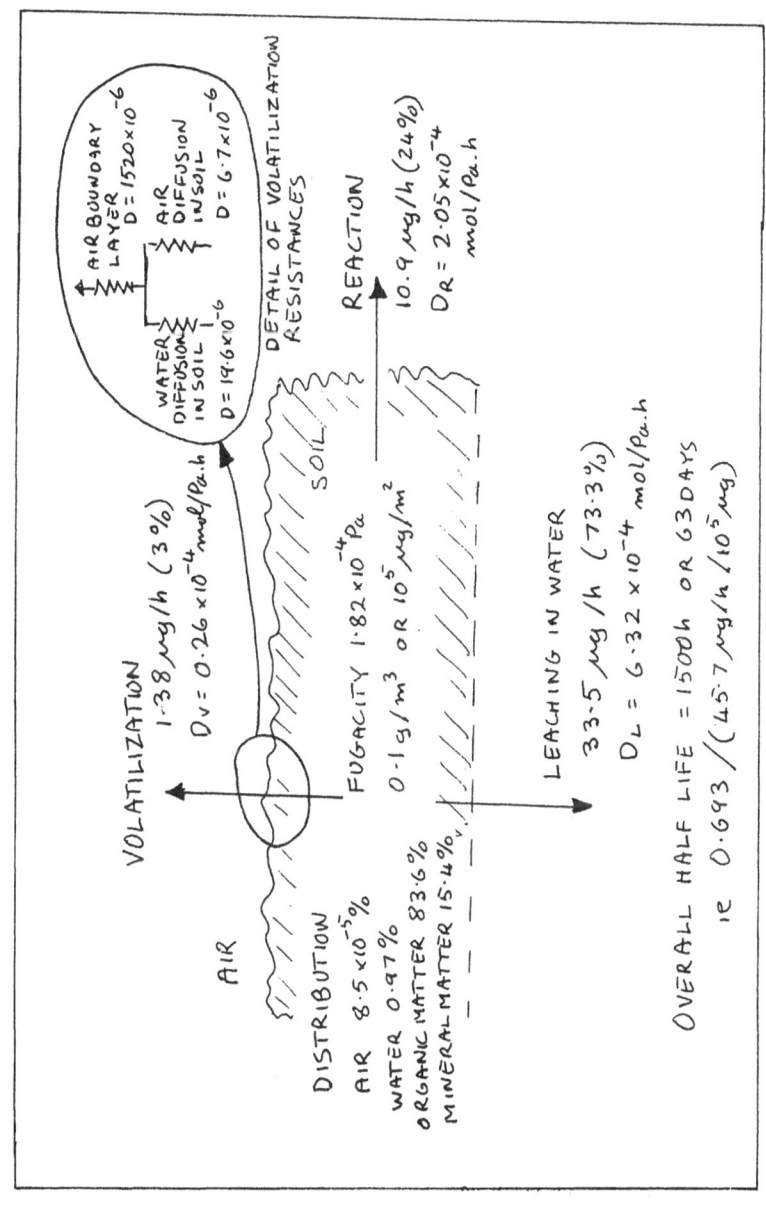

Figure 6: Evaluative Fate Diagram of Lindane in Soil

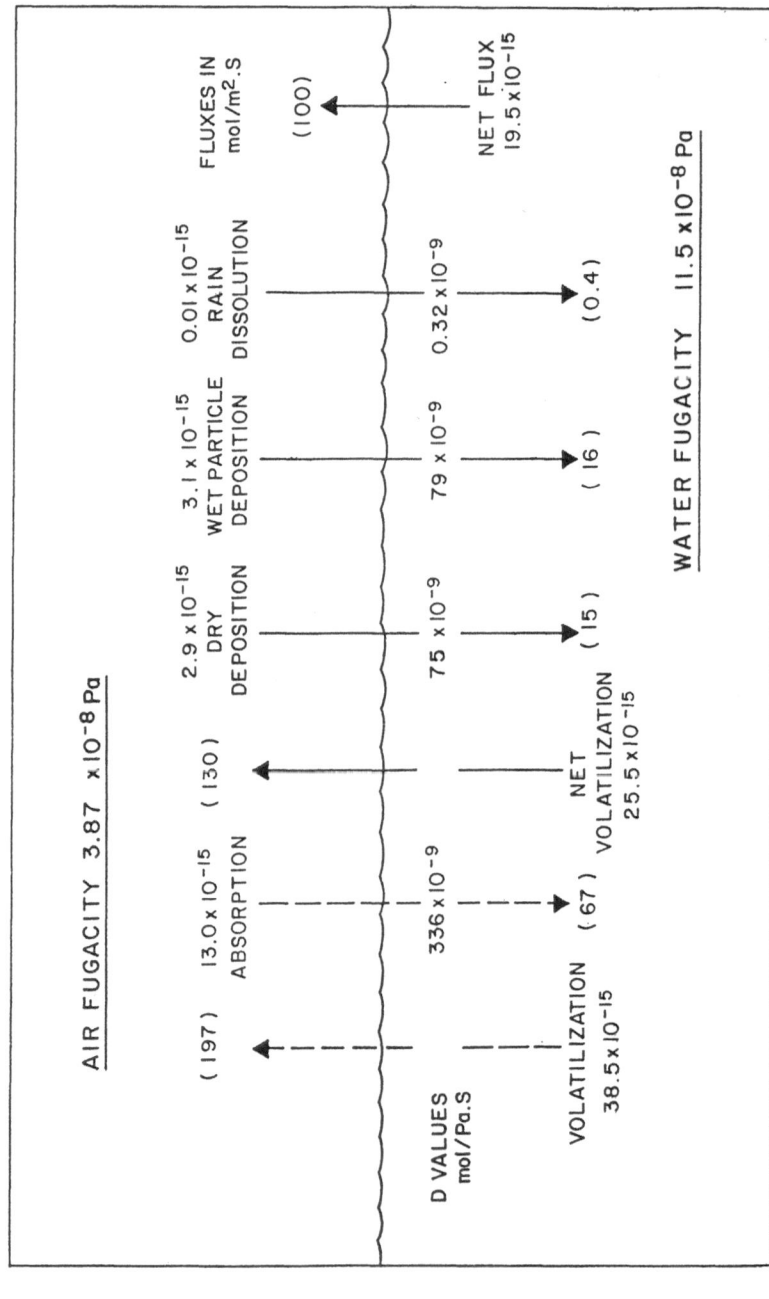

Figure 7: Air-Water Exchange Processes for a PCB like substance. Quantities in parentheses are percentages of the net flux and illustrate the cycling phenomena.

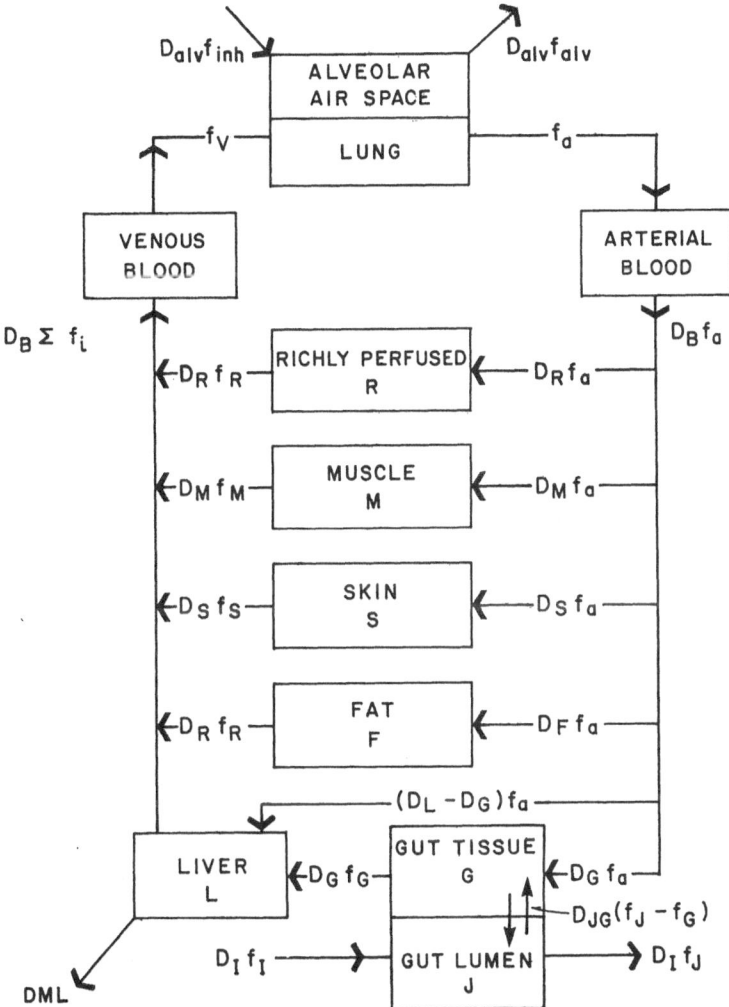

Figure 8:  Steady State Fugacity-Based Pharmacokinetic Model with Multiple
Exposure Routes.

useful step may be to link the real or evaluative environmental calculations to human exposure estimates and to pharmacokinetics in evaluations of chemical fate and effects.

In this way human and other tissues become mathematically linked as additional media to the environmental media and it becomes possible to track toxicant movement from emission to target organ. This is facilitated by the use of fugacity at all stages.

CONCLUSIONS

In this paper we have attempted to show that fugacity has a role to play in multimedia calculations. It simplifies certain operations, it provides additional insights and it facilitates the linking of models. It has an attractive elegance in its capacity to simplify complex situations. It must be appreciated that it does not enable calculations to be done which cannot be done by conventional methods. It suffers from some thermodynamic difficulties when treating inorganic substances, ionizing substances, surfactants and macromolecules. Finally, using it requires the intellectual effort of learning about and understanding a new, (and to some) a strange concept. But it is our conviction that the benefits outweigh these disadvantages and that its use will increase, especially as young environmental scientists become more familiar with the concept.

REFERENCES

Baughman, G. L., and Lassiter, R. R., 1978, in "Estimating the Hazard of Chemical Substances to Aquatic Life", J. Cairns, Jr., K. G. Dickson, A. W. Maki, eds., ASTM Tech. Pub. 657, Philadelphia, PA.

Burns, L. A., Cline, and D. M., Lassiter, R. R., 1981, Exposure analysis modeling system (EXAMS): User manual and system documentation, U.S. EPA Environ. Res. Lab., Athens, GA.

Holysh, M., Paterson, S., Mackay, D., and Bandurraga, M. M., Assessment of the environmental fate of linear alkylbenzenesulphonates, submitted to Chemosphere, 1985.

Jury, W. A., Spencer, W. F., and Farmer, W. F., 1983, Behavior assessment model for trace organics in soil: I model description, J. Environ. Qual., 12:558, also 13:573 and 13:580.

Lewis, G. N., 1901, The law of physico-chemical change, Daedalus, Proc. Am. Acad., 37:49.

Mackay, D., 1979, Finding fugacity feasible, Environ. Sci. & Technol., 13:1218.

Mackay, D., and Paterson, S., 1981, Calculating fugacity, Environ. Sci. & Technol., 15(9):1006.

Mackay, D., and Paterson, S., 1982, Fugacity revisited, Environ. Sci. & Technol, 16:654.

Mackay, D., Joy, M., and Paterson, S., 1983a, A quantitative water, air, sediment interaction (QWASI) fugacity model for describing the fate of chemicals in lakes, Chemosphere, 12:981.

Mackay, D., Paterson, S., and Joy, M., 1983b, A quantitative water, air sediment interaction (QWASI) fugacity model for describing the fate of chemicals in rivers, Chemosphere, 12:1193.

Mackay, D., Paterson, S., Cheung, B., and Neely, W. B., 1985, Evaluating the environmental behavior of chemicals with a level III fugacity model, Chemosphere, 14:335.

Mackay, D., Paterson, S., and Schroeder, W. H., A model describing the rate of transfer processes of organic chemicals between atmosphere and water, submitted to Environ. Sci. Technol., 1985.

Mackay, D., 1982, Correlation of bioconcentration factors, <u>Environ. Sci. & Technol.</u>, 16:274.

Mackay, D., and Hughes, A. I., 1984, A three parameter equation describing the uptake of organic compounds by fish, <u>Environ. Sci. & Technol.</u>, 18:439.

Paterson, S., and Mackay, D., 1985, The fugacity concept in environmental modeling, <u>in</u> The Handbook of Environmental Chemistry, vol. 2, pt. C, O. Hutzinger, ed., Springer-Verlag, Heidelberg.

Paterson, S., and Mackay, D., 1985, A pharmacokinetic model of styrene inhalation using the fugacity approach, <u>Toxicol. Appl. Pharmacol.</u>, (In Press).

Ramsey, J. C., and Andersen, M. E., 1984, A physiologically based description of the inhalation pharmacokinetics of styrene in rats and humans, <u>Toxicol. Appl. Pharmacol.</u>, 73:159.

# A CHEMICAL RUNOFF MODEL

W. Brock Neely and George R. Oliver

The Dow Chemical Company
P.O. Box 1706
Midland, MI  48640

## INTRODUCTION

Extensive surface and groundwater monitoring programs have been installed throughout the country. When a chemical is detected in the discharge from a watershed, modeling is one approach to determine the reason for the presence of the chemical and to predict future trends in chemical levels. Modern environmental models, such as the Hydrological Simulation Program - Fortran (HSPF), are designed to evaluate the fate of chemicals in such situations (1). However, the large number of inputs needed to describe a complex watershed of considerable size require a significant amount of preparation time. This paper describes a technique applicable to early phase review of chemical movement in terrestrial environments. Such a preliminary review will help identify areas that might need additional and more comprehensive analysis. Once the proposed model has been presented, a partial validation will be made based on the extensive field study performed by McCall et. al. (2).

## MODEL

The starting point will be the approach introduced by Mackay (3) in 1979. This model is illustrated in Figure 1 and has been referred to as the ''Unit World'' (4). The name was derived from the observation that the dimensions and properties were designed to simulate important conditions on earth (Table 1). Previous use dealt almost exclusively with the steady state equilibrium situations (5-11). In order to investigate runoff, a dynamic approach becomes mandatory. Before such a model can be designed, it is essential to estimate the transfer coefficients that control the chemical distribution between the compartments in Figure 1.

Prior to discussing these parameters the concept of fugacity (f) will be reviewed. This term has units of pressure and may be visualized as a measure of the ''escaping tendency'' of a molecule from one phase to another (3). At equilibrium fugacities are equal. Consequently, on the approach to equilibrium diffusion processes must be driven by the $\Delta$ f between two phases. Therefore, in a non-equilibrium situation it is necessary to determine the fugacity for each compartment. Once this is accomplished, the concentration (C) is estimated using Equation 1.

$$C = Z \ f \tag{1}$$

where $Z$ = fugacity capacity factor (moles/m$^3$/Pa)

Mackay and Paterson (5, 6) have reviewed the procedure for calculating $Z$ in the different compartments.

In addition to transport being controlled by concentration gradients, there is a second mechanism of interphase transport movement. This occurs when a chemical moves in association with another material like sediment or water. Diffusive transport is reversible while material transport is essentially one way. Both processes will be considered.

For the watershed portion of the Unit World the rate constants in Figure 1 will be discussed. As chemicals are deposited on the ground either intentionally or inadvertently, there are three main driving forces controlling the movement. The first is absorption to the soil particles, which are then subjected to runoff processes and may end up in the water. The second is leaching through the soil causing the chemical to end up in the groundwater. Finally, there is loss due to volatility. Once the chemical has moved into the water compartment, there are two additional diffusive mechanisms that are operating. One controls the loss to the atmosphere while the second is concerned with the water/sediment layer. These five processes will be discussed below.

1. <u>Runoff and Sediment Balance</u>. From Johnson and Moldenhauer (12) it was deduced that for a watershed of 1,000 square miles, the average sediment production was 0.5 acre foot/square mile/year ($2.32 \times 10^{-4}$ m/yr). For a soil with an average surface horizon depth of 0.15 m (4) the rate constant for runoff, $k_{32}$ (see Figure 1 for the nomenclature to be used), becomes $1.55 \times 10^{-3}$/yr. Amemiya (13) reported that 15.6 cm of water at the rate of 6.25 cm/hr (an intense rainfall) applied to a 5% slope with no surface cover caused a soil loss of 12.4 tons/acre ($2,780$ g/m$^2$). An average rainfall of 0.7 m/year, typical of the mid-western states (14), would create about 5 such events. This would be equivalent to a rate constant of $6.2 \times 10^{-2}$/yr. At the other extreme the flat portion of north central Iowa generates a constant of $1.6 \times 10^{-4}$/yr (12). These numbers are partially substantiated by examining the data for the Great Lakes. The average sediment loading rate for each of the lakes is shown in Table 2 (15). The watershed area was taken from a U.S. Lake Survey prepared by the Great Lakes Commission (16). As may be seen, these numbers are close to those estimated from the data of Johnson <u>et. al</u>. (12). A further confirmation is revealed in the 1967 conservation inventory (17) which showed approximately 20% of the cropland averaged more than 8 tons of soil/acre/yr ($8 \times 10^{-3}$/yr) and 50% averaged 3-8 tons/acre/yr ($3$ to $8 \times 10^{-3}$/yr). Thus, values ranging between $2 \times 10^{-2}$ and $2 \times 10^{-4}$ reciprocal years appear to reflect the rate at which sediment is moved from the land to the water. The constants indicated below will be used to simulate different rates of runoff.

|   |   |   |
|---|---|---|
| 1. | High sediment yield | $2 \times 10^{-2}$/yr – high intensity storms |
| 2. | Average sediment yield | $2 \times 10^{-3}$/yr – typical mid-western farmland |
| 3. | Low sediment yield | $2 \times 10^{-4}$/yr – flat farm land or forest ecosytem |

The following example will illustrate how the runoff constant is used to develop the remaining constants (Figure 1) in the model.

a. The average sediment yield for the Unit World (Table 1) is estimated by means of Equation 2.

$$Yield = (2 \times 10^{-3}) \qquad \times (0.15) \quad \times (3 \times 10^5) \times (1.5 \times 10^6) \qquad (2)$$

$$k_{32} \text{ (1/year)} \quad \times \text{Surface} \quad \text{Area of} \quad \text{Density (g/m}^3\text{)}$$
$$\text{Horizon} \quad \text{Watershed}$$
$$\text{(m)} \qquad \text{(m}^2\text{)}$$

$$= 1.35 \times 10^8 \text{ g/yr}$$

b. By assuming a suspended sediment concentration of 5 g/m$^3$ and a surface area of $7 \times 10^5$ m$^2$ for the water compartment, a sediment balance was performed. This is illustrated in Equation 3.

$$k_{24} = Yield/A \ C_{ss} \qquad (3)$$

where Yield = g/yr of sediment input ($1.35 \times 10^8$)

A = Water area for the Unit World

$C_{ss}$ = Concentration of sediments (g/m$^3$)

$k_{24}$ = Water column settling rate constant (38.6 m/yr)

c. The burial rate constant is estimated as follows. In one year (38.6 x 5) or 193 grams of sediment are deposited on 1 m$^2$. It is assumed that an equal mass of soil becomes buried in a year (i.e., removed from the active layer). The total mass of sediment in the top 3 cm layer is given by Equation 4.

$$Mass \ (g/m^2) = \rho \ (1 - \phi) \times depth \qquad (4)$$

where $\phi$ = porosity

$\rho$ = bulk density

Using 0.88 for the average porosity (14) and $1.5 \times 10^6$ g/m$^3$ for the bulk density, the burial rate $k_b$ becomes:

$$193/Mass \text{ or } 0.036/yr \ (t_{1/2} = 19.2 \text{ years})$$

One final note on the sedimentation rate constants. Before using in the equations governing the mass balance, they must be modified by the fraction of chemical that is present in the sediment.

a. $k_{32} \ (1 - FR_3)$ (5)

b. $k_{24} \ (1 - FR_2)$ $\qquad$ FR = 1/($K_d$ $\cdot$ conc of sediment + 1)

c. $k_b \ (1 - FR_4)$

where FR is the fraction of chemical present in the water phase of the particular compartment, and $K_d$ is the soil absorption coefficient (mL/g).

These three constants are self-consistent and bring the sediment into balance. In order to maintain this balance, new soil must be created faster than it is depleted. Eventually, the soil will all end up in the sea--a natural process that is going on in the world.

In addition to the movement of chemical from the ground to the stream _via_ sediment the movement in the actual runoff water must also be considered. Once a chemical has been added to the soil the fraction in the water phase will be given by Equation 5. For the soil compartment the concentration of soil in the field is approximately equal to the bulk density. Thus, it is apparent that even for materials with very low absorption coefficients the fraction of chemical in the sediment is greater than the water phase. It will only be during an intense storm immediately following application that much chemical will be found in the surface water. Consequently, the amount transported in the water phase may be ignored for the screening procedure. This assumption is only true for soil containing an organic fraction, obviously for quartz sand no absorption takes place and all the chemical will be in the water phase.

2. Leaching. McCall _et. al._ (18) demonstrated that the leaching or movement of chemicals through soils followed Equation 6. This

$$\text{cm moved} = \frac{1}{K_d} \left[ \frac{\text{cm of water entering soil}}{(1 - \Phi^{2/3})\ \text{density}} \right] \qquad (6)$$

equation was developed from column studies using disturbed soils and a constant head of water. Such a situation is atypical of what is found in the field; however, it was felt that the equation would correctly predict the relative movement of a series of chemicals. Thus, for an average rainfall of 0.7 m/yr where 90% of the rain enters the soil (14) Equation 6 may be converted to 7.

$$\text{m/yr} = (\text{Rain(m/yr)} \times 0.92)/K_d \qquad (7)$$

The porosity ($\phi$) and density in this case have been assigned values of 0.4 and 1.5 g/cc, respectively. For a chemical to move below the 15-cm depth, the leaching rate constant becomes:

$$k_L(1/\text{yr}) = (\text{Rain(m/yr)}/K_d)\ 6.13 \qquad (8)$$

This coefficient does not have to be modified by $FR_3$ since absorption was included in Equation 6 deduced by McCall _et. al._ (18).

3. Volatility (Ground/Air). Volatility is a diffusion process where the transfer occurs in both directions and is controlled by the fugacity difference that exists between the air and ground compartments. Thus, the transfer in moles/yr between ground and air is given by Equation 9. The direction of transfer depends on the sign associated with the fugacity difference.

$$D_{13}\ (f_1 - f_3) \qquad (9)$$

where $D_{13}$ is the diffusion transfer coefficient
in units of moles/Pa/yr between compartments
1 and 3

$f_1$, $f_3$ are fugacities in units of Pa

The reciprocal of diffusion or resistance is an easier concept to understand. Thus, if the resistance is large the diffusion will be small and very little material will be transferred. The total resistance is the sum of the resistances that occur in the boundary layers between the soil and the water. The resistance in the air has been described previously [3] and is given in Equation 10. The gas phase constant $K_g$ is related to the exchange constant ($k_a$) for

$$R_1 = 1/(K_g \ A \ Z_1) \qquad (10)$$

where $\quad K_g = k_a \ (H_2O^{0.5}/MW^{0.5})m/yr \qquad (11)$

$\qquad A = $ Interfacial area

$\qquad Z_1 = $ Fugacity capacity for water (mol/Pa.m$^3$)

$\qquad MW = $ Molecular weight

water vapor [19]. The value that will be assigned to $k_a$ is 30 m/hr established by Liss and Slater [19].

From the work of Spencer and Cliath [20], it is concluded that chemicals volatilize from plant and soil surfaces only when they are wet. Consequently, the resistance to movement through the soil layer will be confined to the water segment. The expression describing this resistance for unsaturated soil will be similar to the one used by DiToro et. al. [15] and is shown in 12.

$$R_3 = d/(M_d \ \gamma \ \phi A \ Z_2) \qquad (12)$$

where $\quad d = $ Average depth of movement (1/2 vertical depth)

$\qquad M_d = $ Molecular diffusion coefficient

$\qquad \phi = $ Porosity (0.4 for soil)

$\qquad \gamma = $ Fraction of water in the void volume (0.2)

$\qquad A = $ Contact area between the ground and air

$\qquad Z_2 = $ Fugacity capacity of water

The molecular diffusion coefficient was calculated by the procedures described by Tucker and Nelken [21]. The estimating equation is shown in 13. Using the additive volume increments [21] the Le Bas

$$M_d = 13.26 \times 10^{-5}/(\eta \ 1.14 \ V_b{}^{.589}) \ cm^2/s \qquad (13)$$

where $V_b = $ Le Bas Molar Volume

$\qquad \eta = $ Viscosity of water in centipoises

Molar Volume ($V_b$) was determined in units of cm$^3$/mole. Viscosity of water as a function of temperature was taken from the article by Tucker and Nelken [21].

The total resistance to mass transfer is the summation of Equations 11 and 13. The reciprocal of this sum becomes the diffusion transfer coefficient. In most cases $R_3 \gg R_1$ and diffusion will be

$$D_{13} = 1/(R_1 + R_3) \qquad (14)$$

controlled in the soil layer.

4. **Volatility (Water/Air)**. Mackay and Paterson (5, 6) and Mackay (3) have discussed these mechanisms. In essence, the transfer coefficient is the summation of the resistance in the gas phase and in the water phase. The gas phase resistance is illustrated in Equation 10, while the resistance in the water layer is given by Equation 15.

$$R_2 = 1/(K_w \ A \ Z_2) \qquad (15)$$

$$\text{where } K_w = k_w CO_2^{0.5}/MW^{0.5}) \ (m/yr)$$

A = surface area of water compartment

The water phase constant $K_w$ is related to the exchange constant $k_w$ for carbon dioxide. The value of 0.02 m/hr (19) will be assigned to $k_w$. The diffusion transfer coefficient is given by 16.

$$D_{12} = 1/(R_1 + R_2) \qquad (16)$$

5. **Diffusion Between Water and the Bottom Sediments**. From the investigations reported by DiToro et. al. (14), the resistance to mass transfer across the water/bottom sediments is all in the sediment layer. Accordingly, the expression for the resistance will be similar to Equation 12 as shown in 17.

$$R_4 = d/(M_d \ \gamma \ \phi A \ Z_2) \qquad (17)$$

where d = 1/2 the vertical depth of the sediment layer

$\phi$ = Porosity for bottom sediments is 0.88

$\gamma$ = Fraction of water in the void volume (1.0)

A = Contact area between water and sediment layer

The diffusion transfer coefficient $D_{24}$ is the reciprocal of $R_4$.

**Mass Balance Equations**. The mass balance for each compartment may be expressed in words as follows:

$$\text{Input} = \begin{bmatrix} \text{Loss from a compartment} \end{bmatrix} - \begin{bmatrix} \text{Gain from an adjacent compartment} \end{bmatrix} \qquad (18)$$

where Loss is from: reaction, diffusion, advective flow and material transport

Gain is from: diffusion and material transport

Input is from: annual environmental release and advective flows from adjacent compartments

For this screening model advective flows will not be considered. Expressing Equation 18 in mathematical terms generates 19.

$$I_i = \left[V_i Z_i (k_i + k_{ij}) + \Sigma D_{ij}\right] f_i - \left[V_j Z_j k_{ji} + \Sigma D_{ij}\right] f_j \qquad (19)$$

(see Figure 1 for identification of terms)

Assuming that the input can only be added to either ground, air, or water, a mass balance equation for each of the four compartments may be written as follows:

$$I_1 = f_1 X_1 - f_2 D_{12} \quad -f_3 D_{13} \qquad (20)$$

$$I_2 = -f_1 D_{12} + f_2 X_2 \quad -f_3 R \quad -f_4 D_{24} \qquad (21)$$

$$I_3 = -f_1 D_{13} \quad +f_3 X_3 \qquad (22)$$

$$0 = \quad f_2 (D_{24} + S) \quad +f_4 X_4 \qquad (23)$$

where $X_1 = V_1 Z_1 (k_1 + k_s) + D_{12} + D_{13}$

$k_s$ = the rate constant for exit to stratosphere (4).

$X_2 = V_2 Z_2 (k_2 + k_{24}(1 - FR_2)) + D_{12} + D_{24}$

$X_3 = V_3 Z_3 (k_3 + k_{32}(1 - FR_3) + k_L) + D_{13}$

$X_4 = V_4 Z_4 (k_4 + k_b (1 - FR_4)) + D_{24}$

$R = V_3 Z_3 k_{32} (1 - FR_3)$

$S = V_2 Z_2 k_{24} (1 - FR_2)$

Using matrix notation, Equations 20–23 can be illustrated in 24.

$$(I_i) = f_i [A] \qquad (24)$$

where $[A] = \begin{bmatrix} X_1 & -D_{12} & -D_{13} & \\ -D_{12} & X_2 & -R & -D_{24} \\ -D_{13} & & X_3 & \\ & (D_{24}+S) & & X_4 \end{bmatrix}$

By inverting the matrix, the fugacities are given by 25.

$$(f_i) = [A]^{-1} (I_i) \qquad (25)$$

At steady state the moles of chemical in each compartment are

$$M_i = V_i Z_i f_i \qquad (26)$$

and the residence time is given by 27.

$$\text{Residence time} = \Sigma M_i / \sum_j I_j \qquad (27)$$

A dynamic version of the steady state model may be developed by setting up differential equations for each compartment and solving for

$$\frac{df_i}{dt} = Input - Loss + Gain \qquad (28)$$

the individual fugacities as a function of time. Such equations are easily handled using one of the many simulation packages that are available for computers. Agin and Blau (22) have developed SIMUSOLV to run on the IBM mainframe which will be used in this investigation.

VALIDATION

Before such models can be used with any degree of confidence, it is necessary to have some evidence that the predicted results are realistic. The opportunity to test the model presented itself with the extensive field data generated in a runoff study performed by McCall et. al. (2). This study was conducted during the spring and summer of 1982 on a 120-acre farm containing a small pond in Kankakee County, Illinois. Three applications of a pesticide containing chlorpyrifos were made on April 28, May 15, and June 15 avoiding in each case, as much as possible, the pond area. The data for the May 15 surface application will be used for validating the model. This was chosen because 6 hours after the pesticide had been distributed an intense rain occurred which caused measurable runoff. Field measurements indicated that the first application had largely dissipated prior to May 15; consequently, it was possible to associate all of the observed concentrations in the pond water and sediment with the second application. A summary of the studies are given in Tables III and IV. Table V lists the data that will be used in the model. For this analysis the air compartment will be considered as an infinite sink; consequently, once the chemical moves to the air it will be lost to the system. This is a reasonable assumption, since photodegradation and advection will dissipate the chemical as rapidly as it enters. For example, the wind speed at the time of the second application was 6 mph (2.7 m/s) (2). Using a 100 m mixing zone and an average width for the plot of 100 m yields a half-life of 5 minutes a value, which would quickly remove any chlorpyrifos from the air column.

RESULTS

Using an input of 3.82 moles for one day and the parameter values shown in Table V, a simulation was performed using the algorithms discussed above. The equations were solved using the program designed by Agin and Blau (22). The results are illustrated in Figure 2. Considering the assumptions involved, the fit of the predicted values with the experimental data is reasonably good.

McCall et. al. (2) estimated that a constant of 0.4/days (a half-life of 1.7 days) was necessary to describe the data for water. This rapid rate was attributed to volatilization. While the present model incorporates evaporation as a loss mechanism, the construct used may be inadequate to account for the actual site situation. In addition to the discrepancy in the loss rate from water, the present algorithm failed to describe the rapid appearance of chlorpyrifos in the water as seen in Table IV. One reason for the observed deviation is that the model did

not include dissolved chemical in the runoff water. Since an intense rain occurred shortly after application there is a strong probability that the amount in solution will be high, which would contribute to the rapid appearance of chemical in the pond. The CREAMS Model (Chemical Runoff and Erosion from Agricultural Management Systems) used by McCall et. al. (2), with its more definitive description of the site, was capable of making this prediction.

One other area that could be validated dealt with the sediment and water yield. Table III contains the field measurements and indicates a total of 37.4 kg of sediment lost with 109 $m^3$ of rain over a 44-day period. For this time period the runoff constant ($k_{32}$) in the model was assigned a value of $2 \times 10^{-4}$/year, which represents a relatively flat area similar to central Iowa (12). Substituting the proper values into Equation 1 yields 71 kg of sediment in 44 days. CREAMS predicted 53.9 kg of sediment (2). Using 0.7 m/year of average rain and assuming a 10% runoff (14) yields 111 $m^3$ of water in the 44-day period as compared to the measured number of 109 $m^3$ and 118 $m^3$ predicted by CREAMS (2) based on the actual field data.

CONCLUSION

These results indicate that the proposed model is an adequate description of how chemicals are moved and distributed in a terrestrial-aquatic ecosystem. Furthermore, only a minimum data set is required to conduct the analysis. From a field development point of view, the method will permit investigators to rapidly generate a profile of the disposition of a chemical which will target areas that might require further attention.

The derived mass transfer constants may be used in the fugacity model of Mackay (3). This will allow for the development and use of the steady state non-equilibrium and the dynamic version of the model. Both will be a more realistic description of events in the Unit World (4) and provide an improved environmental pattern for the chemical under investigation.

TABLE I.   PROPERTIES OF THE UNIT WORLD (4).

| Compartment | Volume (m$^3$) | Area (m$^2$) |
|---|---|---|
| Air | 6 x 10$^9$ | |
| Water[a] | 7 x 10$^6$ | 7 x 10$^5$ |
| Ground | 4.5 x 10$^4$ | 3 x 10$^5$ |
| B. Sediments | | |

GROUND
| | |
|---|---|
| Bulk density | 1.5 x 10$^6$ g/m$^3$ |
| Depth | 0.15 m |
| Porosity[b] | 0.4 |
| Water content | 0.2 |
| Fraction organic carbon | 0.02 |

BOTTOM SEDIMENTS
| | |
|---|---|
| Depth | 0.03 m |
| Porosity | 0.88 |
| Water content | 1.0 |

(i.e. the void volume is all water)

Exit to Stratosphere          4.1 x 10$^{-5}$/day ($k_s$)

[a]   Water is 70% of surface area of the world.

[b]   Porosity is the fraction of void volume in soil.

TABLE II.   SEDIMENT LOADING IN THE GREAT LAKES.

| Lake | Loading Rate g/yr x 10$^{12}$ | Watershed Area m$^2$ x 10$^{11}$ | Runoff[a] (1/yr) |
|---|---|---|---|
| Superior | 8.11 | 1.25 | 2.88 x 10$^{-4}$ |
| Michigan | 3.66 | 1.18 | 1.38 x 10$^{-4}$ |
| Huron | 4.7 | 1.27 | 1.64 x 10$^{-4}$ |
| Erie | 45.9 | 0.58 | 3.48 x 10$^{-3}$ |
| Ontario | 4.33 | 0.71 | 2.73 x 10$^{-4}$ |

[a]   Runoff = Loading rate/(watershed area x density x depth).
This assumes a depth of 0.15 m and a density of 1.5 x 10$^6$ g/m$^3$
similar to the Unit World (4).

TABLE III.   SUMMARY OF THE FIELD STUDY AS REPORTED IN REFERENCE 2.

1.  The total area of the test watersheds was 3.29 acres (0.64 acres for
    the hillside plus 2.65 acres for the pond site).  Of this total the
    2.65 acre site contributed flow to the pond, and only 1.97 acres was
    actually treated with the chemical.

2.  A broadcast spray application of 1.5 pounds of chlorpyrifos/acre was
    made on May 15, 1982, for a total of 2.95 pounds (3.82 moles/day).

3.  A 1-inch rain occurred 6 hours after application.

4.  The pond was 0.75 acres (3,000 $m^2$) with an average depth of 6 feet
    (1.8 m).  The bottom sediments contained 3% organic carbon and will
    be assigned a depth of 0.03 m.

5.  During the 44 days of the study, the two plots yielded a total of
    37.4 kg of soil along with 0.83 cm (109.23 $m^3$) of rain.

6.  It will be assumed that the average depth of the soil compartment in
    the field plots was 0.15 m with 2% organic carbon and a density of
    1.5 x $10^6$ g/$m^3$ (4).

TABLE IV. CONCENTRATIONS OF CHLORPYRIFOS IN THE WATER AND SEDIMENT OF
THE POND.

| Day | Concentration (ppb)[a] | |
|---|---|---|
| | Water | Bottom Sediments |
| 0 | | |
| 2 | 0.02 | |
| 4 | 0.005 | |
| 5 | | 0.18 |
| 8 | 0.001 | |
| 10 | | 0.28 |
| 15 | | 0.32 |
| 20 | | 0.30 |
| 50 | | 0.18 |
| 75 | | 0.07 |

[a] These concentrations were a result of subtracting the effect of a
previous site contamination as well as contamination of the pond by
drift (2).

143

TABLE V.  PARAMETERS TO BE USED IN THE MODEL.

Field Data:

| | |
|---|---|
| Land area[a] | 7,880 m$^2$ |
| Depth of land | 0.15 m |
| Pond area | 3,000 m$^2$ |
| Depth of pond | 1.8 m |
| Depth of sediment | 0.03 m (4) |
| Runoff ($k_{32}$)[b] | 5.48 x 10$^{-5}$/day |
| Rainfall | 0.7 m/yr. |

Chemical Data (for Chlorpyrifos):

| | |
|---|---|
| Molecular weight | 350 |
| Vapor pressure | 2.5 x 10$^{-3}$ Pa |
| Water solubility | 2 ppm |
| Molecular diffusion coefficient[c] | 4.34 x 10$^{-5}$ m$^2$/day |
| Octanol water partition coefficient | 40,000 |

Rate Constants (Day$^{-1}$):

| | |
|---|---|
| Photolysis in water | 0.02  (2) |
| Hydrolysis in water | 0.23  (23) |
| Biodegradation in bottom sediments | 0.014 (24) |
| Dissipation from soil | 0.28  (2) |

[a] This represents the watershed that ran directly into the pond.

[b] This represents a value for high sediment yield as it would occur during an intense storm.

[c] Calculated using Equation 13 and Reference 20.

[d] Optimized rate constants.

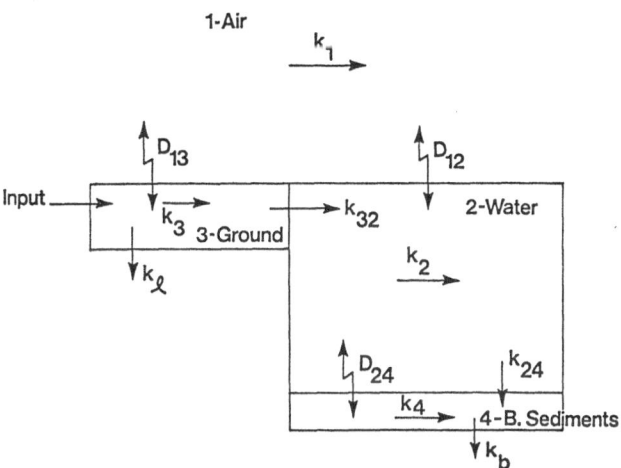

FIGURE 1. A COMPARTMENTAL MODEL FOR THE RUNOFF STUDY. THE DOUBLE-HEADED
ARROWS REPRESENT DIFFUSION. THE SUBSCRIPTS REFER TO COMPART-
MENTS 3 TO COMPARTMENT 2. THE SINGLE NUMBER SUBSCRIPTS
REPRESENT DEGRATION REACTION IN THAT COMPARTMENT, WHILE $k_\ell$ AND
$k_b$ ARE RATE CONSTANTS FOR REMOVAL OF MATERIAL, I.E., LEACHING
AND BURIAL.

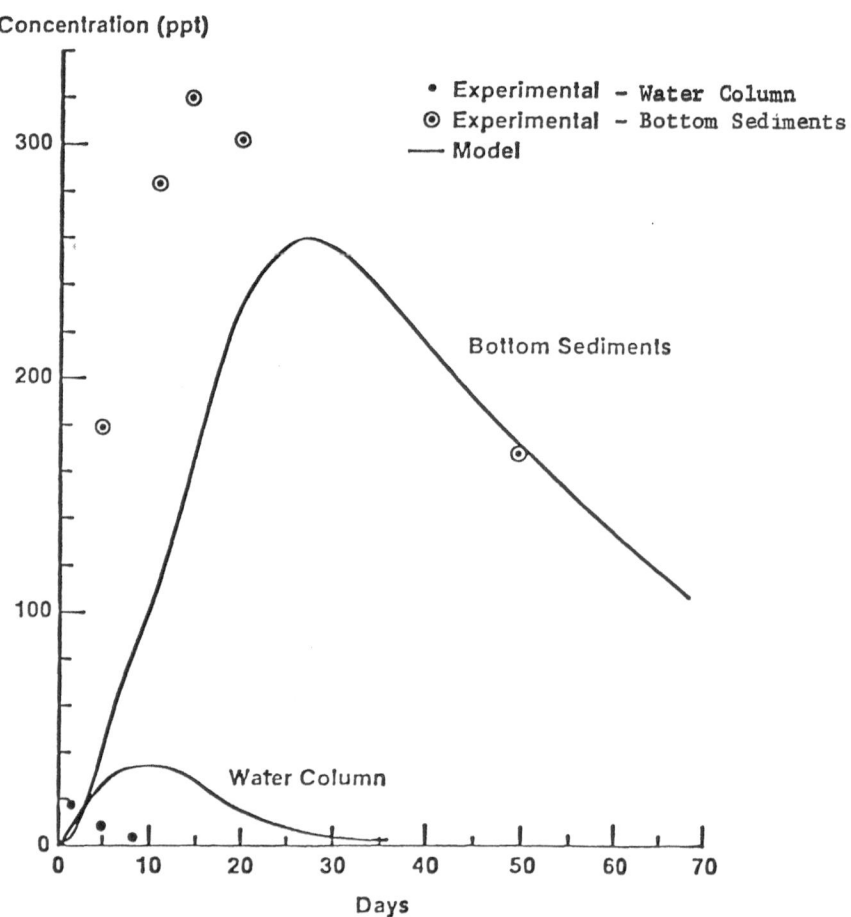

FIGURE 2.  THE RESULTS OF MODELING THE RUNOFF OF CHLORPYRIFOS.  THE
EXPERIMENTAL POINTS WERE FROM THE FIELD STUDY OF McCALL
ET. AL. (2).  THE CONTINUOUS LINE IS THE RESULT OF USING
THE MODEL DISCUSSED IN THIS REPORT.

# REFERENCES

1. Donigian, A. S., J. C. Imhoff, B. R. Bicknell, and J. O. L. Kittle, Guide to the Application of the Hydrological Simulation Program - Fortran (HSPF). EPA Contract No. 68-01-6207 Env. Res. Lab. USEPA, Athens, GA (1982).

2. McCall, P. J., G. R. Oliver, and R. L. McKellar, Modeling the Runoff Potential and Behavior of Chlorpyrifos in a Terrestrial-Aquatic Watershed. Unpublished report from The Dow Chemical Co., Midland, MI (1984).

3. Mackay, D., Environ. Sci. and Technol., 13, 1218 (1979).

4. Neely, W. B., and D. Mackay, Evaluative Model for Estimating Environmental Fate in ''Modelling the Fate of Chemicals in the Aquatic Environment,'' ed. Dickson, K. L., A. W. Maki, and J. Cairns, Ann Arbor Science, Ann Arbor, MI, pg. 127 (1982).

5. Mackay, D., S. Paterson, Environ. Sci. and Technol., 15, 1006 (1982).

6. Mackay, D., S. Paterson, Environ. Sci. and Technol., 16, 654 (1982).

7. Neely, W. B., Environ. Toxicol. and Chem., 1, 259 (1982).

8. Neely, W. B. ''Chemicals in the Environment,'' Marcel Dekker, New York, N.Y. (1980).

9. Mackay, D., S. Paterson, Chemosphere, 12, 143 (1983).

10. McCall, P. J., D. A. Laskowski, R. L. Swann and H. J. Dishburger, Residue Reviews, 85, 231 (1983).

11. Klopffer, W., G. Rippen, and R. Frische, Ecotoxicol. Environ. Safety, 6, 294 (1982).

12. Johnson, H. P. and W. C. Moldenhauer, in ''Agricultural Practices and Water Quality'' ed. by Willrich T. L. and G. E. Smith, Iowa State University Press, Ames, Iowa, pg. 3 (1970).

13. Amemiya M., ibid, pg. 35.

14. Chow, V. T., ''Handbook of Applied Hydrology,'' McGraw Hill Book Co., New York, N.Y. (1964).

15. DiToro, D. M., D. J. O'Conner, R. V. Thoman and J. P. St. John, ''Analysis of the Fate of Chemicals in Receiving Water,'' Hydroqual, Inc., Lethbridge Plaza, Mahwah, NJ (1981).

16. Willford W., personal communication. United States Dept. of the Interior, Great Lakes Fishery Laboratory, Ann Arbor, MI (May, 1984).

17. USDA Agricultural Research Services and USEPA Office of Research and Development, Control of Water Pollution from Cropland. Report No. ARS-H-5-1, U.S. Government Printing Office, Washington, D.C. (1975).

18. McCall, P. J., R. L. Swann, D. A. Laskowski, S. A. Vrona, S. M. Unger, and H. J. Dishburger, Prediction of Chemical Mobility in Soil from Sorption Coefficients in ''Aquatic Toxicology and Hazard Assessment'' ed. by Branson D. R. and K. L. Dickson, ASTM, pg. 49 (1981).

19. Liss, P. S., and P. G. Slater, Nature, 247, 181 (1974).

20. Spencer, W. F., and M. M. Cliath, Vaporization of Chemicals in ''Environmental Dynamics of Pesticides'' eds. Haque, R. and V. H. Freed, Plenum Press Publishers, New York, N.Y. pg. 61 (1975).

21. Tucker, W. A. and L. H. Nelken, Diffusion Coefficients in Air and Water in ''Chemical Property Estimation Methods'' eds. Lyman, W. J., W. F. Reehl, and D. H. Rosenblatt, McGraw Hill Book Co., New York, N.Y., Chapter 17 (1982).

22. Agin, G. L. and G. E. Blau, Applications of DACSL (Dow Advanced Continuous Simulation Language) to the Design and Analysis of Chemical Reactor System,'' AIChE Symposium Series No. 214, Volume 78, pp. 108-118.

23. Neely, W. B., Hydrolysis in ''Environmental Exposure from Chemicals.'' Eds. Neely, W. B., and G. E. Blau, CRC Press, Inc., Boca Raton, FL, Chapter 7, Volume 1 (1985).

24. Klecka, G. M., Biodegradation in ''Environmental Exposure from Chemicals.'' Eds. Neely, W. B. and G. E. Blau, CRC Press, Inc., Boca Raton, FL, Chapter 6, Volume 1 (1985).

STOCHASTIC METHODS FOR INTERMEDIA TRANSPORT ANALYSIS

K. J. Yost

School of Health Sciences
Purdue University
West Lafayette, Indiana   47907

INTRODUCTION

The term "environment" is a generic one which may denote a wide variety
of complex systems and subsystems.  When the chief concern is assessing the
impact of pollutants in the environment, one is typically concerned with a
system comprised of sources (natural and anthropogenic), transport/translo-
cation mechanisms, and sinks.  The analysis of such a system, virtually by
definition, involves "systems analysis".  In order to analyze a system, one
must first characterize it in terms of relationships between component
parts, i.e. system logic.  In this process one typically evolves a set of
parameters which, to some degree of approximation, describes in a quantitative way component interactions.  The credibility of an analysis is thus
heavily dependent upon the accuracy with which the parameters are evaluated.
Of particular concern are (a) the sensitivity of a system to parameter
variations (sensitivity analysis), and (b) measures of estimates of
parameter variabilities.  The combining of (a) and (b) will be termed
uncertainty analysis in the following discussion.

This paper is intended as a review of some elementary methods of
analyses with which these topics can be addressed.  Due to the rudimentary
nature of the subject matter, it was deemed unnecessary to assembly a
comprehensive set of references.  There are, of course, a wide variety of
text and reference books on the first order differential equations,
stochastic/Monte Carlo methods, and numerical modeling techniques available
to the reader to supplement the following discussion.  Where specific applications of the methods to environmental analyses are mentioned, however,
references are provided.

There is a corollary benefit to be derived from an uncertainty analysis. In particular, the aim of most rule making in the occupational and environmental health areas is the protection of especially vulnerable, at-risk populations. The identification of such populations may, in principle, be readily accomplished using uncertainty analysis methods. This is also discussed further on.

It should be said at the outset that more frequently than not, insufficient data are available for establishing the variability of a parameter in a statistically viable way. This fact, however, should not stop the analyst from estimating ranges of likely error based on limited data, intuition, insight, etc. Functional relationships between parameter values and associated probabilities can take many forms, e.g. continuous functions, piecewise continuous functions, frequency histograms (including the one cell version which defines equal probabilities within a range of values), discrete values with assigned probabilities, etc. These options make it possible to incorporate data of many forms and patterns into an analysis.[1]

Finally, and perhaps most importantly, the process of setting up the framework of an uncertainty analysis forces one to approach a problem in a systematic fashion. The formulation of an environmental system frequently leads to additional insight into its' structure, and elucidates the primary data requirements for its adequate characterization. In the latter sense, the analysis can be used as a framework for managing policy-oriented experimental programs.

BASIC CONCEPTS

Many quantities of interest in environmental policy-making are complex functions of time, spatial variables, and the set of parameters which define their relationships to other elements of a system. One can express such a (generic) function as $E(\chi, t, \{p\})$, where $\chi$ denotes spatial variables, t is time, and $\{p\}$ signifies the set ($\{\ \}$) of parameters which give it structure. For the purpose of analyzing uncertainties in E as a function of parameter variability, and to identify at-risk populations where E is related to pollutant exposure, it is useful to develop frequency distributions for E. These consist of plots of probability vs magnitude of E, with the variability related to differing sets of values among individual $p_i$ in the set $\{p\}$[2]. These will be denoted as PDF's (probability distribution functions) in the following discussion. Throughout, sets of $p_i$ ($\{p\}$) will be characterized as points in a "parameter space" which encompasses all possible $p_i$ combinations.

An uncertainty analysis requires that some means of variability be established for each parameter, i.e. parameter PDF's of some form must be developed. Where data availability permits, the PDF may be constructed by statistical curve-fitting to laboratory/field experimental data. Their variability may be embodied in a distribution function ranging in from from the normal, "bell-shaped" distribution to one of arbitrary functionality. The only requirements for the latter are that they be representative of available data, and are normalized. Such an ad-hoc distribution function may be a straight line (horizontal) function where a range of values appear equally likely. Another possibility is a discrete valued function by which a parameter is described by a set of estimates with associated probability measures which sum to unity. We denote all such possible PDF's as $\Phi_i(p_i)$, where

$$\int \Phi_i(p_i)dp_i = 1, \tag{1a}$$

$$\overline{p}_i = \int p_i \Phi_i(p_i)dp_i, \tag{1b}$$

and the integrals extend over the ranges of the $p_i$.

A measure of the probability that a set of parameter values is the "correct" one is given by the expression:

$$W(\{p_i\}) = \prod_i^n \Phi_i(p_i)dp_i, \tag{2}$$

where $\prod_i^n$ denotes the product of n parameters, $\{p_i\}$ is the parameter set, and the $dp_i$ intervals are "centered" about the $p_i$ values. Further, the probability that the correct parameter value set is contained within a domain of parameter space of "volume" $\prod_i^n \Delta p_i$ is given by

$$W(\{\Delta p_i\}_\ell) = \prod_i^n \int_{\Delta p_i} \Phi_i(p_i)dp_i \tag{3}$$

where the $\Delta p_i$ are intervals of integration for the parameter PDF's, the product form of the expression is made possible by the independence of the $p_i$, and the index $\ell\ell$ denotes a parameter space domain.

Assuming the $dp_i$ (differentials) are appropriately small, the value of E related to the small parameter space volume $\prod_i^n dp_i$ is simply $E(\underset{\sim}{\chi}, t, \{p_i\})$. However, the average value of E associated with a finite parameter space domain, $\prod_i^n \Delta p_i$, is given by:

$$\hat{E}\ (\chi,\ t,\ \{\Delta p_i\}_\ell) = \frac{\int_{\Delta p_i} \cdots \int_{\Delta p_i} E(\chi,\ t,\ p_i)\ \prod_i^n \phi_i(p_i)dp_i}{W\ (\{\Delta p_i\}_\ell)} \qquad 4)$$

where the symbol "^" denotes an average over a limited ($\ell$th) domain of the parameter space.

The $\hat{E}\ (\chi,\ t,\ \{\Delta p_i\}_\ell)$ values can be used to construct a histogram representation of the variability of E over all, or a portion of, the parameter space. In particular, a parameter space may be divided into some number, say N, of domains for which sets of $\hat{E}$ values are calculated. A histogram axis can be defined with upper and lower limits for $\hat{E}$, and a bin structure consistent with the $\hat{E}$ values. Each $\hat{E}\ (\chi,\ t,\ \{\Delta p_i\}_\ell)$ is "placed" within the appropriate histogram bin, say the $i$th, by incrementing the bin relative frequency, $I_j$, by an amount:

$$\Delta I_j = \frac{W(\{\Delta p_i\}_\ell)}{\sum\limits_k W(\{\Delta p_i\}_\ell)} N \qquad 5)$$

The average of E over the region, $\overline{E}$ has the form:

$$\overline{E}_\ell = \sum_j E_j I_j \qquad 6)$$

where $E_j$ are the bin midpoints, and the sum extends over all bins. If the N domains encompass all parameter space then $\overline{E}_\ell$ is equivalent to

$$\overline{E} = \int_1 \cdots \cdots \int_n E(\chi,\ t,\{p_i\})\ \prod_i^n \phi_i(p_i)dp_i \qquad 7)$$

where the integrals extend over all parameter space, and $\overline{E}$ is the mean value of E.

The 'sensitivity' of E with respect to the $i$th parameter, $p_i$, will be denoted as $\delta_i E_i\ (\chi,\ t,\ \{p_i\})$. It may be evaluated in the standard fashion by taking the partial derivative of E with respect to $p_i$, with $\chi$, t, and all other parameter values held constant.[3] Thus:

$$\delta_i E(\chi,\ t,\ \{p_i\}) \equiv \frac{dE}{dp_i}\ (\chi,\ t,\ \{p_i\}) \qquad 8)$$

with an increment in $p_i$, $\Delta p_i$, of unit measure implied. Inasmuch as the partial derivative is evaluated for a specific parameter set, $\{p_i\}$, it is valid in principle only for the corresponding point in parameter space. A more general measure of the importance of errors in the values of individual parameters can be generated by averaging the product of $\delta_i E$ and its distribution function, $\Phi_i(p_i)$ over the range of $p_i$. Denoting this quantity as $\Delta_i E$, one has:

$$\overline{\delta_i E} \,(\underset{\sim}{\chi}, \, t, \, \{p_i\}) = \int |\delta_i E(\underset{\sim}{\chi}, \, t, \, \{p_i\})| \, \Phi_i(p_i) dp_i \qquad 9)$$

Comparison between the $\overline{\delta_i E}$ may be used to identify the more important parameters in the Evaluation of E. Usually, parameters with relatively minor impacts on E can be identified, and the analysis may be simplified by concentrating on correspondingly fewer parameters. A measure of the total uncertainty for this parameter space point, $\Delta E_t(\underset{\sim}{\chi}, \, t, \, \{p_i\})$, has the form:

$$\Delta E_t(\underset{\sim}{\chi}, \, t, \, \{p_i\}) = \sum_i^n \, | \, \delta_i E(\underset{\sim}{\chi}, \, t, \, \{p_i\})| \, \sigma_i \qquad 10)$$

where $\sigma_i$ is the standard deviation of the $i^{th}$ parameter. A more meaningful measure of $\Delta E$ involves taking into account the probability that the parameter space point is the "true" one, and then calculating a weighted average, $\overline{\Delta E_t}$, over all parameter space. This expression has the form:

$$\overline{\Delta E_t}(\underset{\sim}{\chi}, \, t) = \int_1 \cdots \cdots \int_n \Delta E_t(\underset{\sim}{\chi}, \, t, \, \{p_i\}) \, \prod_i^n \, \Phi_i(p_i) dp_i \qquad 11)$$

The magnitude of the relative error in E associated with parameter uncertainties is given by the ratio

$$\frac{\overline{\Delta E_t}(\underset{\sim}{\chi}, t)}{\overline{E}(\underset{\sim}{\chi}, t, \{p_i\})} \qquad 12)$$

The foregoing measures of uncertainty/sensitivity can also be generated for parameter sub-space (volumes), normalized with the appropriate $W(\{\Delta p_i\})$.

IMPLEMENTATION

The functions E, as pointed out previously, need not be analytical functions of the parameter variables, i.e. they are not necessarily continuous, piecewise differentiable, readily integrable, etc. They may be single or multiple strings of "transfer coefficients" in analytical and numerical formulations of environmental systems based on first order kinetics. In such models each transfer coefficient may in turn be a complex subsystem with its own set of parameters and related variabilities. This results in a form of "nesting" of probability distribution functions, representing the n-fold PDF product appearing in the methodology discussion.

APPLICATION OF METHODOLOGY

The foregoing methodology will be applied to the multimedia transport and statistical risk assessment factors embodied in dietary exposure of the general population to cadmium (Cd) attendant to the landspreading of municipal sewage sludge and the deposition of airborne Cd particulate. This analysis is based on work supported by the National Science Foundation, the Environmental Protection Agency, and a variety of industry sponsors including the National Association of Metal Finishers, International Lead Zinc Research Organization, Cadmium Council and the Metal Finishers Foundation. Publications evolving from these projects, and associated references, are given in the List of References.

The analysis makes extensive use of the "prototype area" concept. The latter reflects the virtual impossibility of simulating environmental flows of ubiquitous contaminants in an area as large and diverse as the U.S. The alternative is to choose exposure/risk assessment areas a) which are conducive to "conservative" analyses representing maximum credible exposure scenarios for humans or biotic populations, and b) for which data are available sufficient to evaluate flow/exposure model parameters. It is sometimes useful to generate artificial exposure scenarios by superimposing non-resident sources on prototype areas.

The principal Cd environmental flow paths are shown in Fig. 1. We will concentrate on the transport/translocation mechanisms relevant to the two dietary exposure mechanisms cited above as follows:

1) Deposition of airborne Cd
   a) urban areas with subsequent runoff to municipal sewer systems
   b) cropland
2) Landspreading of municipal sewage sludge
   a) effluent discharge from point sources to municipal sewer systems
   b) sludge landspreading rates
   c) sludge Cd concentrations
   d) translocation of Cd from cropsoils to crops

Deposition of Airborne Cd

Chicago was chosen as a prototype area for determining the deposition pattern for Cd particulate in a representative urban-industrial region. The Air Transport Model (ATM) (Culkowski and Patterson, 1976) was used to calculate total airborne and dustfall Cd within the Chicago prototype area. Calculations were obtained for incineration, steel production, and coal combustion point sources of Cd emissions to the atmosphere. National Emissions Data Source (NEDS) printouts obtained from EPA provided locations, process parameters, and particulate emissions for model calculations.

154

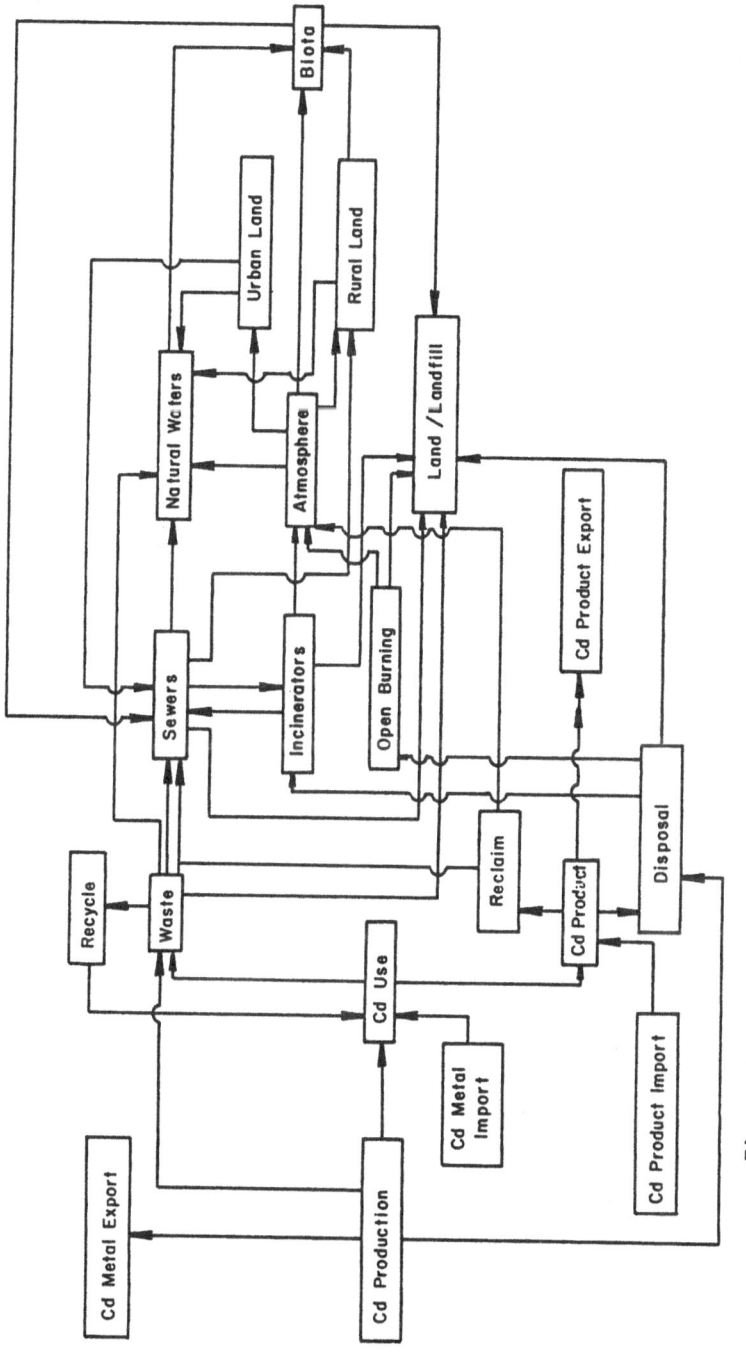

Figure 1.  Primary elements of the U.S. cadmium environmental flow system.

155

Initial ATM results for particulate suspension and deposition were converted to Cd suspension and deposition by use of the following stack particulate Cd concentrations: incineration, 2,250 mg/kg (Jacko and Neuendorf, 1977); coal combustion, 70.5 mg/kg (Jacko and Neuendorf, 1977); open hearth furnace (steel), 367 mg/kg (Jacko and Neuendorf, 1977); electric arc furnace (steel), 851 mg/kg (Dow, 1973); and basic oxygen furnace (steel), 117 mg/kg (Dow, 1973). Total Cd emission rates were scaled as follows:

| Source | Scaling Factor |
| --- | --- |
| Steelmaking | 20% |
| Coal-fired Power Plants | 4% |
| Municipal Incinerators | 4% |

The 4% figure represents a population weighting assuming that the Chicago SMSA includes approximately 4% of the total U.S. population. The 20% factor for steelmaking is based on recent production data from the American Iron and Steel Institute (AISI, 1981). The resulting Cd deposition values were plotted as contour intervals, or isopleths, and are presented in Figure 2.

Forty six percent of the Chicago urban area is sewered, and 54% unsewered (Heanly et al., 1977). Approximately 63% of sewered land is served by combined sanitary/storm systems, and 37% by separate storm sewers. In addition, 58% of the metropolitan area is residential, and 42% industrial or commercial. Using results from Louisville, KY as typical of urban-industrial areas (Terstriep and Stall, 1974); 80% of industrial land area, and 40% of residential land area, are paved. For Chicago as a whole, then, 57% of the land area is paved, and 43% unpaved. If it is assumed that the distribution of sewered and paved areas within the metropolitan area are independent, it follows that 63% of the paved area drains into combined sanitary/storm sewers, and 37% into separate systems. Finally, if it is assumed that urban land has the same erosion potential as rural pasture land, approximately 10% of air-deposited Cd in nonsewered soils will wash into natural waters, and 90% will remain in the soil. This is summarized in Figure 3.

Table 1 gives a summary of the fate of Cd particulate emitted from steelmaking, municipal incinerator and coal-fired power plant facilities. The fraction of Cd-bearing particulate transported out of the urban area is assumed to be deposited uniformly on rural New Jersey soils. Tables 2 and 3 summarize airborne Cd inputs to New Jersey sewage treatment plants and cropsoils, respectively.

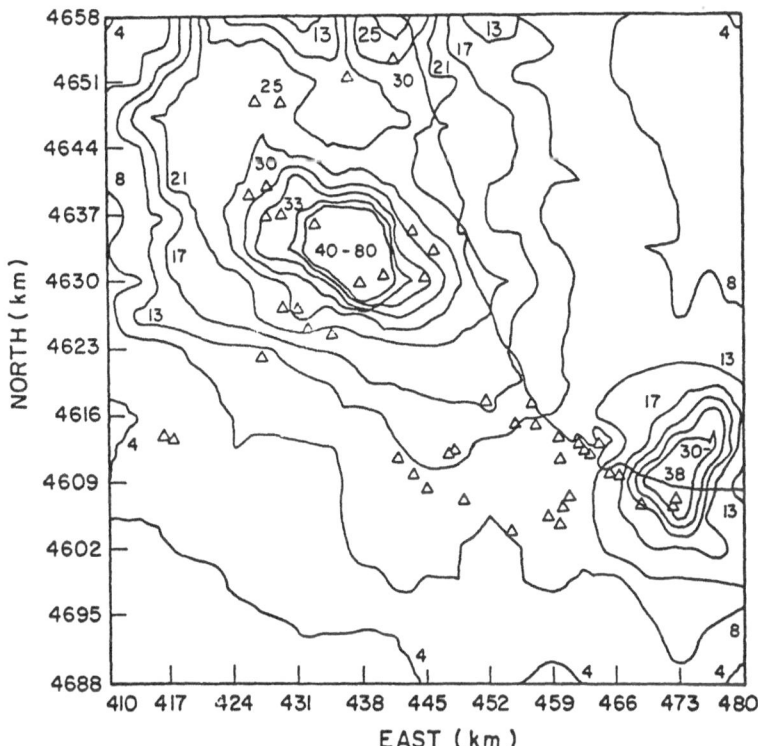

Figure 2.   Annual Cd deposition (g/ha/yr) isopleths for municipal incinerators, steel production, and coal-fired power plants, with source locations (△):  Chicago prototype area; Current Practice.

Table 1.  Summary of ATM computer modeling deposition results for
the Chicago prototype area.

| | Process | | | |
|---|---|---|---|---|
| | Total | Total Steel | Incinerator | Coal |
| Cadmium Emission (MT) | 16.64 | 3.68 | 11.14 | 1.61 |
| Cd Dep (MT) | 7.86 | 1.64 | 5.86 | .364 |
| % Cd Staying Within Area | 47.3 | 44.5 | 52.6 | 22.6 |
| % Cd Leaving Area | 52.7 | 55.5 | 47.4 | 77.4 |

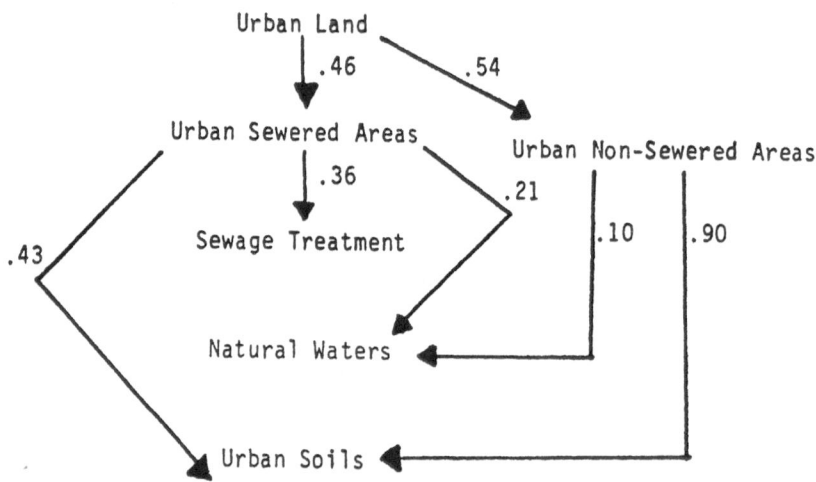

Figure 3.    Fate of air deposited Cd in urban areas.

Table 2. Cd input rates (MT/yr) to New Jersey prototype region sewage tratment plants from airborne particulate deposition.

| Source | Airborne Cd Deposition and Runoff |
|---|---|
| Municipal incineration | .50 |
| Sludge incineration | .01 |
| Steelmaking | .10 |
| Coal-fired power plants | .03 |
| Totals | .64 |

Table 3. Airborne Cd deposition rates on New Jersey cropland.

| Source | Cropsoil Cd Deposition Rates (MT/yr) |
|---|---|
| Municipal Incineration | .27 |
| Sludge incineration | .004 |
| Steelmaking | .075 |
| Coal-fired power plants | .019 |
| Totals | .37 |

## Municipal Sludge Landspreading

Details of the analysis are summarized in two publications cited in the List of References (Yost and Greenkorn, 1983; Yost, 1984). For the sake of brevity, they will not be repeated here. Of relevance to this discussion is the fact that dietary exposure projections involve the evaluation of expressions (e.g. Eq. 4) containing a variety of probability distribution functions (PDF's). They are as follows:

Sewage sludge nitrogen content

Nitrogen-based sludge application rates for six crop categories

Sludge Cd content normalized to New Jersey municipal treatment plant inputs (point source and runoff) and sludge generation rates

Consumption of six crop/food categories by 18-20 year-old U.S. male population

Cd dietary intake for each of six crop categories

Tables 4-9 give examples of the above frequency distributions. Crop-specific sludge application rate histograms are the quotients of crop nitrogen requirements and sludge nitrogen concentrations.

Soil Cd concentration frequency histograms were generated for each crop type by numerically executing the integrals in Eq. 4 with the nitrogen-based sludge application rates (e.g. Table 5) and sludge Cd concentration (Table 6) histograms as the constituent "functions". A resulting frequency histogram representing Cd contamination of leafy vegetable crop soils is given in Table 7. Cd crop uptake factors were then employed to define Cd concentration frequency histograms in the six crop categories factored into the dietary exposure analysis. This represents a simple "mapping" of the soil Cd concentration histograms by multiplying the latter by the appropriate uptake factors. Several statistical assumptions are implied in this process:

a) there is no correlation between sludge nitrogen and Cd concentrations

b) the distribution of sludges for landspreading is independent of Cd concentration and crop

Cd dietary intake frequency histograms for the six crops are generated by executing Eq. 4 with (the foregoing) crop Cd concentration and food consumption (e.g. Table 8) histograms as the constituent PDF's. The latter are derived from the 1965 U.S. Department of Agriculture survey of dietary habits (Household Food Consumption Survey USDA (1966), U.S. Government Printing Office). Table 9 gives a Cd intake frequency histogram associated with the consumption of leafy vegetables. The indicated mean (27 ug/day) is high as compared to the maximum ADI set by the World

Health Organization (70 ug/day). This result should be interpreted with the following in mind:

   a)  The figure applies only to persons consuming leafy vegetables on a given day, i.e. it is not a population average

   b)  The mean is skewed by the small but non-zero contributions of the two largest intake frequencies

The extent to which the latter truly represent Cd exposure "target populations" is not clear. This type of analysis is useful in identifying potential exposure problems which require more thorough elucidation before decisions on regulatory measures can be made.

Table 4.  Sewage sludge nitrogen content frequency histogram (USDA, 1975).

| Nitrogen Content (%) | Relative Frequency |
|---|---|
| 0.25 | 0.063 |
| 0.75 | 0.042 |
| 1.25 | 0.052 |
| 1.75 | 0.11 |
| 2.25 | 0.089 |
| 2.75 | 0.11 |
| 3.25 | 0.079 |
| 3.75 | 0.042 |
| 4.25 | 0.068 |
| 4.75 | 0.073 |
| 5.25 | 0.052 |
| 5.75 | 0.037 |
| 6.25 | 0.058 |
| 6.75 | 0.042 |
| 7.25 | 0.011 |
| 7.75 | 0.016 |
| 8.25 | 0 |
| 8.75 | 0 |
| 9.25 | 0.011 |
| 9.75 | 0.026 |
| 14.0 | 0.026 |

MEAN = 5.11%

Table 5.  Nitrogen sufficiency sludge application rate frequency
histogram for leafy vegetables.

| Sludge Application Rate (MT/ha) | Relative Frequency |
|---|---|
| .812 | .026 |
| 1.17 | .026 |
| 1.23 | .011 |
| 1.30 | 0.00 |
| 1.38 | 0.00 |
| 1.47 | .016 |
| 1.57 | .011 |
| 1.68 | .042 |
| 1.82 | .058 |
| 1.98 | .037 |
| 2.17 | .052 |
| 2.39 | .073 |
| 2.68 | .068 |
| 3.03 | .042 |
| 3.50 | .079 |
| 4.14 | .11 |
| 5.05 | .089 |
| 6.50 | .11 |
| 9.10 | .052 |
| 15.2 | .042 |
| 45.5 | .063 |

MEAN = 6.80 MT/ha

Table 6. Sewage sludge cadmium concentration frequency histogram.

| Cadmium Concentration (ppm) | Relative Frequency |
|---|---|
| 5.50 | .072 |
| 15.5 | .074 |
| 25.5 | .012 |
| 35.5 | .050 |
| 45.5 | .11 |
| 65.5 | .0035 |
| 75.5 | .0037 |
| 95.5 | .0031 |
| 105. | .019 |
| 115. | .0067 |
| 125. | .0034 |
| 145. | .079 |
| 165. | .015 |
| 175. | .11 |
| 195. | .11 |
| 205. | .21 |
| 215. | .0012 |
| 245. | .056 |
| 255. | .015 |
| 265. | .00053 |
| 275. | .0016 |
| 295. | .0057 |
| 375. | .0012 |
| 475. | .015 |
| 525. | .014 |
| 675. | .014 |
| 775. | .0033 |
| 1100. | .0013 |
| 2170. | .0042 |
| 3170. | .0011 |
| 3410. | .0055 |

MEAN = 179 ppm

Table 7. Cd addition to leafy vegetable cropsoils from sludge landspreading to achieve nitrogen sufficiency.

| Cd Addition Rate (g/ha) | Relative Frequency |
|---|---|
| 9.1 | .027 |
| 25 | .062 |
| 72 | .104 |
| 200 | .18 |
| 570 | .36 |
| 1,600 | .18 |
| 4,600 | .045 |
| 13,000 | .039 |
| 55,000 | .004 |

Table 8. Leafy vegetable consumption frequency histogram for U.S. 18-20 year old male population.

| Consumption (g/day) | Relative Frequency |
|---|---|
| 2.0 | .67 |
| 51.5 | .088 |
| 88.3 | .069 |
| 120 | .050 |
| 156 | .034 |
| 188 | .023 |
| 220 | .033 |
| 262 | .007 |
| 292 | .007 |
| 327 | .007 |
| 364 | .0064 |
| 600 | .017 |

MEAN = 91 g/day

Table 9.  Cadmium intake frequency histogram for adults consuming
leafy vegetables grown in sludge-amended soils.

| Cd Intake (ug/day) | Relative Frequency |
|---|---|
| .024 | .003 |
| 0.10 | .024 |
| 0.55 | .122 |
| 2.8 | .258 |
| 14.5 | .422 |
| 74.5 | .142 |
| 380 | .028 |
| 2230 | .00096 |

MEAN = 27 ug/day

# APPLICATION OF MULTIMEDIA POLLUTANT TRANSPORT

# MODELS TO RISK ANALYSIS[*]

Thomas E. McKone[a] and William E. Kastenberg[b]

[a] Environmental Sciences Division
Lawrence Livermore National Laboratory
Livermore, California

[b] Mechanical, Aerospace and Nuclear Engineering Department
University of California
Los Angeles, California

## INTRODUCTION

The impact of a toxic substance on public health depends upon its inherent toxicity, the size and location of the source term, its behavior once released to the environment, the number and type of exposure pathways, and the susceptibility of the exposed population. Because the process of quantifying impact through this sequence of events is inherently uncertain, we refer to the estimated impact using the concept risk. Thus, regulating toxic risks requires consideration of the source, chemical fate, exposure routes, human susceptibility and combined health effects of an ever-increasing list of chemicals. The very magnitude of this list precludes a detailed assessment of each chemical of concern. For this reason there is a demand for comprehensive but not necessarily precise models for quantifying human exposure. Multimedia transport models fill this demand. As Herman Daly (1980) notes, "[i]t is better to deal incompletely with the whole than wholly with the incomplete."

This paper will explore how multimedia models can be used in an environmental health risk analysis. The purpose of risk analysis is to assess and manage the adverse consequences of human activities. Multimedia models offer a way of organizing information about landscape and chemical properties in order to provide a comprehensive picture of the link between source and receptor for toxic chemicals. We begin with an overview of the risk analysis process, including risk assessment and management. This is followed by a description of a multimedia model, called GEOTOX, which was developed for use in evaluating the health risks of trace elements, radionuclides, and organic chemicals. The model uses landscape properties such as run-off, precipitation, evaporation, biomass density, and soil properties and chemical properties such as molecular weight, vapor pressure, and solubility to determine how a substance partitions within the environment. The predicted environmental

---
* Work performed under the auspices of the U.S. Department of Energy by Lawrence Livermore National Laboratory under Contract W-7405-ENG-48.

concentrations are used with exposure models and combined with health-effects data to estimate the potential human health risks associated with a given source. We demonstrate the use of GEOTOX by a sample application to three chemicals. We examine how these models can be used in risk assessment to better define the relationship between contaminant releases and resulting doses to humans. We also examine how these models can provide risk managers and decision makers with a means for determining the value of reducing uncertainties about chemical and landscape properties.

## BACKGROUND ON RISK ANALYSIS

The National Research Council (1982) divides the risk analysis process into two phases ~ risk assessment and risk management. In addition to risk assessment and management, it should be recognized that risk analysis usually begins with a hazard identification. Without the measurement or perception of a hazardous condition, there will be little or no effort to assess and manage risk. The object of risk assessment is to develop models and measures that can be used to determine the magnitude of the risk, parameters that contribute to this magnitude, and the likely uncertainty about the magnitude. Risk management is the process of weighing policy alternatives and selecting an appropriate institutional response. This process should integrate the results of risk assessment with engineering data and with social, economic, and political valuation to reach a decision. Linking these processes is the concurrent effort to evaluate risk. Risk evaluation directs the risk assessment/management process in terms of individual and societal valuations of risk.

Figure 1 provides a view of how the risk analysis process might proceed for a toxic chemical. Each of the major steps requires one or more actions that are listed to the right of the box representing that step. It is typical for the risk-analysis process to begin with efforts to identify the potential hazards associated with a chemical species. We have listed three ways to obtain hazard information. These are epidemiological studies on humans, animal studies, and <u>in vitro</u> testing. One or more of these pieces of information are required to identify a hazard. If a substance has been found to present a significant hazard, a regulator will prescribe a risk assessment. The risk assessor quantifies the relative potency of the substance, estimates exposures to human populations and determines susceptibility of various populations and population subgroups. Once the level of risk has been assessed, the risk is controlled or managed. We list three ways of managing risk. These are (1) public information, (2) exposure control (such as controlling food and water supplies) and (3) source control. The level of risk assessment and management requires that decisions be made which balance the valuation of risk against the cost of reducing risk. For this reason we have listed risk evaluation as a separate box that links risk assessment and management and receives input from social, political and economic institutions.

## Hazard Identification

The process of assessing and managing risk is driven by the identification of an actual or at least perceived hazard. The distinction between risk and hazard is made clear in an example posed by Okrent (1980) of considering the choice between a rowboat and an ocean liner for a trip across an ocean. Both choices provide the hazard of drowning; but the risks (and costs) differ substantially. For toxic chemicals, hazards are usually identified through observations or experiments with tissue or cell cultures, animal species, or human subjects. However, the hazard can sometimes be identified using analysis based on existing physiological and/or toxicological studies.

Analysis based on physiological and toxicological data and laboratory studies using living tissues or cells provide limited but useful input about toxic and carcinogenic materials. Such studies fail to provide definitive answers on the toxic effects of any substance. However, they are a useful method for identifying the existence and likely severity of any effect. These techniques offer a low-cost way of acquiring preliminary data. They are viewed by many as an attractive way to undertake preliminary screening of a large number of substances in order to establish regulatory priorities (Morgan, 1985). However, there is the problem that such studies will fail to screen out a substance that eventually proves toxic to humans.

Animal exposure studies under controlled conditions provide a much more reliable technique for exploring potential human health effects. There are, however, several complications. First, is the question of whether and how observed effects in animals will show up in human populations. Second is a problem of cost: animal studies can be very expensive. In order to reduce costs, much higher exposure levels than those expected for humans are used. This introduces problems involving the shape of the dose-response function and the possible existence of thresholds. Finally, there is an issue of ethics. There are groups and individuals who argue that experiments with animals are cruel and perhaps immoral. These people will suggest that such experiments should be used only if no alternative exists.

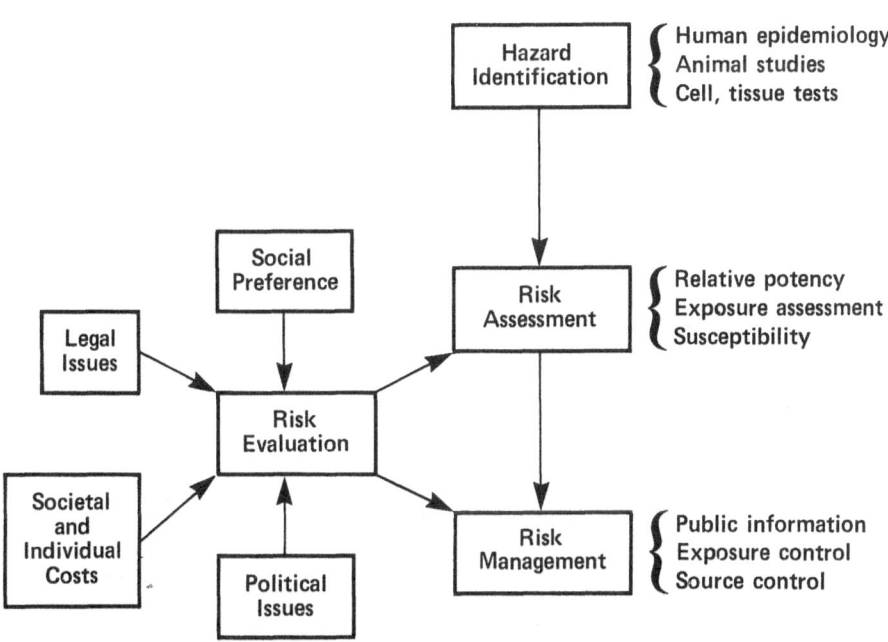

Fig. 1. Schematic diagram of the risk-analysis process.

Studies involving human subjects are based on controlled exposure studies, analysis of special cases of high-level exposure, and studies of population groups undergoing chronic low-level exposure. Controlled human subject studies are clearly the best way of obtaining reliable data. In most cases ethical considerations prohibit such studies. These studies are only used to assess short-term exposures to substances with reversible effects. Accidents or acts of war which have resulted in special high-level or chronic exposures often provide reliable data on human effects. The radiation victims of Hiroshima and Nagasaki and people exposed to toxic clouds from industrial accidents are groups that can be studied to infer long-term hazards. The problem in studying these groups often involves reconstructing the exposure. Finally, many human groups are exposed voluntarily (through occupation or life-style) or involuntarily (from contaminants in drinking water, food, and air) to toxic substances. The effects of such exposures can be studied both prospectively and retrospectively by identifying, tracking, and collecting data on specific individuals.

## Risk Assessment

Risk assessment is the process of determining the adverse consequences that may result from some action (or inaction) such as the use of an industrial process or technology. As applied to a toxic substance, risk assessment includes four principal elements: (1) quantification of releases and environmental levels near the source, (2) prediction of the movement of the toxic substance away from the sources, (3) an estimate of the number of people, plant, and animal species or other environmental elements likely to be exposed to toxic hazards, and (4) the numbers likely to suffer adverse consequences. In addition, risk assessment should provide a process for determining how inadequacies and random variations in data combined with limits and assumptions included in models effect uncertainty about the risk.

Source quantification requires that one identify hazardous inventories, points of release, and release rates and profile the uncertainties associated with these inputs. Description of hazardous inventories should include estimates of the volumes, masses, and chemical form. Quantification of releases rates should specify transfers from the source to adjacent components of the air, water, and soil. Often this quantification requires the enumeration and characterization of a number of potential release scenarios.

Predicting the movement of a toxic substance in the environment involves the use of models that describe partitioning of chemical species among various environmental media. One approach to this process is the use of multi-compartment models. Models of this type have been employed for studying the global fate of toxic elements and radionuclides (Garrels, et al., 1975; Stumm, 1977; McKone, et al., 1983) and for assessing the fate of organic chemicals in soil/lake/river systems (Mackay, 1979; Mackay and Paterson, 1981, 1982; Frische et al., 1984; Cohen and Ryan, 1985). Compartment models are a reasonable starting point because they cover all of the primary media simultaneously and thus provide a "cross-media" perspective. Often the models can be limited to only a few compartments. There are situations in which the level of information needed in the risk assessment requires more complex and mathematically sophisticated models. Nevertheless, the amount of information available for most chemicals limits the analyst to a model consisting of one or more atmospheric compartments, two or three soil zones and land biota, groundwater, and sediment compartments.

Detrimental effects on human health caused by toxic species can be classified as "stochastic" and "nonstochastic" effects (ICRP, 1977). Stochastic effects are those for which the probability of an effect occurring, rather than the severity of effect, is proportional to dose; without threshold. Nonstochastic effects are those for which severity of effect is a function of dose and for which a threshold may exist. The health effects of carcinogens are usually assumed to be stochastic, whereas the health effects of agents such as neurotoxins (like lead and mercury) are nonstochastic. An exposure assessment translates environmental concentrations in air, land, and water into human exposures through deposition, inhalation, and ingestion. A health risk estimate should identify the number of people exposed, the dose per individual, relevant characteristics of those exposed, and the effect that exposure to other hazards might have in modifying the effects of the chemical under consideration. For several chemicals EPA has enough data to adequately estimate cancer risks for individuals using the linearized-multistage model. The lifetime cancer risk (P) is calculated using an expression of the form:

$$P = (B \times d) \quad \text{(for low dose rates)} \qquad (1)$$

In this expression d is dose rate in units of ng/day per kg (body weight). B is the unit cancer risk slope. The exposure d is obtained by multiplying environmental levels of a toxic chemical in air, water, and food by the rate of human uptake and by the effective absorption factor.

There are likely to be many sources of uncertainty in the risk assessment process. The term uncertainty is often used to include two different concepts: random variability in some parameter or measurable quantity and an imprecision in the analyst's knowledge about models, their parameters, or their predictions. An uncertainty analysis is used to determine uncertainty in the overall estimate of risk as a result of uncertainties in each of the submodels and inputs.

## Risk Evaluation

Risk evaluation provides a link between risk assessment and risk management. In the evaluation phase the results of a risk assessment are integrated with social, economic and political considerations to provide input to the risk management phase. A variety of techniques have been proposed and are used to systematically apply one's own values or the values held by others in the evaluation of risks. There is currently no consensus as to which technique is most appropriate for a given set of circumstances. The purpose of this section is to compare and contrast a variety of techniques for risk evaluation. Consideration is given to techniques of risk evaluation including individual or societal preference, science policy, economic methods, and ethics as inputs to risk management.

The decision to expend societal resources in order to identify, assess and manage risk carries with it an implicit valuation of the risk being controlled. Because of the inherent uncertainty of the risk assessment/management process, it is importance to consider how individuals and societies value uncertain adverse consequences. We expect such valuations to be expressed in terms of relative preferences; economic preferences or ethical constraints.

Even though exposure and dose-response assessments are an important part of understanding risk, they do not provide the complete picture on how individuals and societies decide to manage risk. It is also important to understand the processes by which people perceive health

risks and decide how good or bad they are. Recently, the field of
experimental psychology has been examining the problem of risk perception
in the process of risk management. Kahneman et al. (1982) have prepared
a review of this work. The economic problem of individual and societal
valuation of life and health has been considered by Raiffa et al. (1977).

Psychologists have suggested that when people are asked to make
judgements involving uncertainty, they subconsciously adopt a number of
heuristics or simple rules of thumb for decision-making. Briefly, their
argument states that people tend to assess the probability of an event by
the ease with which they can think of previous occurrences of the
event. Usually this heuristic serves people well, but in modern
societies there are ways this process can lead to biases. For example,
deaths from botulism are quite rare, but Americans, through the press,
learn about virtually every one that occurs. Deaths by stroke are fairly
common, but Americans learn about them only when a friend, relative or
famous person is involved. Thus, Americans systematically overestimate
the frequency of death by botulism and underestimate the frequency of
death by stroke.

Economists often measure the value of commodity or action by the
willingness of people to make trade-offs. For example, people who live
in California are willing to trade the higher cost of housing for the
perceived increased value of the location. When human lives and health
effects are among the outcomes at stake, the problem of confronting
trade-offs becomes more difficult. Raiffa et al. (1977) have identified
three kinds of trade-off problems that arise in decisions involving life
and health. These are: (1) trade-offs between economic resources and
human lives and health; (2) trade-offs among mortality, morbidity, and
"quality of life;" and (3) intertemporal trade-offs including the
questions of discounting and future generations. According to Raiffa et
al. (1977):

> The chief difficulty in valuing lives (or health) stems from the fact
> that life has no market price. Under nearly all conceivable
> conditions no sum of money is sufficient to compensate a man for loss
> of his life; lives cannot be bartered and are obviously qualitatively
> different from all other commodities that society produces. In an
> attempt to circumvent this difficulty, two major strategies have been
> employed. The first utilizes an ex post approach; that is, it employs
> one or another technique to determine what dollar value society should
> place on a life that is lost. For such an assessment, the most
> commonly used measure is so-called human capital. The second strategy
> is radically different in that it approaches the problem ex ante; that
> is, it estimates the dollar value that should be assigned to some
> given statistical reduction in the probability of death or illness.
> In effect, it focuses on the value of life-saving rather than the
> value of a life. The most widely-discussed method for obtaining such
> values depends on an assessment of the individual's willingness to pay
> for a public program that offers some statistical improvement in his
> outlook for survival or health.

Risk Management

Once the risks associated with chemical releases have been
identified and quantified, it is necessary to develop a basis for
evaluating these risks and carry out the decision-making process. The
goal of risk management is to establish the significance of the estimated
risk, compare the costs of reducing this risk to benefits gained, compare
the risks to societal benefits derived from incurring the risk, and carry
out the political process of reducing the risk. There is no single,

objectively correct procedure for managing risks. Still, there are a variety of techniques that can increase our understanding of how to systematically apply one's own values to a complex, multidimensional choice and how to analyze the values held by others and incorporate such values in policy decisions.

There are at least four types of studies that can be used in risk management: risk-benefit, cost-benefit, risk-risk, and cost-effectiveness studies. Risk-benefit studies provide a comparison of the risks added by a process or technology to the concurrent benefits (usually economic) provided to the society. A cost-benefit study measures the financial cost (in dollars) of reducing risk relative to the benefit (in equivalent dollars or an appropriate surrogate) gained by reducing the risk. A risk-risk comparison establishes the importance of an estimated risk. Such a study considers the magnitude of an added risk relative to some other acceptable risk such as background cancer incidence, occupational hazards, or natural disasters. Cost-effectiveness studies compare the risk reduction per unit cost among several options for dealing with a risk. For example, at a contaminated waste site, a cost-effectiveness study would compare cancer risk reduction per dollar spent among several options for dealing with the site.

A MULTIMEDIA MODEL FOR USE IN RISK ASSESSMENT

A multimedia model suitable for risk assessment/management studies must meet several requirements. Among these we have identified the following:

1) the ability to handle organic chemicals, trace elements and radionuclides;
2) a model simple enough to allow multiple runs for sensitivity studies;
3) capability of interfacing environmental concentrations with pathways to calculate exposure;
4) a model that can handle steady-state and dynamic calculations; and
5) a modular model that can be easily altered to simulate different types of compartment structures.

We have found no model that fully satisfies all these criteria. However, there are models which address most of these requirements. The model we describe here, referred to as GEOTOX, has been adapted by McKone (1985) from a model first developed by McKone (1981) and McKone et al. (1983) for ranking the health risks of heavy metals and radionuclides in a generic global landscape. GEOTOX has two main components – a chemical fate model and an exposure model. The chemical fate model uses a set of landscape properties and chemical properties to determine the distribution and concentration of chemicals among a set of specified compartments, i.e. air, water, soil, etc. Environmental concentrations are linked to human exposures and cancer risk using an exposure model that accounts for intake through inhalation, consumption of food and water, and dermal absorption.

The general form of the compartment equations used in GEOTOX are:

$$\frac{dN_i^n}{dt} = -\lambda_i^n N_i^n - \sum_{\substack{i=1 \\ i \neq j}}^{m} T_{ij}^n N_i^n - T_{io}^n N_i^n + \sum_{\substack{j=1 \\ j \neq i}}^{m} T_{ji}^n N_j^n + S_i^n \qquad (2)$$

where

$N_i^n$ = time varying inventory of species n in compartment i, moles;

$\lambda_i^n$ = decay constant for species n in compartment i accounting for radioactive decay, chemical decomposition, etc., $y^{-1}$;

$T_{ij}^n$ = transfer rate of species n from compartment i to compartment j, $y^{-1}$;

$T_{io}^n$ = transfer rate of species n from compartment i to some external system, $y^{-1}$;

$S_i^n$ = source term for the introduction of species n to compartment i, mole/y; and

m = number of compartments.

When Eq. 2 is written m times for m compartments, the following matrix equation is obtained

$$\underline{N}(t) = [T]\,\underline{N}(t) + \underline{S}(t) \qquad\qquad (3)$$

in which $\underline{N}$ and $\underline{S}$ are vectors of length m and $[T]$ is a matrix of size m by m. A steady-state solution for this system of equations can be obtained by defining a constant source vector $\underline{S}$. A time-varying solution to this system is set up by allowing the source vector $\underline{S}(t)$ to be a function of time. A transient solution for the vector $\underline{N}(t)$ is obtained using numerical integration. The chemical fate analysis provides estimates for the transfer coefficients $T_{ij}^n$ and decay constants.

The choice of compartment structure requires a balance between two competing concepts, simplicity of the model and the value of the information provided. Because the model is required to provide a framework for assessing the environmental fate of chemical species, it should, even in its most simple form, include at least three compartments: air, water, and soil. However, because much of the exchange of chemicals occurs at interfaces such as soil/water, soil/atmosphere and water/sediment and because of the role of the surface biota as a possible reservoir for chemicals, we choose to use a minimum of eight compartments. These are atmospheric gas and particles, land biota, upper soil layer, lower soil layer, ground water zone, surface water, and surface-water bottom sediments.

Each compartment is composed of from one to as many as three phases: solid, liquid, and/or gas. A compartment is described by its total mass, total volume, solid mass, liquid mass and gas mass. Mass flows among the compartments consist of solid flows and liquid phase flows. Figure 2 illustrates the conceptual layout of the environment as it is viewed by the eight-compartment GEOTOX model. In the paragraphs that follow, we describe how the mass, volume, liquid and solid exchanges among the compartments are obtained.

Atmospheric Gas Phase. We have found it useful to divide the atmosphere into two compartments, gas phase and particulate matter, in order to more accurately trace the fate of chemicals moving between soil, air, and surface waters. For both atmospheric compartments a box model is used to represent the atmosphere. In this model the concentration is assumed to be uniform across a region of along-wind width, $\Delta x$. Chemicals are confined within the mixing height $H_c$. The gas phase compartment is used to handle vapor transport to and from the lower atmosphere. A

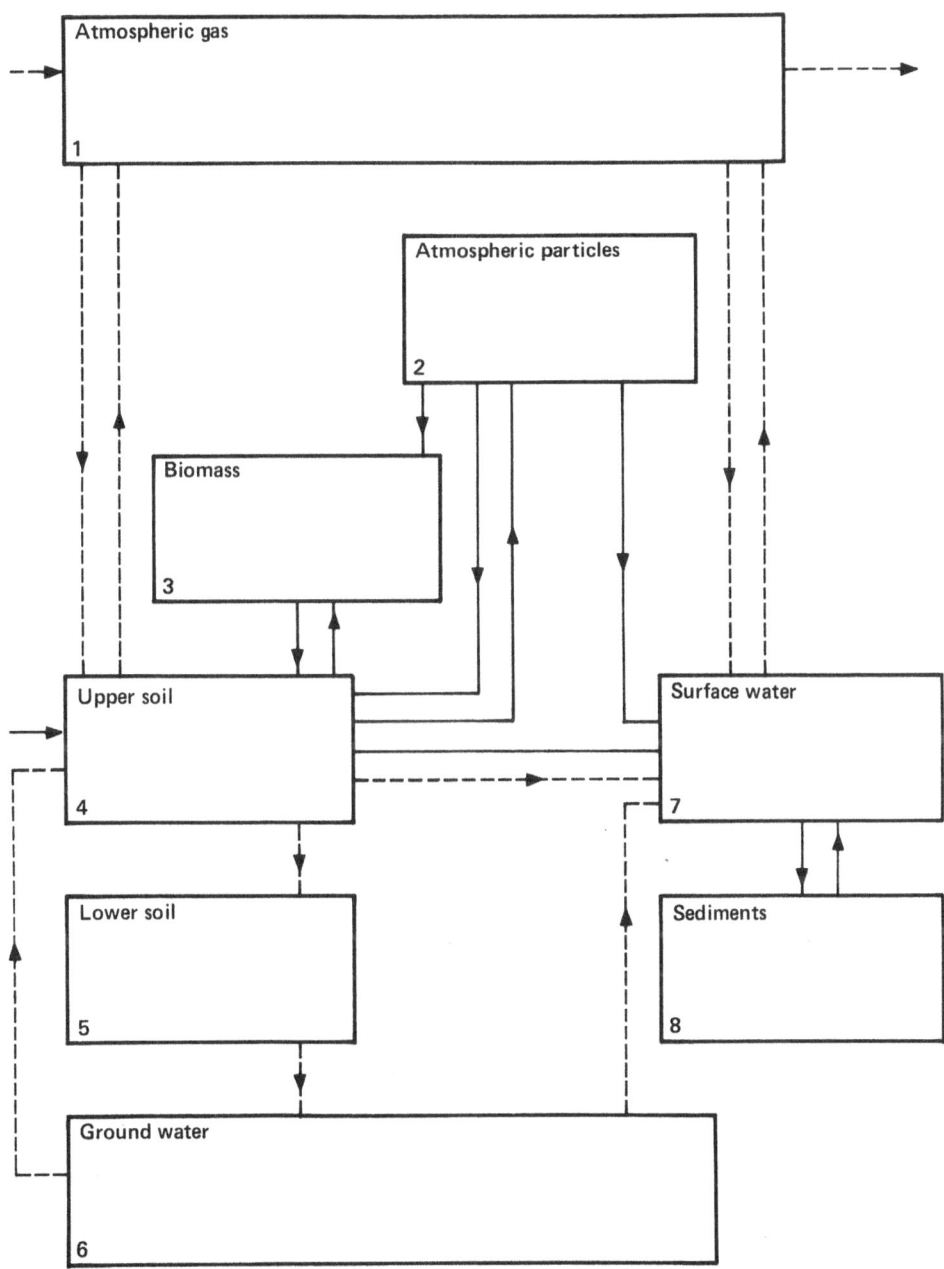

Fig. 2.   Interactions among the eight compartments in the GEOTOX model. Solid lines refer to solid phase flows, dashed lines to liquid phase flows.

chemical enters this compartment directly (from a source) or by diffusion from soil and/or surface waters. The chemical leaves the compartment through dry deposition, wet deposition, advective loss due to wind transport, venting out the top of the box, and chemical decomposition. Each of these processes is characterized by a first-order rate constant, which is the inverse of the effective residence time for that process. The rate constant for dry deposition is set equal to the diffusive capacity of the chemical across the air/soil boundary layer. The rate constant for wet deposition (or washout) is obtained by assuming that the ratio of chemical concentration in precipitation to that in the ambient air is given by Henry's law constant. The removal rate is then the product of precipitation and air concentration divided by the air/water partition coefficient.

Atmospheric Particulate Matter. This compartment is used to account for the movement of soil particles to and from the lower atmosphere as windblown dust with sorbed chemicals. Chemicals enter this compartment attached to the suspended dust and leave as a result of dry and wet deposition, advective loss due to wind transport, and chemical decomposition. We model this compartment as a uniform suspension within a box of mixing height $H_o$. For the GEOTOX model we assumed the average concentration of atmospheric particles to be 100 $\mu g/m^3$. We input 1000 m as the height of the lower atmosphere. This value results in an effective 3 d residence time of particles in the lower atmosphere due to wet and dry deposition.

Land Biota. The GEOTOX model treats chemical turnover in the biosphere in terms of the dry mass production per unit area. It is assumed that the dry mass portion of vegetation (excluding carbon) is derived from soil matter and returns to soil matter as the plant decays.

Upper and Lower Soil Layers. The soil layers are characterized by five parameters: thickness, bulk density, porosity, water content and fraction organic carbon. Typically, the upper soil layer has a lower bulk density, higher porosity, and higher organic carbon content than the lower soil layer. The lower soil layer is the vadose zone between the upper soil layer and the water table. It can be viewed as a filter layer for precipitation passing through the upper soil layer on its way to the groundwater compartment.

Groundwater. This compartment consists of the saturated material of the lowest soil horizon and saturated bedrock. All pore spaces are occupied by water. Water and chemical migration in this zone are typically in a horizontal direction. Groundwater moves in response to gravity from areas of infiltration (recharge) to points of discharge where the water table intersects the land surface. However, transport of substances dissolved in subsurface waters is not at the same rate as groundwater motion. Several processes retard the migration of chemicals in groundwater. Among these processes are absorption, ion exchange, precipitation, colloidal filtration, and irreversible mineralization. Taken together these processes are referred to as "sorption" and accounted for by a partition constant, which expresses the ratio between solid and liquid phase concentrations.

Surface Water and Sediments. These compartments are used to represent the fate of chemicals in lakes, reservoirs, ponds, sloughs, streams and estuaries. The surface water compartment is composed of water, biota, and suspended sediments in chemical equilibrium. Water enters this compartment from run-off, precipitation, and ground water discharge and leaves by evaporation and downstream flow. Solids enter the compartment from atmospheric deposition and surface erosion and leave by deposition to sediments and downstream advection. The sediment compartment exists at the lower boundary of the surface water compartment. This compartment is composed of bottom sediments and water in a ratio that is determined by the porosity of the sediments. Chemical species move into and out of this compartment by deposition from and resuspension to the overlying surface water and by mass diffusion.

Comparison of exposure potential among a number of chemicals requires the definition of a hypothetical environment or "unit world" (Mackay, 1979). The masses, volumes and intercompartmental transfers of liquids and solids in the GEOTOX model depend upon a number of landscape parameters. These parameters and the values used for the comparative analysis in this paper are listed in Table 1. These properties are taken from a review of landscape characteristics in three regions of the U.S. by Layton et al. (1985). They reflect landscape properties that are characteristic of the southeastern United States. Each of the parameters listed in Table 1 is an input to the GEOTOX. This set of parameters is used to define compartment masses and volumes, and to determine the exchange of solids and liquids between compartments. Because the landscape and chemical properties are handled independently by GEOTOX, it can be used to compare several chemicals introduced to a single landscape or to compare the role landscape properties in mitigating the health risks of a set of chemicals in different ecoregions.

## Calculation of Transfer Rates and Decay Constants

Chemical exchange among compartments occurs by advection in the solid phase, advection in the liquid phase and/or diffusion. A given chemical species is assumed to be in chemical equilibrium among the phases of a single compartment. However, there is no requirement for equilibrium between adjacent compartments. As an example of this process consider the upper soil layer. It is composed of three phases: soil water, soil matter, and soil gas. An organic chemical added to the soil distributes itself among these three phases in such a way that chemical and physical equilibrium are achieved. Among the potential chemical pathways from the upper soil compartment are liquid advection as a dissolved component of soil water run-off, solid advection in dust suspensions, and diffusion from soil gas to the atmosphere.

Mathematically, the rate of movement of a chemical from compartment i to compartment j can be represented by the expression:

$$T_{ij}^{n} = \alpha_{ij}^{n} + \beta_{i}^{n\ell} F_{ij}^{\ell} + \beta_{i}^{ns} F_{ij}^{s} \tag{4}$$

where

$\alpha_{ij}^{n}$ = rate constant that relates the molar inventory of chemical n in compartment i to the rate of mass diffusion from compartment i to compartment j, $y^{-1}$:

$\beta_{i}^{n\ell}$ = partition coefficient that expresses the fraction of molar inventory of species n in compartment i that is in the liquid phase, $kg^{-1}$;

Table 1.  Landscape parameters used to determine mass, volume and liquid
and solid transfers of the compartments.

| Compartment | Parameter | Value |
|---|---|---|
| Atmosphere gas and particles | Height | 1000 m |
| | Absolute humidity | 11.4 $g/m^3$ |
| | Precipitation[a] | 100 cm/y |
| | Residence time of dust particles | 3 d |
| | Ambient temperature | 290 K |
| Biota | Dry mass inventory | $3 \times 10^7$ $kg/km^2$ |
| | Dry mass production | $2 \times 10^6$ $kg/km^2$-y |
| Upper soil | Thickness | 0.25 m |
| | Bulk density | 1.5 kg/L |
| | Water content | 0.42 kg/L |
| | Air content | 0.03 L/L |
| | Evapotranspiration | 60 cm/y |
| | Surface run-off | 35 cm/y |
| | Mechanical erosion | $3.0 \times 10^5$ $kg/km^2$-y |
| | Organic carbon fraction | 0.02 |
| Lower soil | Thickness | 2.0 m |
| | Bulk density | 2.0 kg/L |
| | Water content | 0.3 kg/L |
| | Air content | 0.02 L/L |
| | Organic carbon fraction | $1 \times 10^{-3}$ |
| Groundwater | Water inventory | $4.6 \times 10^9$ $L/km^2$ |
| | Rock porosity | 0.1 |
| | Rock density | 2.3 kg/L |
| | Irrigation withdrawal | 2 cm/y |
| | Recharge | 5 cm/y |
| | Organic carbon fraction | $3 \times 10^{-4}$ |
| Surface water | Fraction of the land surface | 0.02 |
| | Average depth | 6 m |
| | Suspended sediment load | 890 $mg/L$ |
| | Sediment-deposition rate | 380 $kg/m^2$-y |
| Sediments | Thickness | 0.05 m |
| | Bulk density | 1.5 kg/L |
| | Porosity | 0.2 |
| | Sediment resuspension rate | 380 $kg/m^2$-y |
| | Organic carbon fraction | 0.02 |

[a] 98% of the total precipitation falls on land and 2% falls on surface
waters.

$F^{\ell}_{ij}$ = rate of mass flow of liquid from compartment i to compartment j, kg/y;

$\beta^{ns}_i$ = partition coefficient that expresses the fraction of species n in compartment i that is in the solid phase, $kg^{-1}$; and

$F^s_{ij}$ = rate of solid flow from compartment i to j, kg/y.

The terms $F^s_{ij}$ and $F^{\ell}_{ij}$ in Eq. 4 are dependent on landscape parameters, whereas the $\alpha$ and $\beta$ terms are dependent on environmental and chemical properties. The assumption of chemical equilibrium among the three phases of a compartment requires that the total molar inventory of a compartment satisfy the following expressions:

$$N^n_i = M^s_i C^{ns}_i + M^{\ell}_i C^{n\ell}_i + V^g_i C^{ng}_i \tag{5}$$

$$N^n_i = \gamma^{ns}_i C^{ns}_i = \gamma^{n\ell}_i C^{n\ell}_i = \gamma^{ng}_i C^{ng}_i \tag{6}$$

where

$M^s_i$, $M^{\ell}_i$ = mass of the solid and liquid phases in compartment i, kg;

$C^{ns}_i$, $C^{n\ell}_i$ = concentration of species n in the solid and liquid phases of compartment i, mole/kg;

$V^g_i$ = volume of the gas phase in compartment i, $m^3$;

$C^{ng}_i$ = concentration of species n in the gas phase of compartment i, $mole/m^3$;

$\gamma^{ns}_i$, $\gamma^{n\ell}_i$ = constants relating the concentration of species n in the solid and liquid phase of compartment i to the molar inventory, kg; and

$\gamma^{ng}_i$ = a constant relating the concentration of species n in the gas phase of compartment i to the molar inventory, $m^3$.

The $\gamma$ constants are derived from chemical partition factors. They can be interpreted as the effective mass or volume capacity of each phase for containing a chemical distributed in equilibrium among all phases. The distribution of a chemical between solid and liquid can be described with a sorption constant, $K^n_d$, that relates the amount of chemical sorbed to soil, rock, or sediment to the amount in the water. For organic compounds the sorption coefficient is the product of the organic carbon fraction of the soil $f_{oc}$ and the organic-carbon partition factor $K_{oc}$ (Hamaker and Thompson, 1972). The distribution of a chemical between water and air is expressed using the Henry's law constant. Combining Eq. 5 and 6 provides a set of expressions for the $\gamma$ factors in terms of compartment masses and the concentration ratios that depend on the sorption constant and Henry's law constant.

The $\beta$ factors are simply the inverse of the corresponding $\gamma$ factors. However, relating the $\alpha$ factor to $\gamma^{ng}$ requires that we consider the net molar flow (in mole/yr) of a species from the gas phase of one compartment to that of another as represented by the expression

$$\text{Molar Flow} = \frac{AD^n_G}{d}(C^{ng}_i - C^{ng}_j) = \alpha^n_{ij} N^n_i - \alpha^n_{ji} N^n_j \tag{7}$$

where

$$A_n = \text{landscape area, } m^2;$$
$$D_G^n = \text{diffusion coefficient for species n in air, } m^2/y; \text{ and}$$
$$d^G = \text{boundary layer thickness, m.}$$

In general, the $\alpha$ terms can be related to A, $D_G^n$, d and $\gamma^{ng}$ using Eq. 6 and 7 so that

$$\alpha_{ij}^n = \frac{A D_G^n}{d \gamma_i^{ng}} \tag{8}$$

A similar approach is used for liquid diffusion from bottom sediments to surface water. The boundary layer thickness, d, was assigned a value of 1 cm for atmospheric diffusion and 2 cm for diffusion from bottom sediments (see Layton et al. [1985] for more details).

Decay and transformation processes in the GEOTOX code are modelled as a first-order removal processes. The decay constant $\lambda_i^n$ in Eq. 2 is used to account for radioactive decay, hydrolysis, photolysis, oxidation, biodegradation, sedimentation and advection losses in the atmosphere. This constant treats each of these processes as first-order irreversible transformations. However, some of these processes are reversible for specific chemicals and cannot be modelled using this approach. There is currently no option in GEOTOX for treating reversible reactions. Another issue regarding transformations is the fate of the transformation product. In some cases the product is more toxic than the parent. However, GEOTOX can account for the accumulation and transport of the decay products by treating them as source terms within the respective compartment.

## Human Exposure and Health Risk Model

The human exposure and health risk model links environmental concentrations to potential exposure pathways and lifetime health risk. Exposure is expressed as daily intake per unit body weight averaged over an individual's lifetime. Two age groups, adult and child, are assumed in the analysis. The general model used to calculate exposure has the form:

$$E = \frac{1}{70 \text{ yr}} \left[ \frac{10 \text{ yr}}{17 \text{ kg}} \sum_{i=1}^{P} I_i^c + \frac{60 \text{ yr}}{60 \text{ kg}} \sum_{i=1}^{P} I_i^a \right] \tag{9}$$

where

$$E = \text{lifetime average exposure, for P pathways, ng/kg-d; and}$$

$$I_i^c, I_i^a = \text{daily intake by pathway i for a child and adult, ng/d.}$$

This formula reflects the use of steady-state environmental concentrations as a basis for exposure. The lifetime average exposure is not intended to represent the exposure to any specific individual, but to an individual who reflects the physiology, life-style and food consumption of the population for screening purposes.

Seven potential pathways are used in the analysis. These are: (1) inhalation; (2) drinking water; (3) ingestion of fruits, vegetables, and grain; (4) ingestion of meat and dairy products; (5) ingestion of fish; (6) ingestion of soil; and (7) dermal absorption. Parameters used to describe the reference adult and child for these exposure pathways are given in Table 2. Intake of a chemical by inhalation is the product of atmospheric concentration and breathing rate. Intake of chemicals through drinking water is the product of water concentration and daily

Table 2. Parameters used in the exposure model[a].

| Parameter | Child | Adult |
|---|---|---|
| Age range (yr) | 0 - 10 | 10 - 70 |
| Average weight (kg) | 17 | 60 |
| Inhalation rate ($m^3$/d) | 10 | 22 |
| Drinking water intake (L/d) | 1.4 | 2.0 |
| Ingestion of fruits, vegetables and grains (kg/d)[b] | 0.15 | 0.13 |
| Ingestion of milk and dairy products (kg/d) | 0.5 | 0.3 |
| Ingestion of meat (kg/d) | 0.1 | 0.3 |
| Ingestion of animal fat (kg/d) | 0.06 | 0.17 |
| Ingestion of fish (kg/d) | $2.0 \times 10^{-3}$ | $6.5 \times 10^{-3}$ |
| Ingestion of soil (kg/d)[a] | $1.0 \times 10^{-4}$ | $6.0 \times 10^{-5}$ |

[a] Values compiled from ICRP (1975) and USNRC (1977) except for soil ingestion, which comes from Layton et al. (1985).
[b] Dry mass ingestion.

water intake. Because roughly half of the water used in the U.S. comes from ground water, we assume the drinking water supplies are split equally between surface and ground water. The intake of a chemical species through fruits, vegetables, and grains is the product of biota dry mass ingestion, the chemical concentration in soil, and the plant/soil partition coefficient $K_{ps}$. The remaining pathways involve more detailed assumptions that are dealt with in the following paragraphs.

The ingestion of meat and dairy products and of fish present the opportunity for bioaccumulation through the food chain. Meat and dairy product exposures are based on the intake of chemicals by cattle. For an organic compound the human intake is the product of daily human intake of animal fat, the rate of intake of the chemical by cattle, and the fat/diet partition factor $K_{fd}$. For other compounds human intake is the product of human ingestion of animal mass, the rate of intake of the chemical by cattle and a meat or milk/diet partition factor $K_{md}$. It is assumed that cattle ingest 15 kg/d of vegetation, 0.72 kg/d soil and 27 L each of surface and ground water (Layton et al., 1985). The intake of chemicals through fish is based on the assumption that they reach chemical equilibrium with their surroundings - either surface waters or bottom sediments. The concentration in fish tissue is the product of surface or bottom water concentrations and a bioconcentration factor, BCF. The daily human intake of fish is assumed to be divided equally between bottom feeders and surface feeders.

Soil ingestion and dermal absorption are direct pathways from soil to humans that are important in the absence of other soil-mediated pathways. Ingestion of soil occurs throughout one's life, but is most significant during childhood. Dermal absorption results from bathing in contaminated water or from the accumulation of contaminated soil on skin. Dermal absorption from contaminated soil on the skin is based on an estimate that both a child and adult will daily absorb the contaminant content of roughly 40 mg of soil (Layton et al., 1985).

From the discussion above, it should be noted that several assumptions were used to produce a set of single-valved parameters for the exposures analysis. For some screening assessments distributions of valves and sampling might be required to define the expected range of potential exposures.

## Treatment of Uncertainty and Sensitivity in Multimedia Models

There is no universally accepted method for the analysis of parameter uncertainty and sensitivity in multimedia-type systems models. However, there are a number of commonly-used approaches. Uncertainty in model output results from the natural variability of real processes, measurement uncertainty associated with input parameters, and the exclusion and aggregation of processes in models. Sensitivity in models often refers to the rate of change of model output with respect to each input parameter.

Two general approaches are available for investigating the propagation of uncertainty in ecosystem-type models. For some (but very few) cases the variability in model output can be estimated analytically. This is the case for models that can be either reduced to or approximated by a series of multiplicative parameters. For most cases the output uncertainty must be estimated numerically using Monte Carlo techniques. Uncertainty studies using the GEOTOX code require this approach. The goal of the uncertainty analysis should be to identify the parameters that are influential in determining the variability in the output. In addition, we would like to determine the extent to which the overall variability is reducible.

The various approaches for sensitivity analysis are characterized by differences in the way they measure the influence of parameters on model output. A measure of sensitivity may be either local or global. A local measure is one that reflects the influence of small changes in the parameter values only in the region near the set of nominal or baseline values. A global measure reflects their influence over the entire range of values that the parameter may assume. A sensitivity analysis can incorporate the effects of interactions among parameters or it may require the assumption that no interactions exist. In determining the acceptability of contaminated food and water supplies we use a global sensitivity analysis applied to the results of the uncertainty analysis. Mapping inputs to outputs using Monte Carlo techniques followed by a simple stepwise regression can be used to correlate the variability of individual input parameters with the variability in model predictions.

## APPLICATIONS

In this section we consider ways in which multimedia models can be applied to risk assessment and management. In risk assessment, these models offer the analyst the potential to prepare a more comprehensive picture of the impact of toxic chemicals. In the risk management area, these models provide a holistic framework, which, although incomplete, does offer decision makers an opportunity to compare a number of options for controlling risk.

## Multimedia Models and Risk Assessment

We have identified four areas in which multimedia models can be used to enhance the risk assessment process. These are:

1) chemical screening
2) source evaluations
3) sensitivity studies
4) uncertainty studies

In the paragraphs below we describe how multimedia models contribute to each of these processes.

Managing environmental health risks requires that we assess the environmental fate, exposure, and health risks of an ever increasing list of pollutants. The very magnitude of this list precludes a detailed experimental evaluation of each pollutant. For this reason, there is a continuing need for toxic screening methods. In reviewing hazard and risk indexes for toxic substances, Smith et al. (1980) found that in most measures, the screening is based on quantity and toxicity. The role of the environment is usually ignored or handled using a subjective score. In a recent paper McKone (1985) has proposed and demonstrated the use of a generic multimedia model in combination with exposure and health-risk estimates as a way of screening toxic substances.

Source evaluation is enhanced by multimedia models because they allow us to compare the public health impacts in terms of the source compartment. Layton et al. (1985) have used a multimedia model to compare the potential impacts of the same toxic waste streams introduced to the soil of three different ecoregions. Currently, one of us (McKone) is using GEOTOX to compare the impact of groundwater versus soil or atmosphere as the source compartment for TCE.

The comprehensive character of multimedia risk assessment models makes them well-suited to sensitivity and uncertainty studies. These models can be used to compare the sensitivity of predicted human health risks to variabilities in chemical properties and environmental transport parameters. Using Monte-Carlo methods, the random variability of input parameters can be propagated through the models to determine the uncertainty in estimates of health risk. A stepwise regression analysis can be used to correlate the uncertainties in risk estimates to the uncertainty or random variability of specifc input parameters.

Multimedia Models and Risk Management

In the paragraphs below we identify three ways that multimedia models contribute to the risk management process. These are:

1) enhanced visualization
2) value of information
3) flexible standards

In each case the test of utility is to ask how the models can help the decision maker select among alternatives for managing risk. Often the decision makers must choose when to stop collecting more information. Because of this, the suitability of multimedia models for value-of-information studies may be their major contribution to the risk management process.

Enhanced visualization is not just a qualitative feature but a quantitative feature of multimedia models. In assisting the regulation of toxic chemicals, these models provide a means for visualizing and quantifying the transfer of chemicals within the environment. For each chemical we can map-out how landscape and chemical properties affect the source/receptor/health-risk sequence. This type of information is quite useful to decision makers in both the public and the private sector.

Value of information refers to the value of increasing the precision of input parameters or the value of expanding models to include more state variables. In a risk assessment the information provided by multimedia models provides the basis for exposure estimates. Exposure estimates using multimedia models are sequential. In the early phase of analysis an incomplete data base is used with simple models to estimate exposure and risk. As the analysis proceeds, more data may be collected and more complex models used. A critical issue in this process is how to determine when to stop refining the analysis and recommend standards. Evans (1985) suggests that a decision maker should collect additional information only when the value of the additional information exceeds the cost. We can estimate the cost of additional information. It is the cost of additional sampling, more chemical and physical analyses and improved models. However, it is not as obvious how to determine the value of additional information. But statistical decision analysis provides an analytical framework for estimating the value of information. Using this framework it is possible to determine when additional information would be beneficial. In general the test of information value is the extent to which new information will affect a decision. This "value-of-information" approach provides a guide to the development and improvement of multimedia models for use in risk assessment.

Flexible standards for exposure to toxic chemicals are risk-based standards that specify acceptable risk instead of specifying acceptable release rates. Such standards are flexible in the sense that the specified release rate varies with the assimilative capacity of the receiving compartment. Flexible standards allow variable treatment of releases based on the character of the ecoregion under consideration. However, standards of this sort could not be implemented without using multimedia models that handle chemical transport, exposure, and health risk in a precise and comprehensive manner.

An Application of GEOTOX to Three Chemicals

In this section we use the transport and exposure models described above to rank the relative health risks of three military explosives residuals--TNT, RDX and benzene. The purpose of this exercise is to demonstrate how GEOTOX can be used to screen compounds in terms of potential health risks. This screening process is not intended to simulate an actual environment or to predict exposures to a specific population group. The aim is, rather, to provide information on the characteristic behavior of each substance and the resulting exposure to an individual living his full life in and receiving all of his air, water, and food from the contaminated landscape. Health risks are compared among the chemicals based on exposures that result from the continuous addition of one $g \cdot mole/km^2$-y to the soil compartment. We refer to the environment transport, chemical fate, and exposure assessment based on this source as the "unit exposure analysis".

The landscape properties used to define the compartment volumes, masses an mass exchange rates are taken from Table 1. The physical and chemical properties that were input to GEOTOX to describe the environmental behavior of the three compounds are listed in Table 3. Figure 3 shows the resulting distribution in the eight compartments based on the steady-state input of one $g \cdot mole/m^2$-y to upper soil.

Table 4 lists the exposures estimated to occur as a result of the concentrations in Fig. 3. These are the lifetime average daily absorbed doses projected using the exposure model described above and assuming that environmental concentrations have achieved steady-state. The absorbed dose for benzene is dominated by inhalation. Whereas, for TNT and RDX the exposure is dominated by water consumption and ingestion of fruits and vegetables.

184

Table 3. Physical and chemical properties of benzene, TNT,[a] and RDX.[b]

| Property | Substance | | |
| --- | --- | --- | --- |
| | Benzene | TNT | RDX |
| Molecular weight (g/mole) | 78.1 | 227.1 | 222.1 |
| Vapor pressure (torr) | 95 | $8 \times 10^{-6}$ | $4.1 \times 10^{-9}$ |
| Solubility in water (mole/L) | $2.3 \times 10^{-2}$ | $5.6 \times 10^{-4}$ | $3.4 \times 10^{-4}$ |
| Henry's law constant | 4130 | $1.4 \times 10^{-2}$ | $1.2 \times 10^{-5}$ |
| Log $K_{ow}$ | 2 | 1.8 | 0.9 |
| $K_{oc}$ | 48 | 30 | 3.8 |
| $K_{sp}$ (soil-plant) | 4 | 1.6 | 0.7 |
| $K_{fd}$ (fat-diet) | 0.01 | 0.009 | 0.003 |
| Bioconcentration factor in fish | 19 | 15 | 3 |
| Diffusion constant in water $(m^2/y)$ | 0.02 | 0.02 | 0.02 |
| Diffusion constant in air $(m^2/y)$ | 157 | 200 | 230 |
| Depletion-rate constant in air $(y^{-1})$ | 81 | 120 | 110 |
| Depletion-rate constant in water $(y^{-1})$ | 42 | 32 | 36 |

[a] TNT = 2,4,6 - trinitrotoluene.
[b] RDX = hexahydro-1,3,5-trinitro-1,3,5-triazine.

Table 4. Daily dose predicted for the seven exposure pathways after the release of 1 mole/km$^2$-y of each contaminant into the upper soil compartment.

| Pathway | Absorbed dose (mg/kg-d) | | |
| --- | --- | --- | --- |
| | Benzene | RDX | TNT |
| Inhalation | $3 \times 10^{-7}$ | $2 \times 10^{-11}$ | $2 \times 10^{-8}$ |
| Water consumption | $5 \times 10^{-10}$ | $1 \times 10^{-5}$ | $1 \times 10^{-5}$ |
| Fruit and vegetable ingestion | $4 \times 10^{-10}$ | $4 \times 10^{-5}$ | $1 \times 10^{-5}$ |
| Meat and dairy ingestion | $2 \times 10^{-10}$ | $6 \times 10^{-8}$ | $7 \times 10^{-8}$ |
| Fish ingestion | $6 \times 10^{-12}$ | $7 \times 10^{-8}$ | $6 \times 10^{-8}$ |
| Soil ingestion | $3 \times 10^{-14}$ | $5 \times 10^{-11}$ | $4 \times 10^{-10}$ |
| Dermal absorption | $2 \times 10^{-14}$ | $3 \times 10^{-11}$ | $2 \times 10^{-10}$ |
| Totals | $3 \times 10^{-7}$ | $5 \times 10^{-5}$ | $2 \times 10^{-5}$ |

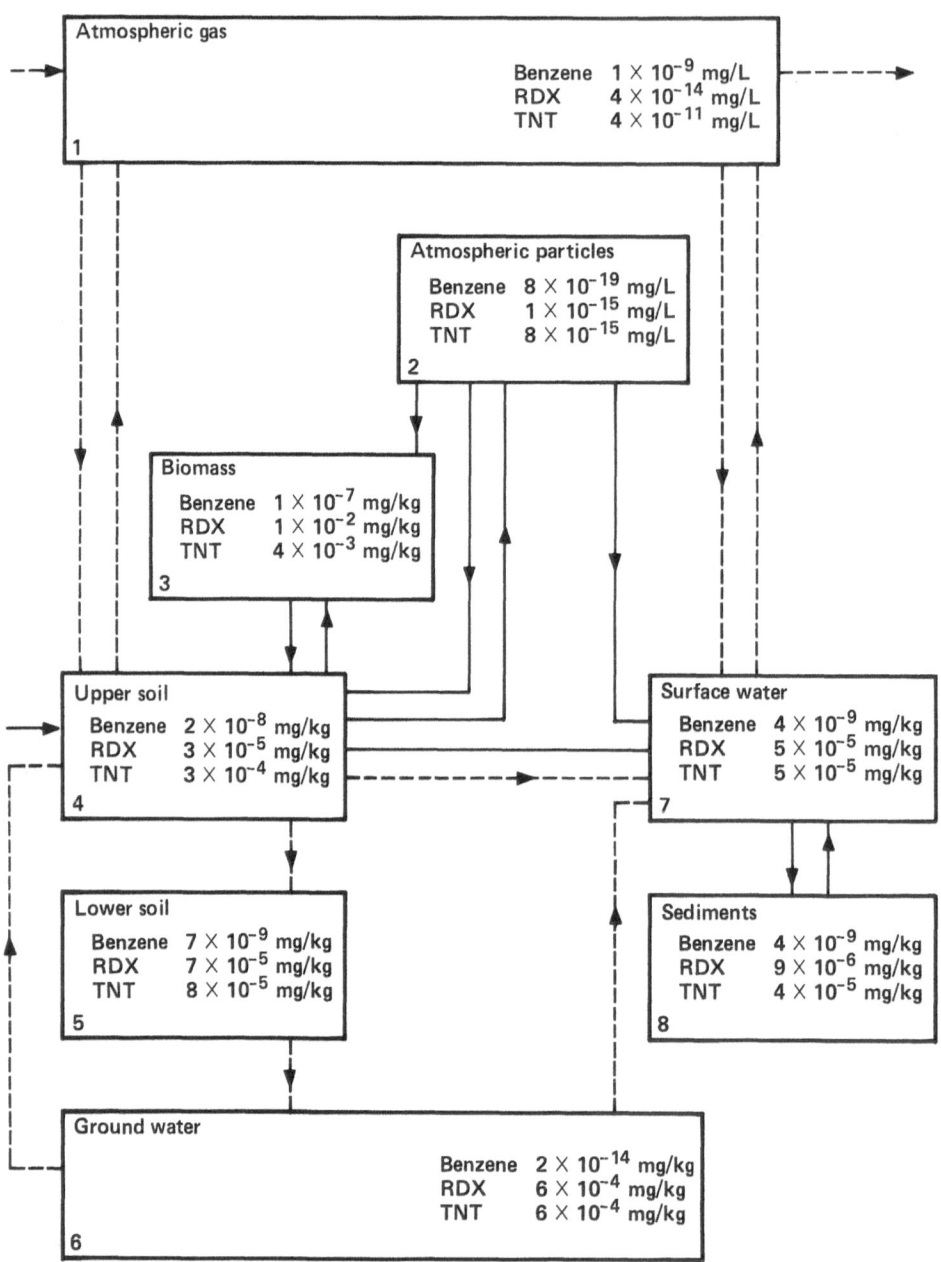

Fig. 3. Steady-state concentration of the contaminants in the eight compartments after the release of 1 mole/km$^2$-y into the upper soil compartment.

In Table 5 we rank the relative risks of the three chemicals by calculating the ratio of the predicted dose rates to virtually safe dose rates (VSDR). The VSDRs are levels that prevent noncarcinogenic effects or levels that correspond to a lifetime cancer risk of $10^{-6}$. The chemicals are listed in Table 5 in order of increasing VSDR. The VSDR for benzene, based on lifetime leukemia risk of $10^{-6}$, is almost two orders of magnitude lower than the VSDR for TNT which is calculated from the drinking-water standard. (TNT has not been identified as a carcinogen). The VSDR for RDX, which is based on a no-observed effects level in animals is an order of magnitude higher than that of TNT. The last two colums of Table 5 indicate that the screening ratio, which includes information on environmental fate and toxicity, does not correlate with the VSDR, which only includes information on toxicity. Benzene, which has the lowest VSDR, ranks second in terms of transport, exposure, and health risk. TNT moves from second to first. RDX ranks third in both columns. Thus, even though benzene is 25 times more toxic than TNT as expressed by the VSDR, benzene's capability for transport, exposure, and toxic risk puts it a factor of three lower than TNT per g-mole released to the upper soil compartment. Most of the differences can be attributed to differences in chemical fate and persistence of benzene versus TNT and RDX.

Table 5. Comparison of estimated contaminant dose rates from all exposure pathways and virtually safe dose rates (VSDR).

| Contaminant | Dose rate mg/kg-d A | VSDR mg/kg-d B | Screening ratio, A/B. | Rank |
|---|---|---|---|---|
| Benzene | $3 \times 10^{-7}$ | $4 \times 10^{-5a}$ | $7.5 \times 10^{-3}$ | 2 |
| TNT | $2 \times 10^{-5}$ | $1 \times 10^{-3b}$ | $2 \times 10^{-2}$ | 1 |
| RDX | $5 \times 10^{-5}$ | $1 \times 10^{-2c}$ | $5.0 \times 10^{-3}$ | 3 |

[a] Based on a lifetime leukemia risk of $10^{-6}$.
[b] Calculated from a drinking-water standard of 0.05 mg/L.
[c] Calculated from a chronic no-observed-effect level of 1 mg/kg-d in laboratory animals and a safety factor of 100.

CONCLUSIONS

Multimedia models can enhance the process of health risk analysis by allowing us to combine knowledge of chemical toxicity with information on environmental transport. In this paper, we have considered the interface between multimedia models and risk analysis in environmental health. We have reviewed the process of risk analysis as it is applied to toxic substances and divided it into four steps - hazard identification, risk assessment, risk evaluation and risk management. We described a model, referred to as GEOTOX, which can handle exposure and health risk assessment in a multimedia context. We have discussed ways that this model can be used to determine the sensitivity and uncertainty in a comprehensive risk assessment. We have shown that there are several ways in which multimedia models can be used to enhance both risk assessment and risk management. In the area of risk assessment, these models provide a means for chemical screening, source evaluation, sensitivity studies and uncertainty analysis. In the risk management arena, these models provide decision makers with a tool for enhanced visualization of chemical risks, value-of-information studies and flexible risk-based standards for controlling chemical health risks. Finally, we have illustrated the use of multimedia models for health-risk screening. In applying the models to three chemicals, we find that the relative rank of the three compounds based on environmental fate and toxicity is quite different from the rank based on toxicity alone.

REFERENCES

Cohen, Y. and Ryan, P. A., 1985, Multimedia modelling of environmental transport: trichloroethylene test case, Environ. Sci. Technol. 19: 412.

Daly, H. E., 1980, Introduction to the steady-state economy, in: "Economics, Ecology, Ethics: Essays Toward a Steady-State Economy", H. E. Daly, ed., W. H. Freeman, San Francisco.

Evans, J. S., 1985, The value of improved exposure estimates: a decision-analytic approach, "Proceedings of the 78th Annual Meeting of the Air Pollution Control Association", Detroit, Michigan.

Frische, R., Klopffer, W., Rippen, G. and Gunther, K-O., 1984, The environmental model segment approach for estimating potential environmental concentrations, Ecotoxicol. Environ. Safety, 8:352.

Garrels, R. M., MacKenzie, F. T. and Hunt, C., 1975, "Chemical Cycles and the Global Environment: Assessing Human Influences," William Kaufmann, Los Altos, California.

Hamaker, J. W. and Thompson, J. M., 1972, Absorption, in: "Organic Chemicals in the Soil Environment," C.A.I. Goring and J. W. Hamaker, eds., Marcel Decker, New York.

International Commission on Radiological Protection (ICRP), 1975, "Report of the Task Group on Reference Man," Report 23, Pergamon Press, New York.

International Commission on Radiological Protection (ICRP), 1977, "Recommendations of the International Commission on Radiological Protection," Report 26, Pergamon Press, New York.

Kahneman, P., Slovic, P. and Tversky, A., 1982, "Judgment Under Uncertainty: Heuristics and Biases," Cambridge University Press, New York.

Layton, D. W., McKone, T. E., Hall, C. H., Nelson, M. A., and Ricker, Y. E., 1985, "Demilitarization of Conventional Ordnance: Priorities for Data-Base Assessments of Environmental Contaminants," Lawrence Livermore National Laboratory, (Draft) UCRL-53620.

Mackay, D. M., 1979, Finding fugacity feasible, Environ. Sci. Technol., 13:1218.

Mackay, D. M. and Paterson, S., 1981, Calculating fugacity, Environ. Sci. Technol., 15:1006.

Mackay, D. M. and Paterson, S., 1982, Fugacity revisited, Environ. Sci. Technol., 16:654A.

McKone, T. E., 1981, "Chemical cycles and health risks of some toxic crustal nuclides," Ph.D. dissertation, University of California, Los Angeles.

McKone, T. E., 1985, The use of environmental health-risk analysis for managing toxic substances, "Proceedings of the 78th Annual Meeting of the Air Pollution Control Association," Detroit, Michigan.

McKone, T. E., Kastenberg, W. E., and Okrent, D., 1983, The use of landscape chemical cycles for indexing the health risk of toxic elements and radionuclides, Risk Analysis, 3:189.

Morgan, M. G., 1985, Risk assessment and risk management decision-making for chemical exposure, in: "Environmental Exposure From Chemicals," Vol. II, W. B. Neely and G. E. Blau, eds., CRC Press, Boca Raton, Florida.

National Research Council, 1982, "Risk and Decision Making: Perspectives and Research", National Academy Press, Washington, D.C.

Okrent, D., 1980, Comment on societal risk, Science 208:372.

Raiffa, H., Schwartz, W. B., and Weinstein, M. C., 1977, Evaluating health effects of societal decisions and programs, in: "Decision Making in the Environmental Protection Agency," Volume II, Committee on Environmental Decision Making, Commission on Natural Resources, National Academy of Sciences, Washington, D.C.

Smith, C. F., Cohen, J. J. and McKone, T. E., 1980, "A Hazard Index for Underground Toxic Material," Technical Report UCRL-52889, Lawrence Livermore National Laboratory, Livermore, CA.

Stumm, W., 1977, "Global Chemical Cycles and Their Alterations by Man," Abakon Verlagsgesellschaft, Berlin.

U.S. Nuclear Regulatory Commission (USNRC), 1977, "Regulatory Guide 1.109: Calculation of Annual Doses to Man from Routine Releases of Reactor Effluents for the Purpose of Evaluating Compliance with 10 CFR Part 50 Appendix I," Washington, D.C.

# OVERVIEW OF THE REMEDIAL ACTION PRIORITY SYSTEM (RAPS)

G. Whelan, B. L. Steelman, D. L. Strenge, and J. G. Droppo

Pacific Northwest Laboratory*
Richland, Washington  99352

ABSTRACT

   All environmental regulations are intended to minimize the risks to
man and his environment resulting from a regulated activity.  Because
lower risk levels are generally accompanied by higher environmental
control costs, optimum management is achieved by balancing risks and
costs.  The U.S. Environmental Protection Agency (EPA) currently employs
the Hazard Ranking System (HRS) to preliminarily assess inactive hazardous waste disposal sites for potential placement on the National Priorities List.  Recently, modifications to HRS have been proposed to more
realistically assess the risks posed by radioactive waste constituents.
These modifications significantly increase the applicability of the HRS
to the U.S. Department of Energy (DOE) hazardous waste disposal sites.

   Although results from applying the modified HRS will be useful for
comparing ranking scores of DOE sites to non-DOE sites, the methodology
is still overly subjective to quantitatively prioritize one site relative
to another site.  To provide DOE with a better management tool for prioritizing funding allocations for further site investigations and possible remediations, Pacific Northwest Laboratory developed a more
objective, physics-based risk assessment methodology called the Remedial
Action Priority System (RAPS).  This methodology uses empirically, analytically, and semianalytically based mathematical algorithms and a pathways analysis to predict the potential for contaminant transport from a
hazardous waste disposal site to local populations.  Four major pathways
for contaminant migration are considered in the RAPS methodology:
groundwater, overland, surface water, and atmospheric.  Using the
predications of contaminant transport, simplified exposure assessments
are performed for important receptors.  The risks associated with the
sites can then be calculated relative to other sites for each pathway and
for all pathways together.

   The RAPS methodology addresses many of the typical limitations
associated with other ranking systems; it considers:  1) more site

---

*The Pacific Northwest Laboratory is operated for the U.S. Department
 of Energy under Contract DE-AC06-76RLO 1830 by Battelle Memorial
 Institute.

information and constituent characteristics associated with the transport pathways; 2) chemical and radioactive wastes; 3) the potential direction of contaminant movement; 4) contaminant retention (e.g., dispersion and decay/degradation), where applicable; 5) population distributions; 6) various routes of exposure (e.g., inhalations, ingestion, and external exposure); 7) contaminant toxicities; 8) duration of exposure of the surrounding population; and 9) contaminant arrival time to sensitive receptors. Because RAPS is based on more site information and constituent characteristics, the scoring system of the RAPS methodology also reduces the subjectivity associated with prioritizing hazardous waste sites.

The RAPS methodology requires minimum user knowledge of risk assessment and a minimum amount of input data. To maximize the utility of the system within DOE and its field offices, RAPS is being designed to operate on a personal computer.


INTRODUCTION

The U.S. Department of Energy's (DOE) inactive hazardous waste disposal sites are currently being evaluated under the Comprehensive Environmental Response, Compensation, and Liability Act of 1980 (CERCLA or Superfund) to determine whether migration of these substances has occurred and whether remediation will be required (DOE, 1985). In response to CERCLA and its associated environmental regulations, the DOE is in the process of locating, identifying, and evaluating potential problems associated with its inactive hazardous and radioactive-mixed waste disposal facilities, and controlling the migration of hazardous substances from such facilities to minimize potential hazards to health, safety, and the environment. Mixed wastes are those wastes that contain both radioactive and nonradioactive constituents.

Limited resources are available to conduct detailed investigations of all identified potentially hazardous and radioactive-mixed waste sites. Therefore, an assessment methodology is required to prioritize waste sites according to risk, based on limited available information, so that detailed site characterizations are performed first on those sites that exhibit the highest potential risks. For identifying sites that may pose significant problems to health, safety, and/or the environment, DOE is using the Hazard Ranking System (HRS), developed by the U.S. Environmental Protection Agency (EPA), to identify chemically contaminated sites for nomination to the National Priorities List (NPL). Because the HRS is not designed to evaluate sites containing radionuclides, a modified Hazard Ranking system (mHRS) was developed by Pacific Northwest Laboratory (PNL) to assess waste sites containing radionuclides (Hawley and Napier, 1984). The HRS and mHRS were not developed to quantify the relative risks of sites or prioritize sites that are nominated to the NPL according to their potential risks (EPA, 1982a). Consequently, PNL has developed a risk assessment methodology called the Remedial Action Priority System (RAPS) to provide DOE with a better management tool for prioritizing funding and other resource allocations for further site investigation. The HRS/mHRS methodology is initially used to evaluate hazardous waste sites; the sites that potentially pose significant problems would subsequently be evaluated using the RAPS methodology to prioritize them for further action.

In addition to presenting the rationale behind the development of the RAPS methodology, this paper briefly reviews the types of assessment methodologies developed for addressing concerns related to the migration, fate, and exposure of contaminants released into the environment and the type of methodology DOE is integrating into its internal program for identifying and evaluating potential problems associated with their inactive mixed radioactive and hazardous waste disposal facilities. These discussions are followed by a description of the structure of RAPS, the various components that compose the system, and the key features and characteristics of the system. Finally, two examples are presented illustrating the application of the RAPS methodology. The examples illustrate the migration, fate, and exposure of the chemical arsenic and the radionuclide strontium-90. The first example simulates the migration and fate of the contaminants from an inactive hazardous waste site through a saturated groundwater aquifer to a nearby municipal drinking-water well. The second example simulates contaminant transport through several layers of a partially saturated aquifer, then through a saturated groundwater aquifer, and finally through a nearby river that serves as a potable water source. In each case, the surrounding population is exposed to the contaminated water, and the respective relative risks to each population are computed.

ASSESSMENT METHODOLOGIES

Assessment methodologies or frameworks have been and are being developed to address concerns related to risks and for the migration, fate, and exposure of contaminants released into the environment. These contaminants can undergo complex processes of transport, degradation and decay, transformation, biological uptake, and intermedia transfer among atmospheric, overland, groundwater, and surface water pathways. The interactions of these various media pathways and linkages to man are illustrated in Figure 1. The assessment frameworks integrate many of these complex components in an attempt to address a complicated environmental setting in a logical, consistent, cogent, objective manner. Each assessment framework is developed to meet a particular objective and, therefore, cannot arbitrarily be applied to all assessment situations. For example, the RAPS methodology is being developed for DOE to prioritize, according to potential risk, inactive hazardous and radioactive-mixed waste disposal sites so the most hazardous sites can be further investigated first (Whelan and Steelman, 1984). The RAPS methodology addresses contaminant migration, fate, exposure, and risk through four major environmental transport pathways (i.e., groundwater, overland, surface water, and atmospheric). Because of the level of sophistication of the RAPS methodology, it can only be employed to rank or prioritize sites according to their relative hazard potential; it cannot be used in a predictive mode to simulate the actual risks at a particular site resulting from the release of contaminants into the environment. The RAPS methodology, therefore, meets the needs of DOE but may not meet the needs of other government agencies or private groups for conducting different types of assessments.

Several computer-based methodologies have been developed to effectively integrate and analyze complex processes involved in the migration and fate of contaminants through various transport pathways. Assessment methodologies can be grouped according to any number of traits. For example, they can be described according to their level of sophistication. At one end, frameworks exist based on simple questionnaires and check lists; at the other end, frameworks exist based on several highly sophisticated computer models. The remaining methodologies exist somewhere between these two extremes. Whelan et al. (1985a) took another approach and divided the various assessment methodologies

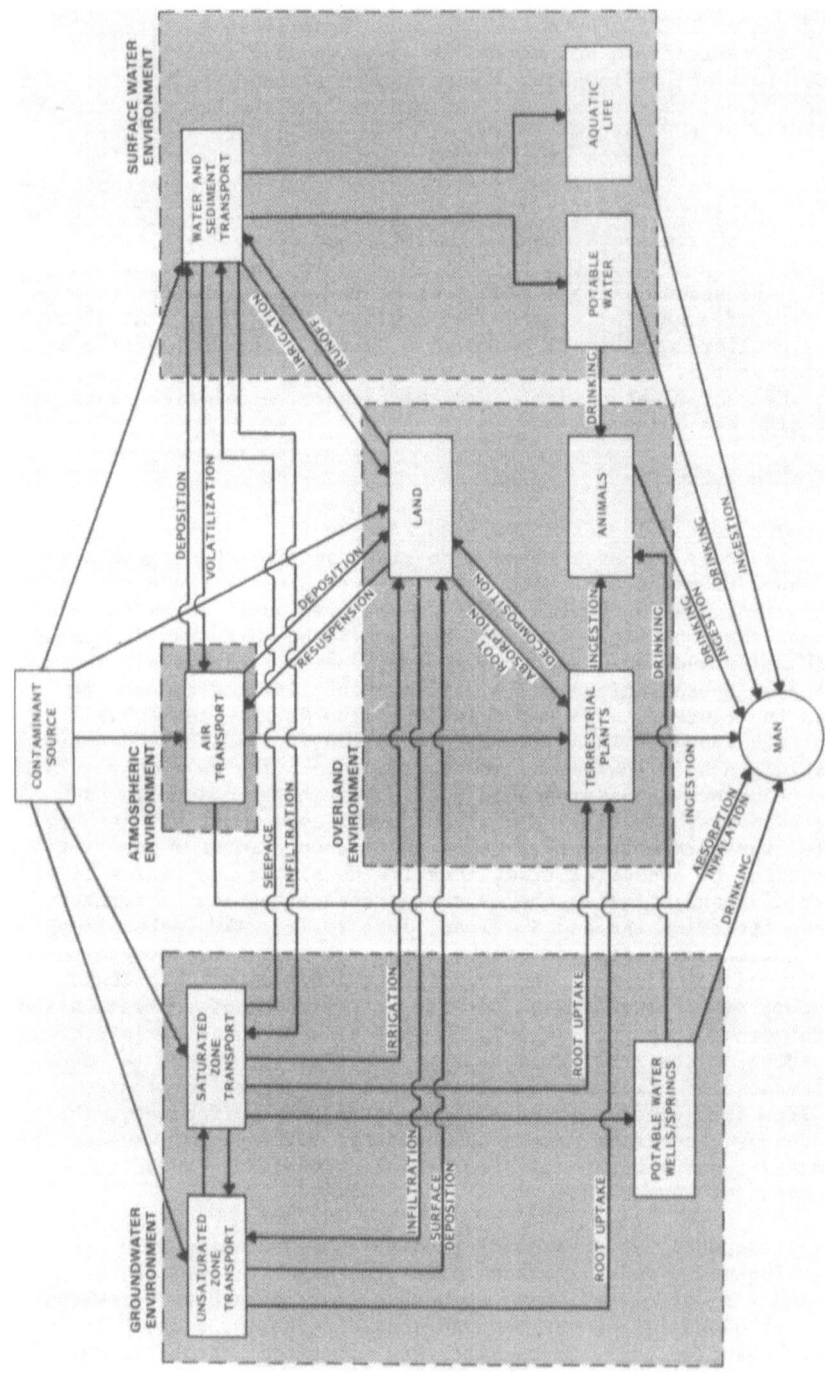

Fig. 1. Schematic Diagram Illustrating the Interactions Between the Various Contaminant Transporting Media and How the Contaminants Affect Man Through His Environment (Whelan et al., 1983).

into three categories according to their structure: check list/questionnaire (CL/Q), fully coupled, and compositely coupled.

Check list or questionnaire methodologies are generally based on a questionnaire that divides site and condition characteristics into pre-determined categories that are each assigned a point value. The user describes the characteristics of the site, waste, demography, etc., by identifying the categories that most closely correspond to those charac-teristics. The points associated with each category are usually totalled, and a score for the site is assigned. This method attempts to provide a simplistic, systematic means of assessing the hazards asso-ciated with waste disposal.

Although the CL/Q frameworks are easily applied, Whelan and Steelman (1984) and JRB* note that they include inherent deficiencies:

● Key parameters, particularly important in describing the migration, fate, exposure, and risks of a contaminant, are usually not directly considered in the assessment. Key parameters usually include, but are not limited to, dispersion coefficients, hydraulic conductiv-ities, degradation rates, modes of exposure (e.g., inhalation, ingestion, dermal contact), and dose-response information.

● The total waste volume is usually assumed to be composed of the most toxic substance present at the site (HWN, 1984), almost totally without regard for that constituent's concentration. This assump-tion even applies to innocuous material, such as soils, that contain relatively low levels of contamination.

● The potential direction of migrating contaminants is not usually addressed (EPA, 1984).

● The site and contaminant characteristics that are employed in the methodology represent oversimplifications of real conditions.

● The time of contaminant arrival to sensitive receptors and the duration of exposure of surrounding populations are not usually addressed.

● All contaminant pathways are usually analyzed using similar, if not identical, questions.

● Little or no information on exposure is included in the CL/Q methodologies. The exposure assessment should include an analysis of the type of exposure (i.e., inhalation, ingestion, dermal con-tact, and external dose), time until exposure (i.e., the arrival time of the contaminant from the waste site to important receptors), and duration of exposure (i.e., the amount of time a population is continually exposed to a contaminant).

● The scoring system is highly subjective; that is, the score assigned is often a matter of personal interpretation of the questionnaire and possible responses to questions.

Typical examples of CL/Q methodologies include the LeGrand model (LeGrand, 1983), Surface Impoundment Assessment (SIA) model (Silka and

---

*JRB. 1982. "The Establishment of Guidelines for Modeling Groundwater Contamination from Hazardous Waste Facilities." Discussion Draft Report. JRB Associates, Inc., McLean, Virginia.

Searingen, 1978, as reported by JRB*), JRB Rating Methodology model (Kufs et al., 1980), Hazard Ranking System (HRS) model (EPA, 1982a), and the modified Hazard Ranking System (mHRS) model (Hawley and Napier, 1984; 1985†).

With a fully coupled approach, each component of the assessment methodology (usually representing a transport pathway or exposure assessment component) is represented by a submodel, with the submodels internally combined into a single code. In effect, each submodel represents a part of the overall multimedia model. Each submodel is usually allowed to interact (i.e., data and information transfer) with other submodels on both a temporal and spacial level. Interfacing and information transfer between submodels are managed by a central executive program, and information transfer readily occurs between pathways. The pathway submodels are chosen a priori (by the original developer of the program), thereby limiting the type of sites and the number of release scenarios that can be addressed at a particular site. The fully coupled approach is intended to allow consistent and unified descriptions of environmental systems. Such methodologies are expected to eliminate user bias and achieve consistency from site to site. Because of the complex phenomena associated with the various pathways, though, a unified model of this type can become extremely large and cumbersome to implement; consequently, simplified models are usually used to reduce the quantity and complexity of information exchanged between components, achieve reasonable computer core size, and obtain more efficient run times. Typical examples of fully coupled assessment methodologies include the RCRA Risk-Cost Analysis model (ICF, 1984), Air Land Water Analysis System (ALWAS)†† model (Tucker et al., 1984), Hydrologic Simulation Program in FORTRAN (HSPF)† model (Johanson et al., 1980; Donigian et al., 1983a), Water Transport Model (WTM) (Fletcher and Dodson, 1971; Fletcher et al., 1973), Simplified Codes for Performance Evaluation (SCOPE) methodology (Petrie et al., 1983), and Unified Transport Model (UTM)† (Patterson, 1986; Baes et al., 1976; Patterson et al., 1974).

With a compositely coupled (i.e., integrated systems) approach, each transport pathway is represented by an independent model. The models are externally coupled by the user to address the appropriate level of detail dictated by the environmental system and the type of assessment required. Therefore, the conceptualization of the modeling scenario is the responsibility of the user; whereas the conceptualization of the modeling scenario using the fully coupled approach is determined before its use. Modeling with a compositely coupled approach occurs in a sequential order; codes for individual pathways do not interact directly between themselves. Interfacing and information transfer occurs by assigning the output file from one pathway model to the input file of the next pathway model. Feedback (i.e., reversing the direction of data transfer) between the pathway models is addressed by the user when assessing the site, and the user decides the specific pathway to address. For example, two codes could be compositely coupled to simulate contaminant movement in a

---

*JRB. 1982. "The Establishment of Guidelines for Modeling Groundwater Contamination from Hazardous Waste Facilities." Discussion Draft Report. JRB Associates, Inc., McLean, Virginia.

†Hawley, K. A., and B. A. Napier. 1985. "A Ranking System for Sites with Mixed-Radioactive and Hazardous Wastes." Prepared for the U.S. Department of Energy, Office of Operational Safety by Pacific Northwest Laboratory, Richland, Washington (Draft).

††Even though the components used in this methodology are fixed, certain components may be more easily updated or changed than with other fully coupled methodologies.

saturated groundwater environment. One code would model the movement of the water (i.e., transporting medium); the other code would model the movement of the contaminant. Each code functions independently and, therefore, can independently be updated or replaced.

The compositely coupled approach allows each component or code to be replaced as the scenario being modeled changes, or as technological advances are made. This approach allows the user to customize frameworks to address specific modeling needs and to allocate resources to optimize the resource requirements of the analysis in relation to the goals of the assessment. Overall, the compositely coupled approach is more flexible than the fully coupled approach because only the necessary pathway models are used for a given problem. Examples of sequential pathway modeling, using the compositely coupled approach, include the Chemical Migration and Risk Assessment (CMRA) methodology (Onishi et al., 1979, 1980, 1981; Parkhurst et al., 1981; Whelan and Parkhurst 1983) and the Multimedia Contaminant Environmental Exposure Assessment (MCEA) methodology (Onishi et al., 1982a, 1982b; Whelan et al., 1982, 1983; Whelan and Onishi, 1983). Bolten et al. (1983) expanded upon the MCEA methodology to include cost and risk analysis components.

INTEGRATED ASSESSMENT METHODOLOGY FOR INACTIVE WASTE SITES

Currently, the EPA uses HRS (a CL/Q methodology) to evaluate hazardous waste sites that fall within the jurisdiction of CERCLA. The HRS is probably the most widely used standardized assessment methodology, although it includes all of the inherent deficiencies listed earlier for CL/Q methodologies. The EPA uses the HRS to identify sites for nomination to the NPL; it is designed as an initial screening tool to discriminate between hazardous wastes that do not and those that are likely to pose significant problems to human health, safety, and/or the environment.

The HRS is limited as a tool for assessing waste sites containing radionuclides. Radioactive sites are likely to receive high scores because of the way the system scores the toxicity and persistence of radionuclides in its waste characteristics section. The HRS waste characteristics scores are based on three criteria: 1) persistence, 2) toxicity or incompatibility/reactivity, and 3) quantity. By HRS definition, all radionuclides potentially cause severe toxic effects such as cancer (EPA, 1984); therefore, radionuclides automatically receive the highest possible toxicity score. In addition, because many radionuclides have relatively long half-lives, most receive a maximum persistence score. By treating most radionuclides alike (i.e., having a maximum toxicity/persistence score), the HRS tends to overestimate the potential hazards of radioactive sites relative to chemical sites and fails to discriminate between the potential risks of sites containing different radionuclides.

Hawley and Napier (1984; 1985*) have proposed modifications to the HRS so radiological hazards would be addressed in a manner consistent with that of chemical wastes in the HRS. This modified Hazard Ranking System (mHRS) was developed by PNL to operate within the existing framework of the HRS, without changing the HRS scoring system. The mHRS

--------------

*Hawley, K. A., and B. A. Napier. 1985. "A Ranking System for Sites with Mixed-Radioactive and Hazardous Wastes." Prepared for the U.S. Department of Energy Office of Operational Safety by Pacific Northwest Laboratory, Richland, Washington (Draft).

allows information to be used in the route characteristics and targets sections common to that of the HRS. The mHRS splits the waste characteristics sections into two subsections--one for nonradioactive or chemical wastes and one for radioactive wastes. At a mixed-waste site (i.e., one containing both radioactive and nonradioactive wastes), the mHRS develops two waste characteristics scores for each exposure route-- one for chemical and one for radioactive wastes. In calculating the migration scores for each route, the higher of the two waste characteristics scores is used in further analyses. For purely chemical sites, the mHRS yields results identical to those of the HRS (Whelan et al., 1985b).

Although these modifications to the HRS alleviate its major limitations in assessing radioactive sites, the mHRS still retains many of the limitations inherent in extremely simple ranking systems. As a relatively subjective ranking tool, the HRS/mHRS cannot be used to quantitatively rank or prioritize sites based on their relative potential hazards. However, as a preliminary screening tool the HRS/mHRS is useful for discriminating between those sites that are essentially innocuous and those that do or may pose significant risk to human health, safety, and/or the environment.

The DOE may be required to use the HRS or the mHRS to determine which sites might be placed on the NPL. However, DOE will be financially and logistically responsible for any additional site investigations and remediations. Because of the potential magnitude of the efforts associated with site investigations and remediations, DOE will have both financial and logistical constraints as to how quickly the work can be completed. Therefore, DOE management must prioritize sites so available monetary and human resources are properly allocated over time (i.e., the sites having the highest potential risk are investigated and remediated first, if necessary). Because of the limitations of HRS/mHRS, PNL has developed the RAPS methodology to help DOE management establish priorities for sites requiring further investigations.

Currently, DOE field offices are locating and identifying those inactive hazardous waste disposal sites that may pose an unacceptable risk to health, safety, and the environment. Two ranking methodologies are used to identify and prioritize, according to risk, waste sites requiring further investigation. For identifying sites that may pose significant risk to health, safety, and/or the environment, DOE is using both the HRS and mHRS methodologies. The sites are then classified into one of two groups: those that may and those that do not pose a potential risk to the surrounding environment. For those sites that do not pose a potential risk to the surrounding environment, no further evaluation is required. Those sites that do pose a potential risk are further evaluated and prioritized according to risk (in a relative sense) by the RAPS methodology. Figure 2 illustrates the utility of integrating the HRS/mHRS methodologies with the RAPS methodology.

REMEDIAL ACTION PRIORITY SYSTEM

The RAPS methodology uses empirically, analytically, and semianalytically based mathematical algorithms and a compositely coupled pathways analysis to predict the potential for contaminant migration from a waste site to important environmental receptors. Four major transport pathways for contaminant migration are considered in RAPS: subsurface (groundwater), overland, surface water, and atmospheric. Using the predictions of contaminant transport, simplified exposure assessments are performed for important receptors. The risks associated with the sites are then calculated relative to other sites for all pathways of concern.

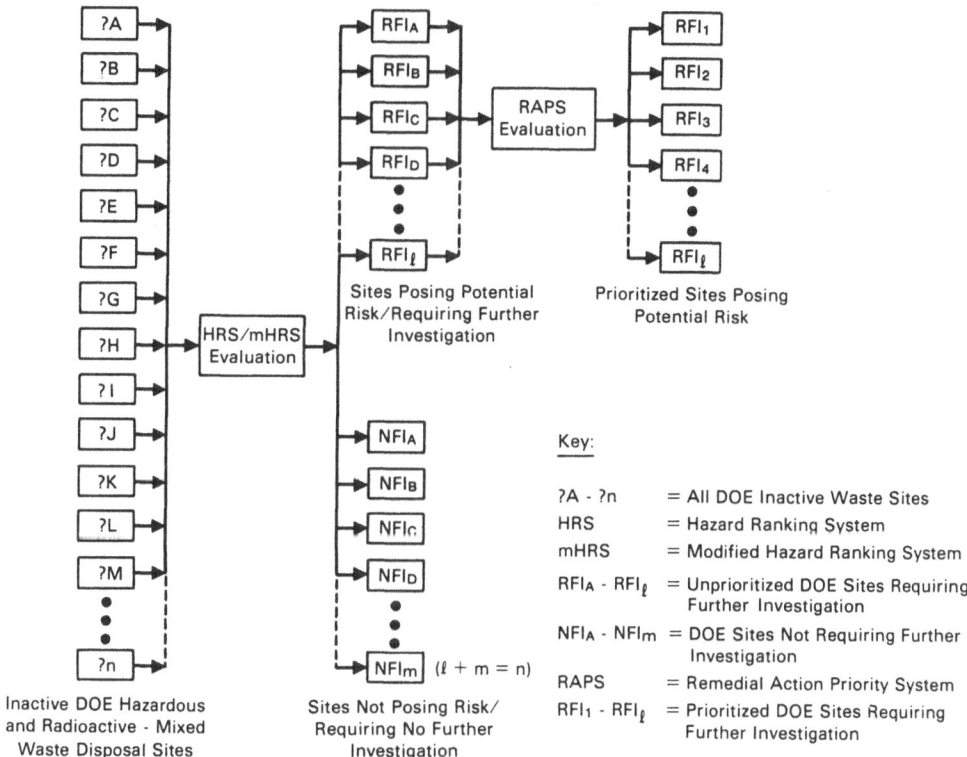

Fig. 2. Utility of the HRS/mHRS and RAPS in Locating, Identifying, and Prioritizing Sites Posing a Potential Risk to the Surrounding Environment (Steelman and DeCarlo, 1985).

Based on input data that are readily available at DOE facilities and require minimum user knowledge of risk assessment, the RAPS methodology addresses many of the typical limitations associated with other simplified ranking systems (e.g., HRS/mHRS). It considers 1) more site information and constituent characteristics associated with the pathways; 2) chemical and radioactive wastes; 3) the potential direction of contaminant movement; 4) contaminant retention (e.g., environmental mobility, dispersion, and decay/degradation), where applicable; 5) contaminant toxicities; 6) population distributions; 7) various routes or types of exposure (e.g., inhalation, ingestion, dermal contact and external dose); 8) time until a population is exposed or exposure begins (i.e., time of contaminant arrival); and 9) duration of exposure (i.e., the length of time a population is continually exposed to a contaminant). Time of contaminant arrival and duration of exposure are critical considerations in a site prioritization; the sooner a population is exposed, the greater the urgency for site characterization and possible remediation. Likewise, the longer a population is exposed to a contaminant the greater the potential severity of that exposure. Both of these factors are absent from the HRS/mHRS evaluations.

Structure of RAPS

Structurally, the RAPS methodology is based on the compositely coupled multimedia modeling approach. Each transport pathway addressed by RAPS has a set of codes that describe the migration and fate of contaminants through each pathway. These transport pathway codes are systematically integrated with an exposure assessment component that considers the type, time, and duration of exposure and the location and size of the population exposed. Figure 3 presents a simplified diagram outlining the various pathways and their interactions considered by the RAPS methodology.

To implement the methodology at a site, the user designates which transport pathways and the path the contaminants may take from the waste site through the various media. The user is then prompted by the computer to supply site and constituent (i.e., contaminant) information. The user-supplied information is then augmented with information from a computerized data base. Based on these data, the migration and fate of the contaminants are simulated from the source through the designated transport pathways to important environmental receptors. The exposure route to the population is integrated into the analysis, and the subsequent risk [i.e., the Hazard Potential Index (HPI)] to the population is computed for the site. The site HPI is compared to HPIs at other sites that have been previously analyzed. The sites are then objectively ranked in order of increasing risk, according to the HPI for each site.

The HPI is a parameter that reflects the results of the multimedia transport calculations, exposure assessment, and the effects of an exposure to a population of concern. It directly considers contaminant levels that reflect persistence and mobility at important receptors, population distributions, contaminant toxicity, routes and levels of exposure, duration of exposure, and the time until a population is exposed. The HPI is used as a relative marker for quantitatively comparing the potential for the migration, fate, and effects of hazardous substances. By itself, an HPI does not indicate the absolute risk at a site but does indicate whether one site potentially presents a higher risk to surrounding receptors of concern than another site. The values of an HPI range between zero and unity per person.

The shaded boxes in Figure 3 illustrate an example application of the RAPS methodology. According to this example, leachate leaves the waste site and enters the groundwater pathway, travels through the partially saturated zone, enters and travels through the saturated zone, leaves the groundwater pathway and enters the surface water pathway, and migrates through a nearby river. At designated usage locations, the population is externally exposed to contaminants of concern and, in addition, ingests a portion of the contaminated river water. An HPI is computed based on the exposure to the population and is compared to HPIs for other sites to prioritize the site, relative to others, based on relative risk.

RAPS Solute Transport Pathways and Exposure Assessment Component

As illustrated in Figure 3, four transport pathways and an exposure component are addressed by the RAPS framework. For each pathway, contaminant retardation is described, where applicable, by an equilibrium (i.e., partition or distribution) coefficient. First-order degradation/ decay is assumed for all contaminants that do not result in decay products (e.g., radionuclides).

For contaminants that decay, the parent contaminants are initially treated as conservative substances (i.e., no decay products or degradation). Upon reaching the environmental receptor, radiological decay is corrected in a separate calculation, and the code subsequently computes the temporal distribution of each decay product. The Bateman equation is then used to calculate the concentrations of all important decay products in the chain (Bateman, 1910 as reported by Codell et al., 1982).

200

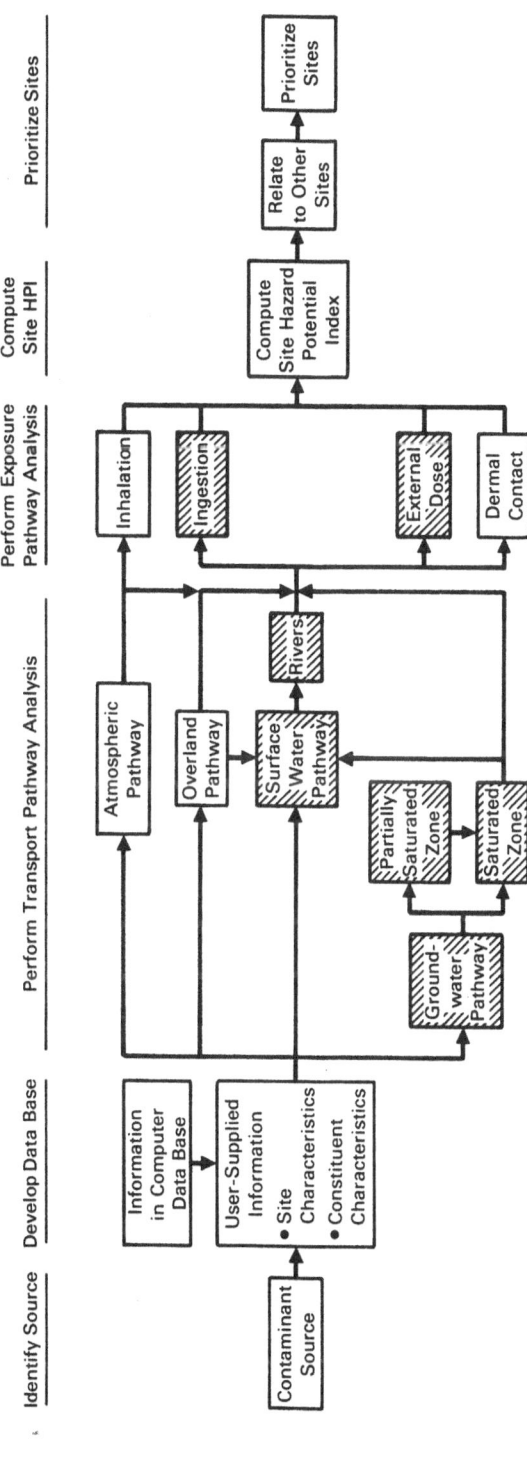

Fig. 3.  Simplified Diagram Outlining the Interactions Between the Transport Pathways and Exposure Assessment Components of the RAPS Methodology.  (Shaded boxes indicate a potential contaminant transport and exposure route using the RAPS methodology.)

$$C_i = \left(\frac{\lambda_i C_1}{\lambda_1}\right)\left(\prod_{j=1}^{i-1} \lambda_j\right) \left[\sum_{j=1}^{i} \left(\frac{e^{-\lambda_j t}}{\prod_{r=j}^{i} \lambda_r - \lambda_j}\right)\right] \qquad (1)$$

where   $C_i$ = decay product concentration (Ci/ml)
$C_1$ = parent concentration (Ci/ml)
$\lambda_i$ = radiological decay coefficient of the $i^{th}$ decay product.

The approach for analyzing each of the pathways considered in RAPS is briefly discussed below.

Groundwater pathway. The quantity of leachate likely to be generated during the operational lifetime of an inactive hazardous waste facility is a major factor controlling the degree to which a site will require analysis. This quantity is controlled by local meteorologic, geologic, and hydrologic conditions and the design and operation of the facility. Given the limited availability of literature data on leachate quantities generated by inactive landfills, available estimation techniques are used to quantify the leachate.

A modified method of that proposed by Dass et al. (1977), Fenn et al. (1975), and Thornthwaite and Mather (1955, 1957) is currently under consideration for computing leachate quantities from inactive landfills. The methodology is based on a water budget analysis; it estimates the quantity of leachate produced at a given landfill site and involves a water-balance calculation using monthly estimates of precipitation, potential evapotranspiration, temperature, and runoff. The principal source of moisture is precipitation (rainfall and snowfall) over the landfill site. Of the precipitation that falls on a landfill site, a portion runs off, some is lost to evapotranspiration, and the remainder percolates through the landfill. Water that percolates through the fill eventually exits as leachate. Simpler methods have been proposed (e.g., Knight et al. 1980); however, these methods are not nearly as precise. More complex methods also have been proposed and developed (e.g., Schroeder et al., 1983; ICF, 1984), but their complexity precludes their use in a preliminary ranking system. At this time, the RAPS methodology assumes that each inactive hazardous waste site has a 70-year (one human lifetime) life expectancy and that the waste leaches at a constant annual average rate from the site.

Contaminants exiting the bottom of the landfill migrate through a partially saturated or saturated groundwater zone. In the partially saturated zone, flow is usually assumed to be in a vertical direction. Because this flow is generally unidirectional, one-dimensional modeling is performed. The RAPS methodology uses a one-dimensional, unsteady, semianalytical code to simulate contaminant leaching and movement through the partially saturated zone. The solution algorithm to the advective-dispersive equation is based on homogeneous and isotropic soil parameters (see Donigian et al., 1983b; Van Genuchten and Alves, 1982). The partially saturated soil beneath the waste site is assumed at a unit potential hydraulic gradient. The moisture content is assumed to fluctuate between field capacity and saturation. The hydraulic conductivity is based on an empirical equation proposed by Gardner (1960) and is expressed as (Hillel, 1980)

$$K(\theta) = K_s \, (\theta/n)^{1/m} \qquad (2)$$

where $K(\theta)$ = hydraulic conductivity (cm/s)
      $\theta$ = moisture content (dimensionless)
      $K_s$ = saturated hydraulic conductivity (i.e., permeability) (cm/s)
      n = total porosity (dimensionless)
      m = empirically based parameter that is a function of soil properties.

Hillel (1980) notes that although attempts have been made to develop theoretically based equations relating hydraulic conductivity to moisture content, the state of the art is such that consistently accurate a priori predictions of $K(\theta)$ from basic soil properties are difficult. If the infiltration rate (leach rate) of water from the waste site is less than the soil transmission rate (i.e., hydraulic conductivity at field capacity), as described by the general equation for liquid flow in the partially saturated zone (see Hillel, 1980; Hanks and Ashcroft, 1980), the water moves through the soil at the infiltration rate. For an infiltration rate equal to or greater than the transmission rate, the leachate is assumed to move at the transmission rate.

The predominant movement of the leachate in the saturated zone is assumed to be in the direction of the groundwater flow. A three-dimensional advective-dispersive equation describes the migrating plume as it disperses and attenuates through the saturated aquifer. Advection represents the transport of solute caused by the mass motion of water, while dispersion represents solute transport by unaccounted variations in the fluid velocity and molecular motion. Dispersion is considered in the longitudinal, lateral, and vertical directions. Soil properties are assumed homogeneous, and the flow is assumed steady and only in the longitudinal direction.

The advective-dispersive equation for solute movement through a porous medium with a constant, steady-state flow velocity forms the basis of all groundwater solution algorithms. For the general case of unidirectional advective transport with three-dimensional dispersion in a homogeneous saturated aquifer, assuming that the diffusion in the nonflowing voids is negligible compared with the dispersion in the following voids, a mass balance on a differential volume gives (Codell et al., 1982)

$$n_e \frac{\partial C}{\partial t} + (n - n_e) \frac{\partial G}{\partial t} + (1 - n) \frac{\partial P}{\partial t} + n_e u \frac{\partial C}{\partial x} = \tag{3}$$

$$n_e \left\{ E_x \frac{\partial^2 C}{\partial x^2} + E_y \frac{\partial^2 C}{\partial y^2} + E_z \frac{\partial^2 C}{\partial z^2} \right\} - n_e \lambda C - (1 - n)\lambda P - (n - n_e)\lambda G$$

where n = total porosity (dimensionless)
    $n_e$ = effective porosity (dimensionless)
    $\tilde{C}$ = dissolved contaminant concentration in the liquid phase in the flowing voids (g/ml or Ci/ml)*
    G = dissolved contaminant concentration in the liquid phase in the nonflowing voids (g/ml or Ci/ml)

---

*When two sets of units are provided, the first refers to chemicals and the second refers to radionuclides.

P = particulate contaminant concentration in the solid phase
(g/g or Ci/g)
u = x-component groundwater (pore water) velocity (cm/s)
$E_x, E_y, E_z$ = dispersion coefficients in the x-, y-, and z-directions, respectively
$\lambda$ = degradation/decay constant [= (ln 2)/(half-life)].

Equation (3) can be streamlined with some simplifying assumptions. One assumption is that the dissolved concentration in the nonflowing voids (G) equals the dissolved concentration in the flowing voids (C). A second assumption is that the contaminant sorption process can be described by a constant (i.e., Kd) representing the ratio between the contaminant adsorbed to the soil matrix (i.e., P) and the contaminant dissolved in solution (C). Using these assumptions, Equation (3) can be rewritten as

$$\frac{\partial C}{\partial t} + u^* \frac{\partial C}{\partial x} = D^*_x \frac{\partial^2 C}{\partial x^2} + D^*_y \frac{\partial^2 C}{\partial x^2} + D^*_z \frac{\partial^2 C}{\partial z^2} - \lambda C \qquad (4)$$

in which

$$D^* = \frac{(E)(n)}{(n_e)(R_f)} \qquad (5)$$

$$R_f = \frac{n}{n_e} + \frac{\beta}{n_e} Kd \qquad (6)$$

$$u^* = u/R_f \qquad (7)$$

where $D^*_x, D^*_y, D^*_z$ = pseudodispersion coefficients in the x-, y-, and z-directions, respectively ($cm^2/s$)
$R_f$ = retardation factor (dimensionless)
$\beta$ = bulk density (g/ml)
Kd = equilibrium (partition or distribution) coefficient ($cm^3/g$).

As written, Equation (4) specifically addresses the general conditions for saturated flow and solute movement. However, Equation (4) can also be applied to the partially saturated zone if minor modifications are made. To apply the equation to the partially saturated zone, the porosities (n and $n_e$) are assumed to be equal to the soil matrix moisture content. In addition, one-dimensional, unidirectional flow and dispersion are assumed only in the vertical (z) direction.

By solving Equation (4) with the appropriate boundary and initial conditions, a set of semianalytical expressions representing concentrations (Ci) for instantaneous contaminant releases at the source is obtained that characterize the transport of contaminants through the partially saturated and saturated groundwater zones. These expressions are based on Green's functions and have been reported by several researchers (e.g., Codell et al. 1982; Selim and Mansell 1976; Yeh 1981; Yeh and Tsai 1976).

Solutions for the advective-dispersive equations for the partially saturated and saturated zones have been formulated in terms of an instantaneous contaminant release (i.e., a pulse release over zero time). The RAPS methodology generalizes these solutions for arbitrary time-varying releases by convoluting response functions (i.e., temporally varying contaminant leach or flow rates) with instantaneous contaminant release solutions.

$$C(\tau) = {_0}\!\int^{\tau} f(t)\ Ci(\tau - t)\ dt \tag{8}$$

where  $\tau$ = the time over which contaminant concentration is computed (s)
  $f(t)$ = source term expressed as a temporally varying contaminant flux (g/s or Ci/s)
   $Ci$ = instantaneous solute concentration for an instantaneous source release ($cm^{-3}$).

Contaminant flux computations are based on contaminated material that crosses an area perpendicular to the flow axis. Using the flow direction (i.e., x-direction) in the saturated zone as an example, the instantaneous contaminant flux perpendicular to the x-direction is described by (based on Codell et al., 1982)

$$\frac{dFi}{dA} = (n_e)\ (uCi - D_x \frac{\partial Ci}{\partial x}) \tag{9}$$

in which

$$D_x = \frac{E_x n}{n_e} \tag{10}$$

where  $Fi$ = instantaneous contaminant flux resulting from an instantaneous contaminant release at the source (g/s or Ci/s)
    $A$ = cross-sectional area perpendicular to the flow direction ($cm^2$)
   $D_x$ = psuedolongitudinal dispersion coefficient ($cm^2/s$).

The total flux across the plane is therefore described by laterally and vertically integrating Equation (9).

$$Fi = n_e\ {_0}\!\int^{h}{_{-\infty}}\!\int^{+\infty} (uCi - D_x \frac{\partial Ci}{\partial x})\ dy\ dz \tag{11}$$

where  h = depth of the saturated aquifer (cm).

The RAPS groundwater component computes contaminant levels at wells and at the edge of streams and calculates solute fluxes from the groundwater environment to the surface water environment. The solution algorithms are based on Green's functions and have been reported by several researchers (e.g., Codell et al., 1982; Yeh, 1981). A review of the mathematical algorithms describing the groundwater pathway is presented by Whelan (1985b, 1985c).

Surface water pathway. Of the many surface water components (e.g., nontidal rivers, estuaries, lakes, open coasts, reservoirs, impoundments, etc.), RAPS is currently capable of addressing nontidal rivers. Nontidal rivers refer to freshwater bodies with unidirectional flow in definable channels. Because the RAPS methodology is compositely coupled, other surface water pathways can be added when deemed necessary.

Contaminant releases to the surface water environment in the RAPS methodology are relatively long term. Because transient solutions for contaminant migration and fate calculations are most applicable for batch and infrequent releases over relatively short periods of time (Codell et al., 1982), steady-state solutions to the advective-dispersive equation are most applicable. The steady-state, vertically integrated mass balance equation for contaminant transport in a riverine environment (where longitudinal advection dominates longitudinal dispersion) can be written as follows.

$$u \frac{\partial C}{\partial x} = E_y \frac{\partial^2 C}{\partial y^2} - \lambda C \tag{12}$$

in which

$$\frac{\partial C}{\partial y} = 0 \text{ at } y = 0 \text{ and } y = B \tag{13}$$

where  C = dissolved instream contaminant concentration (g/ml or Ci/ml)
   u = average stream flow velocity (cm/s)
   $E_y$ = lateral or transverse dispersion coefficient ($cm^2$/s)
   B = width of stream (cm).

For a point source contaminant release from the bank of a stream, solution to Equation (12) is very similar to that outlined by Codell et al. (1982):

$$C = \left(\frac{Qc}{uBh}\right) \exp\left(-\frac{\lambda x}{u}\right)\left[1 + 2 \sum_{n=1}^{\infty} \left[\exp\left(-\frac{n^2 \pi^2 E_y x}{uB^2}\right)\left(\cos \frac{n\pi y}{B}\right)\right]\right] \tag{14}$$

where  Qc = contaminant flux at the source (g/s or Ci/s)
   h = depth of stream (cm)
   x = distance downstream (cm)
   n = index on series expansion (dimensionless)
   y = lateral distance from bank where source release is located (cm)

All other terms retain their previous definitions. The potential interactions between the surface water pathway and the other environmental transport pathways addressed in RAPS is illustrated in Figure 3.

Overland pathway. Overland flow is that portion of precipitation that ultimately appears as flowing water on the ground surface; it occurs primarily because of rainfall or snowmelt in excess of abstraction demands (i.e., interception, evapotranspiration, infiltration, etc.) and/or the emergence of soil water into drainage pathways. The driving mechanism transporting contaminants through the overland pathway is overland flow. Because overland flow controls the distribution of contaminants on land surfaces, its spatial and temporal distribution is simulated for describing solute migration and fate through this environment. Many of the characteristics describing the watershed and hazardous waste sites are used in computing overland water movement and subsequent contaminant transport. If an unlimited supply of contamination was available for transport, then the overland flow rate would control the mass flux of contaminant moving down gradient. As the flow rate increases, the potential for increasing the contaminant mass flux would also rise.

The algorithms describing the overland pathway are based on data that are easily attainable. Estimation techniques are based on the curve number technique of the U.S. Department of Agriculture's Soil Conservation Service (SCS), as presented by SCS (1972), Kent (1973), USBR (1977), and Haun and Barfield (1978). The techniques are also based on the method of characteristics as illustrated by Croley (1978), Eagleson (1970), Hjelmfelt (1976), Whelan (1980), Witinok (1979), and Witinok and Whelan (1980).

The SCS curve number technique incorporates into its computations soil classifications, soil cover, land use treatment or practice, hydrologic condition for infiltration, locale (i.e., location within the United States), initial moisture abstraction, antecedent moisture conditions, and potential maximum moisture retention. The algorithms are empirically based and represent a method of estimating direct runoff volumes from storms. The direct runoff volume for each storm event can be estimated as

$$V = \frac{[(P)(CN) - (0.2)(1000 - (10)(CN))(a)]^2}{(P)(CN)^2 + (0.8)(CN)(1000 - (10)(CN))(a)} \qquad (15)$$

where  V = 24-hr storm event runoff volume per unit area (cm)
      P = effective precipitation (rainfall adjusted for snowmelt) (cm)
   CN = SCS curve number
     a = conversion parameter between centimeters and inches
       (a = 2.54).

The direct runoff inventory computed using the SCS technique is temporally distributed using the method of characteristics with the kinematic wave approximation. The method of characteristics defines the path of wave propagation along which partial differential equations become ordinary differential equations with analytical solutions. Under the kinematic wave approximation, inflow, free surface slope, and inertial terms in the one-dimensional flow equation as described by the equations of motion (i.e., continuity and momentum) are considered negligible as compared with overland slope and friction slope. The governing equations for overland sheet flow can be expressed as (Eagleson, 1970)

$$S_o = S_f \qquad (16)$$

$$\frac{\partial h}{\partial t} + \frac{\partial q}{\partial x} = i \qquad (17)$$

$$q = \alpha h^m \qquad (18)$$

in which
$$\alpha = \frac{a}{n} S_f^{1/2} \tag{19}$$

$$m = 5/3 \tag{20}$$

where  q = overland flow discharge per unit width of overland segment ($cm^3/s/cm$)
$S_o$ = overland slope
$S_f$ = friction slope
h = overland flow depth (cm)
i = excess rainfall rate (i = V/86400) (cm/s)
$\alpha$ = coefficient that is a function of overland roughness characteristics and slope
m = parameter determined by use of Mannings equation
n = Mannings roughness parameter
a = 4.642 (when using units of centimeters)

Based on Equations (16) through (20) and using the definition of total differentials, the set of partial differential equations can be modified such that along characteristic x-t curves they reduce to ordinary differential equations (Eagleson, 1970):

$$dq/dx = i \tag{21}$$

$$dh/dt = i \tag{22}$$

$$dq/dt = i\alpha m h^{m-1} \tag{23}$$

$$dh/dt = i/(\alpha m h^{m-1}) \tag{24}$$

For Equations (21) through (24) to be valid, rainfall excess must be spacially and temporally constant over the overland segment. Based on Equations (21) through (24), a suite of analytical solutions is available describing the temporal and spacial distribution of overland flow. Overland contaminant levels are then computed based on the overland flow hydrographs.

As Figure 3 indicates, the overland transport pathway can interact with the surface water pathway or directly supply the exposure assessment component with contaminant levels for computing the site HPI. A review of the mathematical algorithms describing the overland pathway is presented by Whelan (1985a, 1985b).

Atmospheric pathway. Complex phenomena are associated with the migration and fate of contaminants released to the atmosphere (Cupitt, 1980). The atmospheric component of the RAPS methodology considers release mechanisms and characteristics, dilution and transport, washout by cloud droplets and precipitation, and deposition on the underlying surface cover. The atmospheric pathway model provides a realistic computation of these processes within the constraints of limited site information.

The prediction of contaminant movement through the atmospheric pathway involves the use of models that address atmospheric suspension/emission, transport, diffusion, and deposition. Input to the models include site-specific climatologic information such as wind direction, wind speed, and precipitation. Output from the models consist of average air and surface contaminant levels that are then used as input to the exposure assessment component. Currently, contaminant transport is assumed to occur sufficiently fast that chemical transformations can be neglected. The validity of this assumption needs to be confirmed for the various contaminants that exist at DOE inactive hazardous waste sites.

The atmospheric pathway is modeled in a manner to maximize the validity of comparisons between sites. The suspension/emission rates are based mainly on empirical relationships using site characteristics. The atmospheric transport and dispersion are computed in terms of sector-averaged values using Gaussian dispersion principles similar to that proposed by Busse and Zimmerman (1973) and examined by Culkowski (1984).

The atmospheric pathway computes 1) airborne contaminant levels and 2) overland surface contaminant levels from dry and wet deposition. The airborne concentrations are used in the exposure assessment analysis for computing contaminant inhalation rates; the surface concentrations are used for computing ingestion rates.

For computing airborne contaminant levels, the sector-averaged airborne concentration for one set of wind speed conditions, stability conditions, and elemental contaminant form is given by

$$C_{ijk}(x,\theta,z) = Q_k \, R_k(x,\theta) \, F_{ij}(x,\theta z) \qquad (25)$$

where  $C_{ijk}$ = sector-averaged atmospheric concentration for wind speed i, stability condition j, and contaminant form k ($g/m^3$)

i = index on wind speed (i=1, ...m; m = number of wind speed classes)

j = index on stability conditions (j=1, ...n; n = number of stability conditions)

k = index on elemental contaminant form (k=1, ...p; p = number forms representing (p-1) ranges of particle sizes, and a gaseous state)

x = distance between waste site and population (km)

$\theta$ = direction to the population located at distance x from the waste site

z = height of air concentrations over local ground level (m)

$Q_k$ = contaminant release rate for contaminant form k (g/s)

$R_k$ = deposition and/or decay plume depletion fraction which varies as a function of the position of the plume for contaminant form k (dimensionless)

$F_{ij}$ = dispersion factor for wind speed group i and stability j which varies as a function of downwind distance with the form of the function $F_{ij}$ depending on whether the release is best modeled as a point, line, or area emission (dimensionless).

The average air concentration near the earth's surface [Equation (25) at z = 0] is required for the inhalation component of the exposure assessment analysis. The average air concentration at ground level for a population located at distance x and direction θ from the waste site is computed as

$$\bar{C}(x,\theta,0) = \sum_{i=1}^{n} \sum_{j=1}^{m} \sum_{k=1}^{p} [f_{ij} C_{ijk}(x,\theta,0)] \qquad (26)$$

where $\bar{C}(x,\theta,0)$ = average air concentration near the earth's surface $(g/m^3)$

$f_{ij}$ = frequency of occurrence of the wind speed and stability conditions in the specified wind direction (dimensionless)

The total deposition at a specified location is computed as the sum of wet and dry deposition fluxes to the overland surface,

$$T(x,\theta,0) = \sum_{i=1}^{n} \sum_{j=1}^{m} \sum_{k=1}^{p} f_{ij} [D_{ijk}(x,\theta,0) + W_{ijk}(x,\theta,0)] \qquad (27)$$

where $T(x,\theta,0)$ = total surface contaminant concentration $(g/m^2)$
$D_{ijk}(x,\theta,0)$ = dry deposition flux $(g/m^2)$
$W_{ijk}(x,\theta,0)$ = wet deposition flux $(g/m^2)$.

Dry and wet deposition are computed using the relationship incorporating the sector-averaged concentrations described by Equation (25). Dry deposition is based directly on the computed near-surface air concentrations using

$$D_{ijk}(x,\theta,0) = \frac{C_{ijk}(x,\theta,0)}{B_{ijk}} \qquad (28)$$

where $B_{ijk}$ = dry deposition resistance (s/m).

Wet deposition is given by

$$W_{ijk}(x,\theta,z) = \int_{o}^{Z} \Lambda C_{ijk}(x,\theta,z) \, dz \qquad (29)$$

where $\Lambda$ is a scavenging coefficient (1/s) and Z is the depth of the wetted plume layer (m). The details of the empirical deposition algorithms are described in Van Voris et al. (1984). A review of the mathematical algorithms describing the atmospheric pathway is presented in Droppo (1986).*

The relative importance of the atmospheric pathway between different sites is controlled by a combination of geographic and climatic influences. Distances, directions, winds, and atmospheric stability are controlling parameters. The dispersion relationships used in the atmospheric component depend on local site characteristics (Pasquill and Smith, 1983). Because the dispersion is a strong function of downwind distance from the source, the physical distances between the contaminant sites and population centers are of prime importance. The relative proximity of sites and population centers are important in terms of the local frequencies of wind directions, particularly in areas with topographic channeling of winds. The relative rates of atmospheric dilution between the sites are mainly a function of local wind speeds and atmospheric stability parameters.

In the operational mode, the atmospheric pathway component computes contaminant levels as a function of the direction and distance that coincides with population centers surrounding the site. Inhalation represents the major route of exposure to contaminants via the atmospheric pathway. RAPS also considers the ingestion route of exposure from wet and dry deposition on vegetation and subsequent ingestion of contaminated food materials derived from the soils. In addition, external dose can be addressed, though its effects are usually insignificant as compared to the inhalation exposure route. The interaction and coupling between the atmospheric pathway and exposure assessment components of the RAPS methodology is illustrated in Figure 3.

Exposure assessment component. Results from each of the four transport pathways are used in the exposure assessment component to calculate the HPI for each important waste site contaminant. The exposure assessment component considers potential exposure of the surrounding population through the following exposure routes: 1) external dermal contact to chemicals; 2) external dose from radiation; 3) inhalation of airborne contaminants; and 4) ingestion of contaminated drinking water, soil, crops, animal products, and aquatic foods. In evaluating the HPI values, the important exposure routes and populations at risk are first defined, then based on the air, water, and soil contaminant levels provided by the transport pathway analyses, an estimate is made of the average daily exposure to each contaminant. Estimation of the daily exposure is based on simple multiplicative models describing the transfer of pollutants from air, water, or soil to humans. The daily exposure rate is evaluated for each exposure route using a general expression as follows:

$$D = \sum_{p=1}^{P} U_p C_p F_p \tag{30}$$

---

*Droppo, J. G. 1986. "Development of the Remedial Action Priority System (RAPS): Mathematical Formulations for the Atmospheric Environment." Letter Report. Prepared for the Office of Operational Safety, U.S. Department of Energy by Pacific Northwest Laboratory, Richland, Washington (Draft).

where $D$ = daily exposure rate to an individual (mg/kg/d or rem/d)

$P$ = number of exposure routes contributing to the individual's dose

$p$ = index on exposure route

$U_p$ = average daily usage rate for exposure route p (kg/d food ingested, $m^3$/d air inhaled, or hr/d exposed)

$C_p$ = concentration of contaminant in the media of exposure for route p (mg/kg or pCi/kg for food, $mg/m^3$ or $pCi/m^3$ for air, and $pCi/m^2$ for external exposure to contaminated ground surfaces)

$F_p$ = conversion factor from concentration to dose for exposure pathway p ($kg^{-1}$ or rem/pCi for radionuclide inhalation or ingestion and rem/hr per $pCi/m^2$ for external exposure to ground surface).

The above expression is evaluated for each contaminant of interest.

The daily average exposure is next converted to an average individual risk factor using mathematical models for radionuclides, carcinogenic chemicals, and noncarcinogenic chemicals. The risk factor is intended to indicate the level of potential risk to an average member of the exposed population. For radionuclides, the risk factor is based on cancer risk estimates of the National Academy of Sciences Committee on the Biological Effects of Ionizing Radiation (NAS, 1980). Based on NAS (1980), Buhl and Hansen (1984) have recommended methods for estimating health effects related to low-levels of radiation exposure. Using their recommendations, the risk factor for radioactive contaminants is calculated as

$$R_r = (2.7 \times 10^{-4})\ (D)(70)(365) \tag{31}$$

where $R_r$ = risk factor for a radionuclide (dimensionless)

$D$ = radiation dose attributed to one day of exposure (rem/d)

$2.7 \times 10^{-4}$ = risk conversion factor (effects per rem for lifetime exposure)

70 = number of years per lifetime

365 = number of days per year.

The risk from chemical carcinogens are currently based on cancer potency factors defined by the EPA (1982b). The cancer potency factors relate the average daily intake per unit body mass to the risk of developing cancer. The risk factor for chemical carcinogens is estimated as

$$R_c = 1 - \exp(-I\ q) \tag{32}$$

where $R_c$ = risk factor for a chemical carcinogen (dimensionless)

$I$ = average daily intake rate of a contaminant (mg/kg/d)

$q$ = cancer potency factor for a contaminant ($(mg/kg/d)^{-1}$).

Risk estimates for noncarcinogenic chemicals are difficult to make; potential methods to estimate these risk factors are currently being investigated.

212

One of the key features of the exposure assessment component is the estimation of the average exposure. The exposure modes included in RAPS are as follows:

- Drinking water ingestion: for groundwater, overland, and surface water transport pathways. Factors may be applied to the water concentration to account for purification of the water in a treatment plant.

- Aquatic food ingestion (fish and invertebrates): for overland, surface water, and groundwater transport pathways. Average daily intake is estimated using bioconcentration factors and average daily ingestion rates for aquatic foods.

- Crops: for all transport pathways. Crops may be contaminated from irrigation with contaminated water or by direct deposition onto plants and soil. Two crop types are considered: leafy vegetables, with the edible portion subject to direct deposition and other crops such as root and pod vegetables and fruit. Crop concentrations are estimated using soil to plant transfer factors and air to edible plant transfer factors. Average daily intake is estimated using average daily ingestion rates for vegetables and leafy vegetables.

- Animal product: for all transport pathways. Contaminated animal products result from animal ingestion of contaminated water and contaminated feed. Feed contamination may occur from direct deposition onto feed crops or pasture from air or through use of contaminated irrigation water. Use of contaminated animal drinking water is only considered for the three water transport pathways. The concentration of contaminant in animal meat and milk is estimated using animal ingestion to animal product transfer factors. Average daily intake of exposed individuals is estimated using average daily ingestion rates for meat and milk.

- Water immersion (domestic bathing and swimming): for groundwater and surface water transport pathways. Dermal contact (for chemicals) and radiation exposure are included for domestic bathing for both water transport pathways. Exposure from swimming in contaminated water is also considered for the surface water pathway. For chemicals, an equivalent daily intake amount is estimated based on dermal contact time and absorption characteristics of the chemical pollutant. For radiation exposures, the dose from immersion in water is estimated using dose conversion factors. A contribution to radiation dose may also be included for recreational boating and shoreline fishing.

- Soil ingestion: for the atmospheric transport pathway. Contaminated soil is assumed to be ingested each day with the ingestion rate based on a lifetime average.

- Inhalation: atmospheric transport pathway. The daily average intake is estimated using an average inhalation rate for the exposed population.

A review of the mathematical algorithms describing the exposure assessment component is presented in Strenge and Whelan (1986).*

EXAMPLE APPLICATIONS OF RAPS

To illustrate its application, the results of applying the RAPS methodology to two hypothetical case studies are presented in this section. These simplified case studies are presented to demonstrate the utility of RAPS in simulating the migration and fate of hazardous and radioactive-mixed wastes through and between various environmental media, and their interaction with the media. The case studies are simplified for illustration and comparison. The subsurface information is partially based on data presented by Perlmutter and Lieber (1970) and Anderson (1979), as noted by Codell et al. (1982), and is referenced by the latter document. Whelan et al. (1985a), EPA (1985, 1980), ICRP (1977, 1979-1982), Napier et al. (1980), NRC (1977), Mualem (1976), and Israelsen and Hansen (1962) also supply data pertaining to the transport pathways and constituent characteristics. Each reference is cited accordingly.

Example Case Study Assumptions

The two example case studies addressed by the RAPS methodology assume that identical hazardous and radioactive-mixed waste sites are located above an unconfined, unconsolidated aquifer. The waste is represented by a mixture of chemical and radioactive sludge wastes and soil aggregate. For each case study, first-order degradation/decay is assumed for each potentially important contaminant, with constituent-soil matrix interaction described by an equilibrium coefficient. Climatic and meteorologic conditions for each case study are assumed to be equivalent. Although RAPS is capable of computing a leach rate from the waste site, it is assumed for the purpose of simplification that the leach rate is equivalent to 20% of the average annual precipitation.

Case 1. Under Case 1, the contaminated waste is stored in an unlined landfill with a bottom that coincides with the water table surface. The waste is assumed to leach at a constant average annual rate from the site over a 70-year period (i.e., one human lifetime) and enter an alluvial saturated aquifer. The leachate then travels 6.0 km (3.7 mi) through the saturated aquifer to the nearby municipal drinking-water well, Well X. Well X supplies drinking water to the surrounding population of Town X and represents the only municipal drinking-water supply. For simplicity in this example, it is assumed that ingestion of drinking water represents the only route of exposure to these contaminants. Dermal contact during bathing is not considered a significant pathway for the contaminants, as compared to direct ingestion of water. The well water is assumed to be used directly, without purification. It is also assumed that no other populations are exposed to these contaminated waters. Under Case 1, only one transport pathway, the saturated zone of the groundwater environment, is considered in the analysis. The Case 1 scenario is illustrated in Figure 4.

---

*Strenge, D. L., and G. Whelan. 1986. "Development of the Remedial Action Priority System (RAPS): Mathematical Formulations for the Exposure Assessment Component." Letter Report. Prepared for the Office of Operational Safety, U.S. Department of Energy by Pacific Northwest Laboratory, Richland, Washington (Draft).

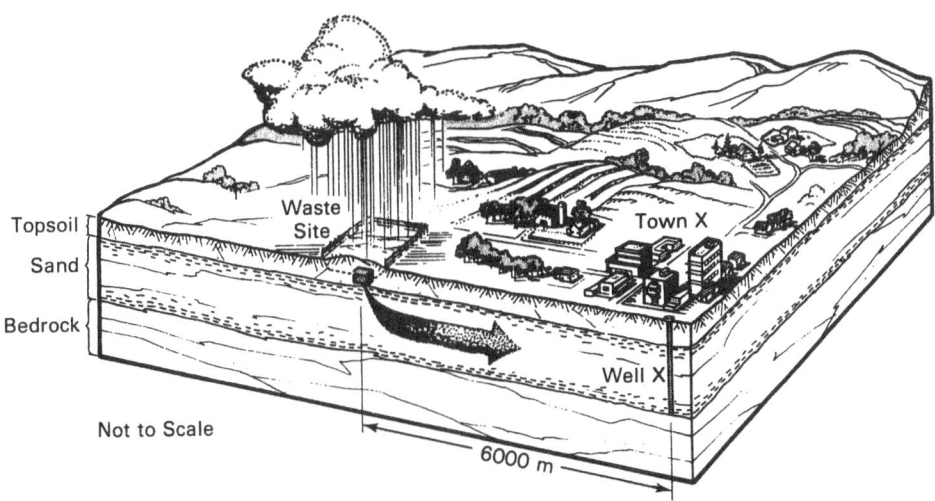

Fig. 4. Case 1 Scenario: Contaminated Wastes Leaching from the Disposal
Site and Migrating Toward Well X and Town X.

Case 2. Under Case 2, contaminated waste is stored in a landfill
that is situated on 6 m of partially saturated soil. This partially
saturated zone is composed of two distinct layers. The topmost layer
(i.e., the one touching the bottom of the landfill) is 1-m thick and is
composed of clay-like material. The soil layer beneath the clay layer
consists of alluvial material and is 5-m thick. The bottom of this layer
coincides with the water table surface of a saturated aquifer composed of
the same alluvial material. The saturated aquifer supplies groundwater
to a nearby river (River Y), 500 m from the waste site. The waste is
assumed to leach from the site over a 70-year period and migrate through
four consecutive transport pathways: two partially saturated zones, one
saturated zone, and one river. The RAPS methodology computes contaminant
fluxes between transport pathways and contaminant levels in the final
transporting medium (i.e., River Y) for use by the exposure assessment
component.

The migration and fate of the contaminants leaching from the waste
site under Case 2 is explained as follows. The leachate initially infil-
trates into and through the clay layer, then into and through the
alluvial layer of the partially saturated zone. Because the infiltration
rate (i.e., leach rate) is less than the maximum transmission rate of
either soil layer, the flux of water through each layer is equivalent to
the infiltration rate. Although the fluid movement through the clay and
sand layers is not hindered by the respective characteristics of each
soil layer, the characteristics of each soil layer affects the
adsorption-desorption process and affects the rates of contaminant transport through each layer. Due to its higher affinity for adsorbing contaminants (assuming an infinite supply of adsorption sites), the clay
layer has a greater impact on impeding solute movement than does the
alluvial layer. Upon leaving the partially saturated zone, the waste
enters the saturated alluvium below and travels, adsorbing to the surrounding soil matrix and dispersing as it migrates, towards River Y. The
contaminants eventually enter the river and are transported to a surface
water intake structure where drinking water supplies for Town Y and crop
irrigated stock-water are collected. River Y supplies drinking water to
Town Y and represents the only municipal drinking water supply. The
drinking water though is assumed to pass through a treatment plant that
removes different fractions of each contaminant. In addition, it is

assumed that the water from River Y is used to irrigate nearby agricultural lands that supply food crops to meat and milk producing arrivals;
these food crops and animal products are assumed to be harvested for
human consumption. It is further assumed that no populations other than
Town Y are exposed to these contaminated waters. For simplicity, this
scenario limits the routes of exposure to the surrounding population to
ingestion. The Case 2 scenario is illustrated in Figure 5.

Site, constituents, and exposure assessment characteristics.
Arsenic and strontium-90 represent the important respective chemical and
radioactive contaminants leaching from the waste site. In these examples, arsenic represents a persistent and relatively immobile substance;
strontium-90 represents a mobile and decaying substance. Contaminant
levels of arsenic and strontium-90 contained in the waste site were
chosen such that if an individual ingested the leachate leaving the site,
arsenic and strontium-90 would pose an equal risk to the individual. By
choosing the waste concentrations such that they initially pose equivalent risk, these examples can better illustrate how RAPS can discriminate
relative risks to important receptors based on contaminant persistence
and mobility and the effects of dilution. For computing the relative
risk for both Cases 1 and 2, arsenic is assumed to be a carcinogen with
a cancer potency factor of 14 $(mg/kg/d)^{-1}$ (EPA, 1985*) and the radio-
logic dose for strontium-90 is based on an ingestion dose factor of
$1.2 \times 10^{-8}$ rem per pCi ingested and a risk factor of $2 \times 10^{-4}$ per person-
rem (ICRP 1977, 1979-1982).

Site, constituent, and exposure assessment characteristics for
Cases 1 and 2 are presented in Tables 1 through 6. Table 1 presents
general information pertaining to both example cases. Table 2 presents
data associated with the saturated zone of the groundwater pathway for

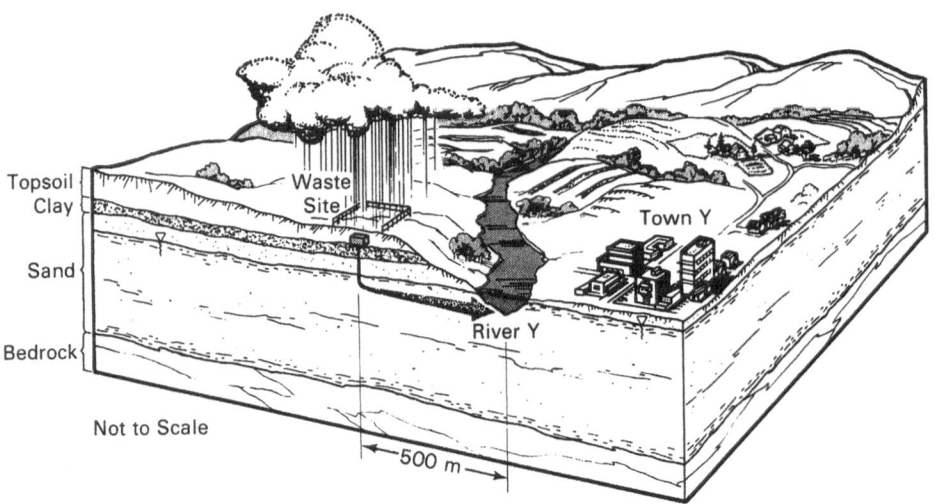

Fig. 5. Case 2 Scenario: Contaminated Wastes Leaching from the Disposal
Site and Migrating Toward River Y and Town Y.

---

*The cancer potency factor for arsenic was obtained from a soon-to-be
published EPA draft document. At the time of this writing, the document
was unavailable for citation.

Table 1. General Information Pertaining to Cases 1 and 2

| Parameter | Value |
|---|---|
| Precipitation Infiltration Rate[a,b] | $8.0 \times 10^{-7}$ cm/s ($2.3 \times 10^{-3}$ ft/day) |
| Area of Square Disposal Site (Codell et al., 1982) | 2916 m$^2$ (31,390 ft$^2$) |
| Life of Disposal Site | 1 Human Lifetime |
| Human Life Span (NRC 1977) | 70 yr |
| Half-Life of Arsenic (Whelan et al., 1985a) | $\infty$ yr |
| Half-Life of Strontium-90 (Codell et al., 1982) | 28.5 yr |
| Population of Town X[b] | 500 People |
| Population of Town Y[b] | 500 People |

[a]Assumed equivalent to 20% of the average annual rainfall.
[b]Assumed for descriptive purposes.

Table 2. Information Pertaining to the Saturated Zone for Cases 1 and 2

| Parameter | Value |
|---|---|
| Depth (Codell et al., 1982) | 43 m (140 ft) |
| Soil Porosity (Codell et al., 1982) | 35.0% |
| Bulk Density (Codell et al., 1982) | 1.92 g/cm$^3$ (120 lb/ft$^3$) |
| Distance Between Waste Site and Well X for Case 1[a] | 6.0 km (3.7 mi) |
| Distance Between Waste Site and River Y for Case 2[a] | 0.5 km (0.3 mi) |
| Groundwater Pore Water Velocity (Codell et al., 1982) | 43.2 cm/day (1.42 ft/day) |
| x-Direction Dispersivity (Codell et al., 1982) | 2130 cm (69.9 ft) |
| y-Direction Dispersivity (Codell et al., 1982) | 427 cm (14.0 ft) |
| z-Direction Dispersivity[b] | 427 cm (14.0 ft) |
| Equilibrium Coefficient for Arsenic (Whelan et al., 1985a) | 50 ml/g |
| Equilibrium Coefficient for Strontium-90 (Codell et al., 1982) | 3.4 ml/g |

[a]Assumed for descriptive purposes.
[b]Assumed equivalent to y-direction dispersivity.

both Cases 1 and 2. Tables 3 and 4 list data associated with the partially saturated zone for Case 2, while Table 5 presents the Case 2 hydraulic information assumed for River Y. Finally, Table 6 lists the information related to the exposure assessment component; much of this information is based on standard parameter values provided by NRC (1977).

Table 3. Information Pertaining to the Top Layer (Clay) of Partially Saturated Zone for Case 2

| Parameter | Value |
| --- | --- |
| Depth[a] | 1.0 m (3.3 ft) |
| Soil Porosity (Mualem, 1976) | 49.5% |
| Bulk Density (Mualem, 1976) | 1.32 $g/cm^3$ (82.5 $lb/ft^3$) |
| Field Capacity (Israelsen and Hansen, 1962) | 28.0% |
| Hydraulic Conductivity at Field Capacity (Mualem, 1976, Israelsen and Hansen, 1962) | $6.9 \times 10^{-8}$ cm/s ($2.0 \times 10^{-4}$ ft/day) |
| Hydraulic Conductivity at Saturation (Mualem, 1976) | $1.2 \times 10^{-5}$ cm/s ($3.4 \times 10^{-2}$ ft/day) |
| z-Direction Dispersivity[b] | 427 cm (14.0 ft) |
| Equilibrium Coefficient for Arsenic[c] | 500 ml/g |
| Equilibrium Coefficient for Strontium-90[c] | 34 ml/g |

[a]Assumed for descriptive purposes.
[b]Assumed equivalent to z-direction dispersivity in saturated zone.
[c]For descriptive purposes, the equilibrium coefficients for the clay layer were assumed to be one-order of magnitude higher than that for the alluvial layer.

Table 4. Information Pertaining to the Bottom Layer (Sand) of Partially Saturated Zone for Case 2

| Parameter | Value |
| --- | --- |
| Depth[a] | 5.0 m (16 ft) |
| Soil Porosity (Codell et al., 1982) | 35% |
| Bulk Density (Codell et al., 1982) | 1.92 $g/cm^3$ (120 $lb/ft^3$) |
| Field Capacity (Israelsen and Hansen, 1962) | 8.5% |
| Hydraulic Conductivity at Field Capacity (Mualem 1976, Israelsen and Hansen, 1962) | $4.9 \times 10^{-4}$ cm/s (1.4 ft/day) |
| Hydraulic Conductivity at Saturation (Mualem, 1976) | $1.5 \times 10^{-2}$ cm/s (42.5 ft/day) |
| z-Direction Dispersivity[b] | 427 cm (14.0 ft) |
| Equilibrium Coefficient for Arsenic (Whelan et al., 1985a) | 50 ml/g |
| Equilibrium Coefficient for Strontium-90 (Codell et al., 1982) | 3.4 ml/g |

[a]Assumed for descriptive purposes.
[b]Assumed equivalent to the z-direction dispersivity in the saturated zone.

Table 5. Information Pertaining to River Y for Case 2

| Parameter | Value |
|---|---|
| Flow Velocity[a] | 0.30 m/s (1.0 ft/s) |
| Flow Depth[a] | 1.5 m (5.0 ft) |
| Width[a] | 91.4 m (300 ft) |
| Distance Downstream to Usage Location[a] | 1000 m (3281 ft) |

[a]Assumed for descriptive purposes.

Table 6. Exposure Assessment Data for Cases 1 and 2

| Parameter | Value | |
|---|---|---|
| | Case 1 | Case 2 |
| Drinking Water?[a] | yes | yes |
| Water Purification?[a] | no | yes |
| Purification Factor for Arsenic (Napier et al., 1980) | --- | 0.7 |
| Purification Factor for Strontium-90 (Napier et al. 1980) | --- | 0.2 |
| Water Consumption Rate (EPA, 1980) | 2.0 L/d (0.53 gal/d) | 2.0 L/d (0.53 gal/d) |
| Crop Ingestion[a] | no | yes |
| Animal Product Ingestion[a] | no | yes |
| Consumption Rate of Vegetables (NRC, 1977) | --- | 0.52 kg/d (1.15 lb/d) |
| Consumption Rate of Leafy Vegetables (NRC, 1977) | --- | 0.08 kg/d (0.18 lb/d) |
| Consumption Rate of Meat (NRC, 1977) | --- | 0.26 kg/d (0.57 lb/d) |
| Consumption Rate of Milk (NRC, 1977) | --- | 0.30 L/d (0.08 gal/d) |
| Crop Growing Period[a] | --- | 120 days |
| Irrigation Rate[a] | --- | 100 (4.7 ft$^3$/day/ac) |

| Transfer Factor | Arsenic | Strontium-90 |
|---|---|---|
| Soil-to-Plant (NRC, 1977) | 0.01 | 0.20 |
| Water/Feed-to-Meat (NRC, 1977) | $1.5 \times 10^{-3}$ d/kg ($6.8 \times 10^{-4}$ d/lb) | $3.0 \times 10^{-4}$ d/kg ($1.4 \times 10^{-4}$ d/lb) |
| Water/Feed-to-Milk (NRC, 1977) | $3.0 \times 10^{-3}$ d/L ($1.1 \times 10^{-2}$ d/gal) | $1.5 \times 10^{-4}$ d/L ($5.7 \times 10^{-4}$ d/gal) |

[a]Assumed for descriptive purposes.

Example Case Study Results

Two simplified examples, demonstrating the application of the RAPS methodology, have been constructed to illustrate the effects of certain constituent characteristics and various environmental media on contaminant transport and eventual exposure of a surrounding population. For the examples presented, five factors help determine the relative importance of arsenic and strontium-90: mobility (adsorption-desorption effects), persistence (degradation/decay), duration of exposure of a population to a contaminant, arrival time of the contaminant, and dilution effects. The example problems have been structured to neutralize other important characteristics such as toxicity, climate, site geometry, etc. The results of each example application are briefly described as follows.

Case 1. Because strontium-90 is approximately 14 times more mobile in the environment than arsenic, it arrives at Well X first. Although strontium-90 is more mobile, it undergoes considerable decay (decay half-life of 28.5 years) prior to reaching Well X [6.0 km (3.7 mi) from the waste site] and, subsequently, the population of Town X; in contrast, arsenic does not degrade. Because arsenic is more persistent in the environment and because of its mobility (i.e., attenuation characteristics), the population of Town X is exposed to arsenic ten times longer than it is to strontium-90. Arsenic potentially poses a greater hazard to Town X, as indicated by its HPI ($2 \times 10^{-5}$), as opposed to the hazard presented by strontium-90 ($1 \times 10^{-13}$). The results of the HPI evaluation for Case 1 are presented in Table 7. Whelan et al. (1985b) present similar results for a similar problem.

Case 2. Although the clay layer in the partially saturated zone does not affect the transmission rate of fluid through the layer, it does effectively reduce the transport rate of arsenic and strontium-90. The peak flux of arsenic is reduced by approximately 38 times, while the peak flux for strontium-90 is reduced by approximately 11 times. Even though strontium-90 decays, it is approximately 14 times more mobile (as in Case 1) than arsenic. Because of the proximity of River Y to the waste site [i.e., 0.5 km (0.3 mi) away] and its relative mobility, strontium-90 does not decay to insignificant levels prior to reaching River Y. Arsenic, on the other hand, is relatively immobile, but because of its persistence, Town Y is exposed over ten times longer to arsenic than strontium-90. Case 2 results indicate that the potential hazard to Town Y associated with arsenic (HPI = $4 \times 10^{-7}$) and strontium-90 (HPI = $4 \times 10^{-7}$) are on the same order of magnitude. The results of the HPI evaluation are presented in Table 7. In this case, the risk posed by a mobile, nonpersistent contaminant (strontium-90) is approximately 45%

Table 7.  Hazard Potential Index Values for Sample Cases

| Case | Arsenic | Strontium-90 |
|------|---------|--------------|
| 1 | $2 \times 10^{-5a}$ | $1 \times 10^{-13}$ |
| 2 | $4 \times 10^{-7}$ | $5 \times 10^{-7}$ |

[a]Each parameter is rounded off to one significant digit.

higher than the risk posed by a significantly less mobile, persistent
contaminant (arsenic) that exposes a population for a longer duration of
time.

Comparison between Cases 1 and 2. The results of applying the RAPS
methodology to Cases 1 and 2 indicate that Case 1, with a maximum HPI
based on arsenic of $2 \times 10^{-5}$, poses a greater potential hazard to the
surrounding population than that of Case 2, which has a maximum HPI of
$5 \times 10^{-7}$. Although Well X in Case 1 is 12 times farther from the waste
site than River Y in Case 2, the attenuation and retardation characteristics inherent to the clay layer and the dilution effects of contaminant
mixing in River Y in Case 2 significantly reduce the adverse effects
associated with the proximity of River Y to the waste site. Based on
these results alone, a further detailed site characterization of Site X
would occur prior to that of Site Y.


SUMMARY

The RAPS is being developed to provide DOE with a better management
tool for prioritizing inactive hazardous and radioactive-mixed waste
sites according to their relative risks to surrounding populations and
for funding allocations for further site investigations and possible
remediation. The RAPS methodology addresses many of the typical limitations associated with other simplified ranking systems (e.g., HRS/mHRS)
as it considers: 1) more site information and constituent characteristics associated with potential contaminant transport pathways; 2) chemical and radioactive wastes; 3) the potential direction of contaminant
movement; 4) contaminant mobility, persistence, and toxicity; 5) time
until a population is exposed (i.e., contaminant arrival time); and
6) duration of exposure.

The risk to a population surrounding a hazardous waste site depends
on each contaminant's mobility, persistence (degradation/decay), transport direction, time of population exposure, and duration of exposure. A
contaminant that is more mobile represents a greater threat to a surrounding population than an equally hazardous contaminant that is immobile. Likewise, a contaminant that degrades/decays is less of a risk
than one that does not, assuming equal environmental conditions. The
numerical algorithms forming the basis of the RAPS methodology consider
each of these conditions when assessing a hazardous waste site.

Two simplified case studies are presented illustrating the use of
the RAPS methodology in assessing the migration, fate, exposure, and
relative risks associated with the release of a chemical (arsenic)
and radionuclide (strontium-90) to a multimedia environment. The
chemical arsenic is an immobile, persistent, suspected carcinogen, and
strontium-90 is a relatively mobile, nonpersistent, known carcinogen.
Case 1 illustrates the transport of arsenic and strontium-90 from a waste
site through one environmental medium (i.e., a saturated alluvial aquifer
of a groundwater system) to a well (Well X) that supplies drinking water
to a nearby town (Town X). Case 2 illustrates arsenic and strontium-90
movement through four environmental media [i.e., two partially saturated
soil layers (one composed of clay and one composed of sand) beneath the
waste site, one saturated alluvial aquifer, and one river] to a water
intake structure in a river (River Y) that supplies the only drinking and
irrigation stock water to a nearby town (Town Y).

Under the Case 1 analysis, arsenic represents the dominant constituent of concern to the surrounding population. Even though
strontium-90 is more mobile than arsenic, the transport time to the

exposure location (i.e., Well X) is so long that strontium-90 decays to very small levels. On the other hand, the persistence of arsenic (i.e., no degradation) causes it to be a long-term potential hazard.

For the Case 2 analysis, the results indicate that strontium-90 is slightly more dominant than arsenic. Because River Y is relatively close to the waste site (River Y is 12 times closer to the waste site in Case 2 than Well X is to its waste site in Case 1), strontium-90 does not decay to insignificant levels. Case 2 illustrates the tradeoff between strontium-90's mobility and arsenic's persistence. In its analysis of Cases 1 and 2, the RAPS methodology indicates that the waste site as described by Case 1 is potentially more dangerous to the surrounding population than that of Case 2. Under Case 2, the clay layer beneath the site significantly reduces the contaminants transport rate to other environmental media; in addition, dilution due to mixing in the river also reduces contaminant levels with respect to Case 2.

These simplified examples illustrate the application of the RAPS methodology to a multimedia environment. The RAPS methodology is structured such that it requires minimum user knowledge of contaminant transfer between various environmental media, risk assessment, and the least possible amount of input data. To maximize the utility of the system, RAPS is being designed to operate on a personal computer.

## ACKNOWLEDGMENTS

Many of the concepts associated with the compositely coupled assessment methodology were developed in concert with Y. Onishi at PNL. The authors would like to thank him for reviewing the document and would like to thank V. J. DeCarlo of DOE for his comments on the development of the RAPS methodology. Thanks are extended to K. A. Borgeson of PNL for editing the document and to PNL word processing for typing it. This work is supported by the Office of Operational Safety, U.S. Department of Energy under Contract DE-AC06-76RLO 1830.

## REFERENCES

Anderson, M. P., 1979, Using Models to Simulate the Movement of Contaminants through Groundwater Flow System. CRC Critical Review in Environment Control, No. 2, pp. 97-156.

Baes, C. F., Begovich, C. L., Culkowski, W. M., Dixon, K. R., Fields, D. E., Holdeman, J. T., Huff, D. D., Jackson, D. R., Larson, N. M., Luxmoore, R. J., Munro, J. K., Patterson, M. R., Raridon, R. J., Reeves, M., Stein, O. C., Stolzy, J. L., and Tucker, T. C., 1976, The Unified Transport Model, in: "Ecology and Analysis of Trace Contaminants Progress Report October 1974 - December 1975," R. I. Van Hook and W. D. Shults, eds., ORNL/NSF/EATC-22, pp. 13-62.

Bateman, H., Proc. Cambridge Phil. Soc., 16:423, 1910.

Bolten, J. G., Morrison, P. F., and Soloman, K. A., 1983, "Risk-Cost Assessment Methodology for Toxic Pollutants from Coal-Fired Power Plants," WD-1589 EPRI RP1826-5, prepared for the Electric Power Research Institute by Rand Corporation, Santa Monica, California.

Buhl, T. E., and Hansen, W. R., 1984, "Estimating the Risks of Cancer Mortality and Genetic Defects Resulting from Exposures to Low Levels of Ionizing Radiation," LA-9893-MS, Los Alamos National Laboratory, Los Alamos, New Mexico.

Busse, A. D., and Zimmerman, J. R., 1973, "User's Guide for the Climatological Dispersion Model," EPA-RA-73-024, Environmental Protection Agency, Research Triangle Park, North Carolina.

Codell, R. B., Key, K. T., and Whelan, G., 1982, "A Collection of Mathematical Models for Dispersion in Surface Water and Groundwater," NUREG-0868, prepared by Pacific Northwest Laboratory for the Office of Nuclear Reactor Regulation, U.S. Nuclear Regulatory Commission, Washington, D.C.

Croley, T. E., II, 1978, "Notes on Hydrologic Computations," Iowa Institute of Hydraulic Research," The University of Iowa, Iowa City, Iowa.

Culkowski, W. M., 1984, "An Initial Review of Several Meteorological Models Suitable for Low-Level Waste Disposal Facilities," NUREG/CR-3838, U.S. Nuclear Regulatory Commission, Washington, D.C.

Cupitt, L. T., 1980, "Fate of Toxic and Hazardous Materials in the Air Environment," EPA-600/3-80-084, U.S. Environmental Protection Agency, Research Triangle Park, North Carolina.

Dass, P., Tamke, G. R., and Stoffel, C. M., 1977, "Leachate Production at Sanitary Landfills." Journal of the Environmental Engineering Division. Proceeding of the American Society of Civil Engineers, Vol. 103, No. EE6.

DOE, 1985, "Comprehensive Environmental Response Compensation, and Liability Act Program," DOE Order 5480.14, U.S. Department of Energy, Washington, D.C.

Donigian, Jr., A. S., Imhoff, J. C., Bicknell, B. R., Baker, J. L., Haith, D. A., and Walter, M. F., 1983a, "Application of Hydrologic Simulation Program - FORTRAN (HSPF) in Iowa Agricultural Watersheds," EPA-600/S3-83-069, U.S. Environmental Protection Agency, Athens, Georgia.

Donigian, Jr., A. S., T. Y. R. Lo, and E. W. Shanahan, 1983b, Rapid Assessment of Potential Groundwater Contamination under Emergency Response Conditions, Prepared by Anderson-Nichols and Co., Inc. for the U.S. Environmental Protection Agency. Athens, Georgia.

Eagleson, P. S., 1970, "Dynamic Hydrology," McGraw-Hill Book Co., New York, New York.

EPA, 1980, "Appendix C - Guidelines and Methodology in the Preparation of Health Effect Assessment Chapters of the Consent Decree Water Criteria Documents," U.S. Environmental Protection Agency, Federal Register, Vol. 47, pp. 79347-79357.

EPA, 1982a, "Appendix A - Uncontrolled Hazardous Waste Site Ranking System: A User's Manual," U.S. Environmental Protection Agency, Federal Register, Vol. 37, No. 137, Friday, July 16, 1982, pp. 31219-31243.

EPA, 1982b, "Health Effects Assessment Summary for 300 Hazardous Organic Constituents," Environmental Criteria and Assessment Office, U.S. Environmental Protection Agency, Cincinnati, Ohio.

EPA, 1984, "Uncontrolled Hazardous Waste Site Ranking System, A User's Manual (HW-10)," U.S. Environmental Protection Agency, Washington, D.C.

Fenn, D. G., Hanley, K. J., and DeGeare, T. V., 1975, "Use of the Water Balance Method for Predicting Leachate Generation from Solid Waste Disposal Sites," Washington, D.C.: U.S. Environmental Protection Agency, SW-168.

Fletcher, J. F., and Dodson, W. L., eds., 1971, "Hermes-A Digital Computer Code for Estimating Regional Radiological Effects from the Nuclear Power Industry," USAEC Rep. HEDL-TME-71-1968.

Fletcher, J. F., Dodson, W. L., Peterson, D. E., and Betson, R. P., 1973, Modeling the Regional Transport of Radionuclide in a Major United States River Basin, in: "Environmental Behavior of Radionuclide Released in the Nuclear Industry," IAEA Vienna.

Gardner, W. R., 1960, Soil Water Relations in Arid and Semi-Arid Conditions, UNESCO, 15:37-61.

Hanks, R. J., and Ashcroft, G. L., 1980, "Applied Soil Physics," Springer Verlag. New York, New York.

Haun, C. F., and Barfield, B. J., 1978, "Hydrology and Sedimentology of Surface Mined-Lands," University of Kentucky, Lexington, Kentucky.

Hawley, K. A., and Napier, B. A., 1984, A Ranking System for Mixed Radioactive and Hazardous Waste Sites," in: "Proceedings of the Fifth DOE Environmental Protection Information Meeting," CONF-841187, U.S. Department of Energy, Washington, D. C.

HWN, 1984, Hazardous Waste News, Silver Spring, Maryland: Business Publishers, Inc., September 17, 1984. Vol. 6, No. 37. p. 290.

Hillel, D., 1980, "Fundamentals of Soil Physics," Academic Press, New York, New York.

Hjelmfelt, A. T., Jr., 1976, "Modeling of Soil Movement Across a Watershed," Completion Report for Project A-076-MO, Missouri Water Resources Center, University of Missouri, Columbia, Missouri.

ICF, 1984, "The Risk-Cost Analysis Model: Phase III Report," Prepared for the Office of Solid Waste, Economic Analysis Branch, U.S. Environmental Protection Agency by ICF, Incorporated.

ICRP, 1977, "Recommendations of the International Commission on Radiological Protection," International Commission on Radiological Protection, ICRP Publication 26, Pergamon Press, New York.

ICRP, 1979-1982, "Limits for Intakes of Radionuclides by Workers," International Commission on Radiological Protection, ICRP Publication 30, Part 1 (and subsequent parts and supplements), Vol. 2, No. 3/4 through Vol. 8, No. 4, Pergamon Press, New York.

Israelsen, O. W., and Hansen, V., E., 1962, "Irrigation Principles and Practices," John Wiley and Sons, Inc., New York.

Johanson, R. C., Imhoff, J. C., and Davis, Jr., H. H., 1980, "User's Manual for Hydrologic Simulation Program - FORTRAN (HSPF)," EPA-600/9-80-015, U.S. Environmental Protection Agency, Athens, Georgia.

Kent, K. M., 1973, "A Method for Estimating Volume and Rate of Runoff in Small Watersheds," SCS-TP-149, U.S. Department of Agriculture, Soil Conservation Service, Washington, D.C.

Knight, R. G., Rothfuss, E. H., and Yard, K. D., 1980, "FGD Sludge Disposal Manual," Second Edition, Palo Alto, California: Electric Power Research Institute, EPRI CS-1515.

Kufs, C., Twedell, D., Paige, S., Wetzel, R., Spooner, P., Colonna, R., and Kilpatrick, M., 1980, "Rating the Hazard Potential of Waste Disposal Facilities," in: "Proceedings on the Management of Uncontrolled Hazardous Waste Sites," U.S. Environmental Protection Agency National Conference. Hazardous Materials Control Research Institute, Washington, D.C., pp. 30-41.

LeGrand, H. E., 1983, "A Standardized System for Evaluating Waste-Disposal Sites," 2nd Edition, National Water Well Association, Worthington, Ohio.

Mualem, Y., 1976, "A Catalogue of the Hydraulic Properties of Unsaturated Soils," Research Project 442," Technion Israel Institute of Technology, Haifa, Israel.

Napier, B. A., Kennedy, Jr., W. E., and Soldat, J. K., 1980, "PABLM - A Computer Program to Calculate Accumulated Radiation Doses from Radionuclides in the Environment," PNL-3209, Pacific Northwest Laboratory, Richland, Washington.

NAS, 1980, "The Effects on Populations of Exposure to Low Levels of Ionizing Radiation," National Academy of Sciences Committee on the Biological Effects of Ionizing Radiations, National Research Council, Washington, D.C.

NRC, 1977, "Calculation of Annual Doses to Man from Routine Releases of Reactor Effluents for the Purpose of Evaluating Compliance with 10 CFR Part 50, Appendix I," Regulatory Guide 1.109, U.S. Nuclear Regulatory Commission, Washington, D.C.

Onishi, Y., Brown, S. M., Olsen, A. R., Parkhurst, M. A., and Wise, S. E., 1979, "Assessment Methodology for Overland and Instream Migration and Risk Assessment of Pesticides," Prepared for the U.S. Environmental Protection Agency by Battelle, Pacific Northwest Laboratories, Richland, Washington.

Onishi, Y., Whelan, G., Parkhurst, M. A., Olsen, A. R., and Gutknecht, P. J., 1980, "Preliminary Assessment of Toxaphene Migration and Risk in the Yazoo River Basin, Mississippi," Battelle, Pacific Northwest Laboratories, Richland, Washington.

Onishi, Y., Brown, S. M., Olsen, A. R., and Parkhurst, M. A., 1981, Chemical Migration and Assessment Methodology, in: "Proceedings of the Conference on Environmental Engineering," American Society of Civil Engineers, Atlanta, Georgia, pp. 165-172.

Onishi, Y., Whelan, G., and Skaggs, R. L., 1982a, "Development of a Multimedia Radionuclide Exposure Model for Low-Level Waste Management," PNL-3370, Pacific Northwest Laboratory, Richland, Washington.

Onishi, Y., Yabusaki, S. B., Cole, C. R., Davis, W. E., and Whelan, G., 1982b, "Multimedia Contaminant Environmental Exposure Assessment (MCEA) Methodology for Coal-Fired Power Plants," Volumes I and II," Prepared for Rand Corporation by Battelle, Pacific Northwest Laboratories, Richland, Washington.

Parkhurst, M. A., Whelan, G., Onishi, Y., and Olsen, A. R., 1982, "Simulation of the Migration, Fate, and Effects of Diazinon in Two Monticello Stream Channels," Battelle, Pacific Northwest Laboratories, Richland, Washington.

Pasquill, F., and Smith, F. B., 1983, "Atmospheric Diffusion," Third Edition, John Wiley and Sons, New York.

Patterson, M. R., 1986, "Unified Transport Model for Organics," In: Proceedings of the First Workshop on Pollutant Transport and Accumulation in a Multimedia Environment, Los Angeles, California, January 22-24, 1986.

Patterson, M. R., Munro, J. K., Fields, D. E., Ellison, R. D., Brooks, A. A., and Huff, D. D., 1974, "A User's Manual for the Fortran IV Version of the Wisconsin Hydrologic Transport Model," ORNL-NSF-EATC-7.

Perlmutter, N. M., and Lieber, M., 1970, "Disposal of Plating Waste and Sewer Contaminants in Ground Water Surface Water, South Farmingdale-Massapequa Area, Nassau County, New York," U.S. Geological Survey Water-Supply Paper 1979-G.

Petrie, G. M., Sneider, S. C., Napier, B. A., and Barnard, J. C., 1983, "Simplified Codes for Performance Evaluation (SCOPE) of Radionuclide Transport: Version 1.0," PNL-4737, Pacific Northwest Laboratory, Richland, Washington.

Schroeder, P. R., Gibson, A. C., and Smolen, M. D., 1984, "The Hydrologic Evaluation of Landfill Performance (HELP) Model," U.S. Environmental Protection Agency, Cincinnati, Ohio.

SCS, 1972, "Hydrology Guide for Use in Watershed Planning," SCS National Engineering Handbook, Section 4, Hydrology, Supplement A, U.S. Department of Agriculture, Soil Conservation Service, Washington, D.C.

SCS, 1982, "SCS National Engineering Handbook, Section 4, Hydrology, 1982 Update, U.S. Department of Agriculture, Soil Conservation Service, Washington, D.C.

Selim, H. M. and Mansell, R. S., 1976, Analytical Solution of the Equation for Transport of Reactive Solutes Through Soils, Water Resources Research, 12(3):528-532.

Silka, L. R. and Swearingen, T. L., 1978, "A Manual for Evaluating Contamination Potential of Surface Impoundments," U.S. Environmental Protection Agency, Groundwater Protection Branch, EPA 570/9-78-003, p. 73.

Steelman, B. L., and DeCarlo, V. J., 1985, Management and Ranking of Department of Energy Radioactive Mixed Waste Sites, PNL-SA-132705, in: "Transactions of the American Nuclear Society 1985 Winter Meeting," American Nuclear Society, La Grange Park, Illinois, November 10-14, 1985.

Thornthwaite, C. W., and Mather, J. R., 1955, "The Water Balance", Publications in Climatology, Vol. VIII, No. 1, Drexel Institute of Technology, Laboratory of Climatology, Centerton, New Jersey.

Thornthwaite, C. W., and Mather, J. R., 1957, "Instructions and Tables for Computing Potential Evapotranspiration and the Water Balance," Publications in Climatology, Vol. X, No. 3, Drexel Institute of Technology, Laboratory of Climatology, Centerton, New Jersey, p. 311.

Tucker, W. A., Eschenroeder, A. Q., and Magill, G. C., 1984, "Air Land Water Analysis System (ALWAS): A Multi-Media Model for Toxic Substances," EPA-600/S3-84-052, NTIS PB 84-171 743, U.S. Environmental Protection Agency, Athens, Georgia.

USBR, 1977, "Design of Small Dams," U.S. Department of the Interior, Bureau of Reclamation, U.S. Government Printing Office, Washington, D.C.

Van Genuchten, M. T., and Alves, W. J., 1982, "Analytical Solutions of the One-Dimensional Convective-Dispersive Solute Transport Equation," U.S. Department of Agriculture, Technical Bulletin No. 1661, p. 151.

Van Voris, P., Page, T. L., Rickard, W. H., Droppo, J. G., and Vaugahan, B. E., 1984, "Environmental Implications of Trace Element Releases from Canadian Coal-Fired Generating Stations," Phase II Final Report, Volume II, Appendix B and Appendix C, Contract No. 001G194, Battelle, Pacific Northwest Laboratories, Richland, Washington.

Whelan, G., 1980, "Distributed Model for Sediment Yield." Master's Thesis in Mechanics and Hydraulics, Iowa Institute of Hydraulic Research, University of Iowa, Iowa City, Iowa.

Whelan, G., 1985a, "Development of the Remedial Action Priority System (RAPS): Mathematical Formulations for the Overland Environment. Letter Report," Prepared for the Office of Operational Safety, U.S. Department of Energy by Pacific Northwest Laboratory, Richland, Washington.

Whelan, G., 1985b, Development of the Remedial Action Priority Systems (RAPS): Mathematical Formulations for the Quantification of the Source Term," Letter Report, Prepared for the Office of Operational Safety, U.S. Department of Energy by Pacific Northwest Laboratory, Richland, Washington.

Whelan, G., 1985c, "Development of the Remedial Action Priority System (RAPS): Mathematical Formulations for the Groundwater and Surface Water Environments," Letter Report, Prepared for the Office of Operational Safety, U.S. Department of Energy by Pacific Northwest Laboratory, Richland, Washington.

Whelan, G., and Onishi, Y., 1983, "In-Stream Contaminant Interaction and Transport," Presented at the Tenth International Symposium on Urban Hydrology, Hydraulics, and Sediment Control, University of Kentucky, July 25-28, 1983, Lexington, Kentucky.

Whelan, G., and Parkhurst, M. A., 1983, "Simulation of the Migration, Fate and Effects of Diazinon In-Stream," BN-SA-1384, D. B. Simons Symposium on Erosion and Sedimentation, July 27-29, 1983, Colorado Sate University, Fort Collins, Colorado.

Whelan, G., and Steelman, B. L., 1984, Development of Improved Risk Assessment Tools for Prioritizing Hazardous and Radioactive-Mixed Waste Disposal Sites," in: "Proceedings of the Fifth DOE Environmental Protection Information Meeting," CONF-841187, U.S. Department of Energy, Washington, D.C.

Whelan, G., Onishi, Y., Simmons, C. A., Horst, T. W., Gupta, S. K., Orgill, M. M., and Newbill, C. A., 1982, "Multimedia Radionuclide Exposure Assessment Modeling," PNL-4545, Pacific Northwest Laboratory, Richland, Washington.

Whelan, G., Thompson, F. L., and Yabusaki, S. B., 1983, "Multimedia Contaminant Environment Exposure Assessment Methodology as Applied to Los Alamos, New Mexico," PNL-4546, Pacific Northwest Laboratory, Richland, Washington.

Whelan, G., Brown, S. M., Strenge, D. L., Schwab, A. P., and Mitchell, P. J., 1985a, "Contaminant Assessment Modeling Under the Resource Conservation and Recovery Act," EPRI Project RP2070-01, prepared by Battelle, Pacific Northwest Laboratories, for the Electric Power Research Institute, Palo Alto, California (in press).

Whelan, G., Steelman, B. L., Strenge, D. L., and Hawley, K. A., 1985b, Development of the Remedial Action Priority System: An Improved Risk Assessment Tool for Prioritizing Hazardous and Radioactive-Mixed Waste Disposal Sites, in: "Proceedings of Management of Uncontrolled Hazardous Waste Sites," Hazardous Materials Control Research Institute, Silver Spring, Maryland, November 4-6, 1985, pp. 432-437.

Witinok, P. M., 1979, "Distributed Watershed and Sedimentation Model," Master's Thesis, University of Iowa, Iowa City, Iowa.

Witinok, P. M., and Whelan, G., 1980, "Distributed Parameter Sedimentation Model," Proc. Iowa Acad. Sci. 87(3)103-111.

Yeh, G. T., 1981, AT123D: Analytical Transient One-, Two-, and Three-Dimensional Simulation of Waste Transport in the Aquifer System," Publication No. 1439, ORNL-5602, Oak Ridge National Laboratory, Oak Ridge, Tennessee.

Yeh, G. T., and Tsai, Y. J., 1976, "Analytical Three-Dimensional Transient Modeling of Effluent Discharges," Water Resources Research, Vol. 12, No. 3, June 1976, 553-540.

ESTIMATION OF MULTIMEDIA EXPOSURES RELATED TO

HAZARDOUS WASTE FACILITIES

Seong T. Hwang and James W. Falco

Exposure Assessment Group
U.S. Environmental Protection Agency
Washington, D.C.  20460

INTRODUCTION

Soil contamination often results from improper land disposal, storage, treatment, and spills of hazardous waste in liquid or solid form.  In order to properly address the levels of pollutants up to which contaminated soil needs to be cleaned, it is necessary to estimate the magnitude of human exposure for each likely exposure pathway.  The procedure developed in this paper for estimating the level of pollutant cleanup from soil relates to 1) identification of exposure pathways, 2) exposure assessment to estimate pollutant intakes by humans from each pathway, and 3) the use of health effects data to estimate the acceptable exposure levels.

Dacre et al. (1980) presented an approach to predict preliminary pollutant limit values based on the use of equilibrium partitioning of a pollutant between environmental media.  This approach will be inapplicable for some processes including, to name a few, rate of volatilization from soil, dispersion of air contaminants, and migration of leachate contaminants in the unsaturated and groundwater media, because they represent dynamic processes.  In addition, these exposure pathways were addressed poorly or not at all by Dacre et al. (1980).  However, it was concluded that the treatments represent a conceptual framework for examination of single exposure pathways, and only a beginning for examination of other critical areas.

Exposure assessment should include short-term and long-term periods of exposure, since acute as well as chronic health consequences need to be considered. Acute effects are normally noncancer effects, while chronic effects can relate to noncancer and oncogenic effects. Depending upon the circumstances under which human exposure occurs and the type of pollutant being considered, exhaustive exposure evaluation for all likely pathways may be avoided, with only the most controlling pathways included in the determination of acceptable contaminant cleanup level in soil.

This paper will present a method intended primarily to allow the estimation of permissible residual levels of contaminant associated with the cleanup of contaminated soils to meet acceptable health effects criteria. It can also be applicable to assessing the limits of pollutant that can be allowed in mixtures of soil and waste being landfilled.

After initial cleanup, the concentration profile in soil continues to change as time progresses, due to volatilization. This change in concentration results in changes in exposures. The transient volatilization rate is estimated by solving a partial differential equation which incorporates soil-air partition coefficient and diffusion. Exposures need to be averaged over the time period of concern, including short-term and long-term periods.

EXPOSURE PATHWAYS

Contaminants originally present in soil can be distributed in various environmental media. The soil leaching process is a major concern for groundwater contamination. Bioaccumulation in fish, in animals used for food, or in garden vegetables results from contaminated surface water or from contaminated soil on which animals graze or plants are grown. Volatile constituents in soil can be released into the atmosphere because their partial pressure in equilibrium with the concentrations in soil provides the driving force for diffusion along the soil column. Windblown particulate matter can entrain toxic constituents into the air which people breathe. People can work in a garden where there is contaminated soil, and be subjected to dermal contact with pollutants that can be absorbed through the skin.

The presence of contaminants in environmental media being distributed from wastes disposed of in soil thus can pose a potential health risk to humans from the following sources of intake:

- Ingestion of contaminated soil;

- Inhalation of air containing contaminants in vapor or particulate form;

- Absorption of contaminants through the skin because of contact with contaminated soil;

- Drinking of contaminated water; and

- Ingestion of foods with equilibrated or bioaccumulated contaminants.

Other exposure pathways, such as phytotoxicity to plants, may also need to be considered.

Under normal conditions, significant soil ingestion is limited to children (Lepow, 1975). Although very limited information is available on the ranges of age subject to soil ingestion, one investigation presented a case study of an adult with a history of habitual eating of garden soil, which may have been associated with a pica illness (Wedeen et al., 1978). The fraction of soil contaminant absorbed by humans is dependent upon the type of compound and its soil contaminant adsorption characteristics, and is generally smaller than that which can be expected when contaminants are present in food or drinking water.

When exposure locations are distant from the contaminated site, dilution, biodegradation, dispersion, or chemical reactions may have a pronounced effect on exposures. Examples include dilution of contaminants by winds, dispersion in groundwater, and biodegradation in surface water. It is possible to incorporate such processes in exposure evaluations. The significance of exposures to inhaled or dermally-contacted contaminants will also be dependent upon the absorption rate of such contaminants in the lungs or on human skin. The smaller the absorption rate, the less significant the exposure would be.

Once exposure pathways are identified, exposure evaluation requires information on the levels of concentration to which a given target population may be exposed. Each pathway may require route-specific evaluation, as described in the following sections.

ESTIMATION OF EXPOSURES RELATED TO CONTAMINATED SOIL

The combined human intake of contaminants from all exposure pathways should not exceed the acceptable intake (AI, in mg/day) needed for preventing adverse effects from short-term and lifetime exposures. Refer to the Nomenclature section for the meaning of symbols used. The intake from an individual route when soil is contaminated can be expressed quantitatively as follows:

i)  Intake by soil ingestion (mg/day):

$$I_1 = (C_s)(IR \times 10^{-3})(GI)(SM)(F) \tag{1}$$

The term $(C_s)(IR)$ in Eq. (1) represents the daily amount of a contaminant ingested resulting from soil ingestion, in $\mu g/day$, because $C_s$ represents the contaminant concentration in soil ($\mu g/g$) and IR is the soil ingestion rate (g/day). The factor $10^{-3}$ is needed to convert the unit from $\mu g/day$ to mg/day. The fraction of the ingested contaminants that will enter human organs and systems to cause toxicity is given as GI. If the contaminant undergoes biological or chemical degradation in soil, and follows first-order kinetics in its disappearance under isothermal conditions, the contaminant concentration will change as a function of time according to $C_0 e^{-kt}$, where $C_0$ is the initial contaminant concentration in soil. The concentration in soil will also be affected by transport processes, which

232

are incorporated in the individual pathway analysis. An individual may not always be present on the contaminated site over his lifetime. The frequency factor of exposure over a lifetime, SM, represents the fraction of a lifetime that an individual will be exposed to the contaminants under consideration. The factor F is necessary because soil ingestion only occurs during childhood (1 to 6 years of age), and the weight of the human body changes from childhood to adulthood.

ii) Intake by air inhalation of volatilized contaminants (mg/day):

$$I_2 = (K_{as})(C_s)(D)(IH \times 10^3)(ABA)(SM) \tag{2}$$

In Eq. (2), $K_{as}$, D, and IH represent the soil-air partition coefficient, the extent of dilution of the vapor concentration in equilibrium with the contaminant concentration on soil after being dispersed by winds in the atmosphere, and the average daily inhalation rate of ambient air, respectively. The term $(K_{as})(C_s)(D)(10^3)$ represents the ambient air concentration of a pollutant at the exposure location in $mg/m^3$ when $K_{as}$ and $C_s$ are given in $g$ soil/$cm^3$ and mg/kg, respectively.

The determination of the ambient air concentration at an exposure location, $(K_{as})(C_s)(D)(10^3)$, or the dilution factor in the term, requires the estimation of transient emission rate, and the use of dispersion modeling. The emission rate will not only be transient, but it will also be retarded by soil or equivalent cover material. This phase of the problem requires solution of a partial differential equation, as described in the section entitled "Inhalation of Volatilized Contaminants."

iii) Intake by dermal absorption (mg/day):

$$I_3 = (C_s)(CR \times 10^{-3})(ABS)(SM) \tag{3}$$

There are many occasions when children playing in the yard or adults working in the garden will come in direct contact with contaminated soil. Dermal contact does not necessarily constitute adverse exposure. The contaminant needs to be systemic to be absorbed into the human body and to exert toxicity. In Eq. (3), the term $(C_s)(CR)$ represents the contaminant contact rate with skin in μg/day since $C_s$ is in μg/g (=ppm) and CR is the dermal contact rate of soil in g/day. The factor $10^{-3}$ is used to convert the contact rate from μg/day to mg/day, and SM will be 1 when the short- or longer-term (10-day) exposure is estimated, and will be between 0 and 1 when the lifetime exposure is estimated.

iv) Intake by drinking water (mg/day):

$$I_4 = (C_w)(IW)(SM)$$

(4)

In Eq. (4) it is assumed that the contaminant in drinking water is completely absorbed into the human body at the average daily water consumption rate of IW or the absorption fraction is 1. In order to relate the contaminant concentration in groundwater, $C_w$, to the contaminant concentration in soil, $C_s$, a fate and transport model can be used to estimate the concentration in the leachate entering groundwater, or

$$C_w = C_L/f_g, \text{ mg/L}$$

(5)

where $f_g$ represents a functional relationship describing contaminant transport in groundwater. This function should be selected to suit the most appropriate conditions for the system. The leachate concentration, $C_L$, referring to the contaminant concentration in liquids just before entering groundwater, should not be confused with the contaminant concentration in groundwater, which results from mixing of the leachate with groundwater. Also, care should be exercised in using groundwater transport models, be-

234

cause some models will treat the leachate concentration as a boundary condition, while others require the contaminant concentration in groundwater as a boundary condition, which should be obtained by groundwater monitoring. When the units of $C_w$ and $C_L$ are all in mg/L, then the function $f_g$ becomes dimensionless. Most leachate from hazardous waste land disposal sites may enter groundwater over a finite surface area, favoring area source models for simulating pollutant transport in groundwater.

There is no reliable method of predicting the leachate concentration from the contaminant concentration in soil, or vice versa. For the exposure evaluation, an equilibrium relationship vetween soil and leachate will provide a first approximation. Monitoring data can also be used relating the concentrations between leachate and soil. An equilibrium condition can be written as

$$C_s = (K_{Ls})(C_L), \text{ mg/kg}$$

(6)

where $K_{Ls}$ is a partition coefficient in (mg/kg)/(mg/L). Eqs. (4), (5), and (6) are combined to get

$$I_4 = \frac{C_s}{(f_g)(K_{LS})} (IW)(SM)$$

(7)

When the equilibrium condition is not appropriate, it can be modified to include transport processes between the soil and leachate.

v) Intake by fish ingestion (mg/day):

At the average daily fish consumption rate of IF (kg/day), and under the assumption of complete absorption of the contaminant associated with the consumption of fish, the exposure can be estimated as

$$I_5 = (C_F)(IF)(SM)$$

(8)

where $C_F$ is the contaminant concentration in fish.  The use of the bioconcentration factor BCF, (mg/kg fish)/(mg/L water), to relate pollutant concentrations in fish and water, gives

$$I_5 = (BCF)(C_W)(IF)(SM)$$

(9)

where it is assumed that contaminants are present in water in dissolved form and that bottom sediments or benthal deposits on which pollutants may be adsorbed are not directly swallowed by fish.  Under the condition of equilibrium between the pollutant-containing soil and leachate which is generated from the soil, substitution of Eqs. (5) and (6) into Eq. (9) results in

$$I_5 = \frac{C_s}{(f_g)(K_{LS})} (BCF)(IF)(SM)$$

(10)

The transport functions, $f_g$, in Eqs. (7) and (10) may assume distinct mathematical descriptions, because one pertains to transport in groundwater and the other to that in surface water.

vi)  Intake by inhalation of contaminants adsorbed on particulates (mg/day) may be expressed by

$$I_6 = (C_p)(IH)(C_s \times 10^{-9})(ABP)(SM)$$

(11)

Contaminant-containing soil can be airborne by blowing winds.  In addition, toxic substances volatilized from contaminated soil can be adsorbed on particulate matter present in the ambient air.  Exposure to contaminants occurs because of inhalation of air containing these particulates. The exposure location could be distant from the source of emission, or in the vicinity of the emission source.  Exposure concentrations will change accordingly.  Another form of exposure relates to inhalation of air con-

taining particulate matter on which volatile constituents are adsorbed.
The intake rate can be estimated based on the concentration of contaminants in wind-dispersed soil or on particulate matter, $C_s$ µg/g (=ppm),
and the concentration of the particulates in the ambient air, $C_p$ µg/m$^3$,
as shown in Eq. (11). The absorption fraction, ABP, is used because
contaminants present in or on soil (or particulate matter) may be bound on
the solid material, reducing the contaminant's absorption rate. Finally,
the factor $10^{-9}$ is a conversion factor to make the units consistent.

vii)  Intake by ingestion of vegetables (mg/day):

The intake rate due to ingesting IV kg/day of vegetables, plants, or
agricultural products containing $C_V$ mg/L of contaminants will be

$$I_7 = (C_V)(IV)(SM)$$
(12)

If it is assumed that equilibrium is established between the contaminant
concentrations in plant and soil, then the exposure can be modified as:

$$I_7 = (K_{sv})(C_s)(IV)(SM)$$
(13)

where $K_{sv}$ is a partition coefficient defined as contaminant concentration in
plant/total contaminant concentration in soil (mg/kg plant)/(mg/kg soil).

viii)  Intake by ingestion of food meat:

The contaminant intake at consumption rate of IM (kg/day) of meat containing $C_m$ (mg/kg) of pollutant is

$$I_8 = (C_m)(IM)(SM)$$
(14)

Here again, an equilibrium relationship is assumed between the contaminant
concentrations in the animal body and plants. Therefore, the intake rate
due to meat consumption is

$$I_8 = (K_{vm})(C_v)(IM)(SM)$$

$$= (K_{vm})(K_{sv})(C_s)(IM)(SM) \tag{15}$$

where $K_{vm}$ and $K_{sv}$ are the partition coefficients used to describe pollutant distribution between meat and vegetables, and the partition between
vegetables and soil, respectively.

Similar expressions can be written for other routes of exposure. In
Eq. (4), it is assumed that the groundwater concentration of a pollutant
at an exposure location can be determined (by using monitoring data or
fate and transport models) from the knowledge of leachate concentration as
given by a function $f_g$, and that the contaminated soil and its leachate
are in equilibrium as given by $K_{LS} = C_s/C_L$. There are a number of functional relationships which can be used in place of $f_g$, as reviewed by
Hwang (1984). Also, equilibrium relationships are assumed between fish
and water for bioaccumulation, between soil and vegetables, and between
food animals and plants consumed by the animals. In Eqs. (7), (10), (13),
and (15), contaminants consumed by humans because they are present in drinking water, fish, and other food sources, are taken as completely absorbed
by the human body. If data on the absorption rate for these pathways are
available, such data should be used.

## DETERMINATION OF PERMISSIBLE POLLUTANT LEVEL IN SOIL

The total intake from all possible exposure pathways is set equal to
the acceptable intake (AI) for short-term and chronic health effects; or

$$AI = I_1 + I_2 + I_3 + \cdots \tag{16}$$

Eq. (16) can be solved for permissible contaminant levels in soil corresponding to each acceptable intake. It is possible that some exposure pathways

occur independently of others. For example, a residence which is located on a contaminated site may use drinking water from a clean public water treatment system, and may thus be free of contaminants found on the site. It is also possible that domestic animals are not raised for food consumption on the contaminated site under consideration. Under such circumstances, all exposure pathways need not be considered. If exposure pathways of significant concern are related to soil ingestion, inhalation of contaminated air, or dermal contact with soil, Eqs. (1), (2), and (3) can be added to solve for $C_s$,

$$C_s = \frac{(AI)(1000)}{[(IR)(GI)(F) + (K_{as})(D)(IH)(ABA \times 10^6) + (CR)(ABS)]SM}$$

(17)

The emission rate is limited by the air phase concentration in equilibrium with the contaminant concentration in soil. Once the contaminant soil concentration reaches the level at which the vapor phase concentration in equilibrium with the soil is at the vapor pressure concentration, a further increase in contaminant concentration in soil ($C_s > C_{sm}$) does not increase the emission rate. At or above this concentration, the ambient air concentration remains constant regardless of the concentration of the contaminant in soil. Under such conditions, $C_s$ in Eq. (2) is no longer a variable, and therefore Eq. (17) does not apply. This situation can be remedied by considering the intakes by the individual route of exposure at a constant value of $C_s$ [$C_s = C_{sm}$ in Eq. (2)] for inhalation exposure, and solving for $C_s$. The form of the equation will be slightly different from that for Eq. (17).

$$C_s = \frac{(AI)(1000) - (K_{as})(C_{sm})(D)(IH \times 10^3)(ABA)(SM)}{[(IR)(GI)(F) + (CR)(ABS)]SM}$$

(18)

INCORPORATION OF TIME-VARYING PARAMETERS

The body weight of a human constantly changes until maturity. The calculation of AIs from the safe dose level (SL) given in mg/kg·day requires requires an assumption of body weight. For rigorous treatment, the estimation of lifetime exposure should take into account changes in body weight. In this case, it is convenient to work with SL instead of AI for exposure calculations. For carcinogens with a potency value at POT $(mg/kg·day)^{-1}$, the equivalent SL at an assumed risk level, R (such as $10^{-6}$, etc.), can be obtained by

$$(SL)eq. = \frac{R}{POT} , \quad mg/kg·day$$

The risk level shown represents an upper-bound estimate. An upper-bound estimate of risk of $10^{-6}$, for example, means that upon lifetime exposure to a contaminant, a person experiences an increased maximum risk of developing cancer in a probability of 1 in one million.

Snyder (1975) presented data on the change of body weight as a function of age. A regression analysis on Snyder's data for average male weight provides the following relationship.

$$BW = 3.14 + 3.52 \text{ (age), kg for age } 0 - 18 \text{ yr} \qquad (19)$$

$$BW = 70, \text{ kg for age greater than 18 yr} \qquad (20)$$

To obtain the daily exposure averaged over an individual's lifetime, intake rates given by Eqs. (1) - (3), (7), (10), (11), (13), and (15) should be divided by the body weight, and the daily intake per unit body weight should be averaged by summing the total intake per unit body weight over the period during which exposure occurs and dividing the result by LT. For purposes of illustration, Eqs. (1) and (2) are repeated below:

i) The average daily exposure by soil ingestion per unit body weight in mg/kg·day can be determined as

$$\frac{I_1}{BW} = \sum_{1\ day}^{25550\ days} \frac{(C_o e^{-kt})(IR \times 10^3)(GI)(SM)}{(BW)(LT)} \tag{21}$$

Again, in Eq. (21) [also in in Eq. (22)], the contaminant present in soil is assumed to disappear by biodegradation and other reactions, according to first-order kinetics. Other processes affecting the concentration in soil are considered in the exposure analyses for individual pathways.

ii) The average daily exposure by inhalation of volatilized contaminants in mg/kg·day is calculated from

$$\frac{I_2}{BW} = \sum_{1\ day}^{25550\ days} \frac{(K_{as})(C_o e^{-kt})(D)(IH \times 10^3)(ABA)(SM)}{(BW)(LT)} \tag{22}$$

Similar expressions can be written for other exposure pathways. For conservative contaminants, the term $C_o e^{-kt}$ in Eqs. (21) and (22) can be replaced by $C_s$. The total dose from all exposures should not exceed SL, or $(SL)_{eq.}$

$$SL = \frac{I_1}{BW} + \frac{I_2}{BW} + \frac{I_3}{BW} + \ldots \qquad \text{for noncarcinogenic effects} \tag{23}$$

$$(SL)_{eq.} = \frac{I_1}{BW} + \frac{I_2}{BW} + \frac{I_3}{BW} + \qquad \text{for carcinogenic effects} \tag{24}$$

As before, Eq. (23) or (24) can be solved for the permissible concentrations in soil, $C_s$. From Eqs. (1) and (21), one can solve for the factor F for use in Eq. (1). The use of LT = 25550 days, and the assumption that soil ingestion occurs during ages 1 through 5 (t = 365 to 1825 days), yield F = 0.323. The factor F does not depend on the soil ingestion rate. Eqs. (1) and (21) use Eqs. (20) and (19), respectively, for BW.


INHALATION OF VOLATILIZED CONTAMINANTS

Exposure occurs at or near contaminated sites through inhalation of ambient air contaminated with toxic vapors or particulate matter on which a contaminant is absorbed. Volatile constitutents emitted from contaminated soil will be diluted by the action of winds before a person inhales the ambient air. When a contaminant is adsorbed on soil, the vapor pressure of the contaminant above the soil surface will be always smaller than the pure component vapor pressure. In other words, the adsorption phenomenon depresses the vapor pressure that can exist under saturated conditions. When adsorption reaches its saturation capacity on soil, the partial pressure will be equal to the pure component vapor pressure.

In calculating ambient air concentrations of contaminants, the first task is to estimate emission rates from the bulk of soil contaminated at various concentrations. The emission rate calculations for steady-state conditions can be performed using the methods presented by Farmer et al. (1980), and subsequently summarized by Hwang (1982). Thibodeaux and Hwang (1982) presented the transient emission rate from land treated facilities. A new mathematical treatment is needed to estimate transient rates of emissions from contaminated soil. The transient emission rate is obtained by solving a partial differential equation based on a material balance around the soil column, as described in this section.

A mass balance over an infinitesimal vertical element of soil for vapor phase diffusion and soil phase adsorption can be written as follows (Hwang, 1985):

$$\frac{\partial C}{\partial t} + \bar{\rho}_s \frac{\partial C_s}{\partial t} = \frac{\partial}{\partial z} \left( \varepsilon \, D_{ei} \, \frac{\partial C}{\partial z} \right) \tag{25}$$

Since changes in soil and vapor phase concentrations of contaminants occur slowly, it can be assumed that the vapor phase and soil-solid phase concentrations of a contaminant are in local equilibrium. The contaminant concentrations in soil and in interstitial vapors are assumed to be at equilibrium, and they are related by the following equation:

$$C_s = \frac{K_d}{41\ H}\ C$$

(26)

where the conversion factor 41 is needed for consistency of the units given in the Nomenclature.

Rearranging Eq. (25) and substituting Eq. (26) into the resulting relationship yields

$$D_{ei}\ \frac{\partial^2 C}{\partial z^2} = \left(1 + \frac{(\bar{\rho}_s)(K_d)}{(\varepsilon)(H)}\right)\frac{\partial C}{\partial t}$$

(27)

or

$$\alpha\ \frac{\partial^2 C}{\partial z^2} = \frac{\partial C}{\partial t}$$

(28)

where

$$\alpha = \frac{(D_{ei})(\varepsilon)}{[\varepsilon + (\rho_s)(1-\varepsilon)(K_d)/H]}$$

(29)

$\alpha$ can also be defined as

$$\alpha = \frac{D_{ei}}{1 + (K)(S)}$$

(30)

where

$$K = \frac{K_d}{H} \, \rho_s$$

$$S = (1-\varepsilon)/\varepsilon$$

Eq. (27) can be solved to estimate contaminant concentration in soil-solids, vapor phase concentration, and emission rate into air above soil for the following two cases:

Case 1.    Surface is exposed to the atmosphere.  The initial and boundary conditions are

        1. I.C.     $C = (H/K_d)C_{SO}$, at $t=0$, $z \geq 0$

        2. B.C.     $C = (H/K_d)C_{SO}$, at $z=\infty$, $t>0$

        3. B.C.     $C = 0$,               at $z=0$, $t>0$

where $C_{SO}$ is the initial concentration of contaminant in soil.  The solution to Eq. (27) for the above initial and boundary conditions is (Crank, 1985)

$$C = (H/K_d)C_{SO} \cdot \text{erf} \left( \frac{z}{2\sqrt{\alpha t}} \right) \tag{31}$$

where

$$\text{erf}(n) = \text{error function} = \frac{2}{\sqrt{\pi}} \int_{o}^{n} \exp(-n^2)dn$$

The flux rate at the soil-air interface ($N_A$) can be estimated as a function of time from Eq. (31) by using the concentration gradient as follows:

$$N_A = -(\varepsilon)(D_{ei}) \left. \frac{\partial C}{\partial z} \right|_{z=o} = \frac{(\varepsilon)(D_{ei})(\frac{H}{K_d})(C_{SO})}{\sqrt{\pi \alpha t}} \tag{32}$$

244

The average flux rate, $N_A$, over an exposure interval, T, is calculated by time-averaging Eq. (32) to yield

$$\overline{N}_A(T) = 2\ N_A(T)$$

(33)

The total average emission rate, Q, can be calculated from

$$Q = (A)(\overline{N}_A)$$

(34)

Case 2. The contaminated surface is covered with contaminant-free soil material. Let $\ell$ = thickness of cover (cm), and L = the depth of contamination measured from the top of cover material (cm). The initial and boundary conditions become:

1. I.C.     $C = 0$,     $0 \leqslant z \leqslant \ell$ ,     at $t = 0$

2. I.C.     $C = C_0$     $\ell < z \leqslant L$,     at $t = 0$

3. B.C.     $C = 0$,     $z = 0$,     at $t > 0$

4. B.C.     $\dfrac{\partial C}{\partial z} = 0$,     $z = L$,     at $t > 0$

where $C_0$ is the initial contaminant concentration in the vapor phase in equilibrium with the solid phase $[C_0 = (H/K_d)C_{SO}]$. Eq. (27), with these initial and boundary conditions, can be solved using the Fourier Series techniques. The solution is (Crank, 1985)

$$C = \frac{4C_0}{\pi} \sum_{n=0}^{\infty} e^{-\dfrac{\alpha(2n+1)^2 \pi^2 t}{4 \cdot L^2}} \cdot \sin[\frac{(2n+1)}{2} \frac{\pi z}{L}]\ \cos[\frac{(2n+1)}{2} \frac{\pi \ell}{L}]$$

(35)

The flux rate at the soil-air interface ($N_A$) can be estimated as a function of time from Eq. (35).

$$N_A = \varepsilon \, D_{ei} \left. \frac{\partial C}{\partial z} \right|_{z=0} = \frac{2C_o \, \varepsilon \, D_{ei}}{L} \sum_{n=0}^{\infty} e^{-\frac{\alpha(2n+1)^2 \pi^2 t}{4 \cdot L^2}} \cos \left[ \frac{(2n+1)}{2} \frac{\pi \ell}{L} \right]$$

(36)

The average emission rate over a time period, T, can be obtained by integrating Eq. (36). The result is

$$\overline{N_A} = \frac{8(H/K_d)(C_{SO})(\varepsilon)(D_{ei})(L)}{\alpha \pi^2 \, T} \sum_{n=0}^{\infty} \frac{1}{(2n+1)^2} \, [e^{-\frac{\alpha(2n+1)^2 \pi^2 t_1}{4 \cdot L^2}}$$

$$- e^{-\frac{\alpha(2n+1)^2 \pi^2 t_2}{4 \cdot L^2}} ] \cdot \cos \left[ \frac{(2n+1)}{2} \frac{\pi \ell}{L} \right]$$

(37)

The time interval $t_2 - t_1$ in Eq. (37) represents the exposure interval. It should be noted that when the value of the expression $\frac{\alpha(2n+1)^2 \pi^2}{4L^2}$ in the exponential term of Eq. (37) is small or considerably less than 1, the average of the exponential term over a time, t, will be close to 1. In this situation, averaging of the exponential term of Eq. (36) by the integration formulae given by Eq. (37) may easily result in an erroneous answer because one has to evaluate very precise numbers of many decimal points for the values of the exponential term. It is more practical to numerically average Eq. (36) than to obtain the average value by using the integration formulae given by Eq. (37):

$$\overline{N_A} = \frac{2(H/K_d)C_{SO}(\varepsilon)(D_{ei})}{L} \frac{1}{T} \sum_{n=0}^{\infty} [\int_{t_1}^{t_2} e^{(-\frac{\alpha(2n+1)^2 \pi^2}{4 \cdot L^2})} ] \, dt \cdot \cos \left[ \frac{(2n+1)}{2} \frac{\pi \ell}{L} \right]$$

(38)

Dispersion models (Bruce, 1969; Wark and Warner, 1981) estimate the ambient air concentrations to which populations may be exposed on a daily or annual basis. While these dispersion models can be used, at any distance x from the site, to estimate ambient air concentrations, they cannot be used on-site. On-site ambient air concentrations can be estimated using the following equation:

$$C_a = \frac{Q}{(LS)(V)(MH)} \tag{39}$$

The dilution factor needed in using Eq. (2) or (17) can be obtained by:

$$D = C_a/C_{as} \tag{40}$$

The value of $C_{as}$ increases continuously as the contaminant concentration in soil increases, until the concentration of contaminant in the air phase corresponds to that of contaminant vapor pressure. Once the condition of vapor saturation is reached, a further increase in contaminant concentration in soil will have a minimal effect on the volatilization rate.

VALUES FOR EXPOSURE ASSESSMENT PARAMETERS

Soil Ingestion

Limited information is available on the likely rate of soil ingestion by children and adults. The situations from which the information is derived differ from study to study. Lepow et al. (1975) studied the mouthing behavior of ten 2- to 6-year-old children in connection with investigations into the principal cause for the excessive lead accumulation observed in the children. The total soil ingestion rate for a 2-year-old child, based on the average amount of street dirt, house dust, and soil ingested by the child as a result of putting his hands and fingers in his mouth, can be summed as 0.6 g of soil per day.

Using the levels of blood lead concentration and the concentration of lead in the soil analyzed, Wedeen et al. (1978) estimated the amount of lead a woman had habitually consumed each year from her garden soil. Based on this estimate, the soil ingestion rate is estimated to have been between 1.96 and 3.9 g/day, with an average value of about 3 g/day. The lead concentration in the dried garden soil was reported to be between 690 and 700 micrograms per gram of soil.

Investigators at the Centers for Disease Control (CDC) presented lifetime rates of ingestion of contaminated soil according to age group (Kimbrough et al., 1984), and for children used in the experimental soil ingestion study (Binder, 1985). The Kimbrough et al. paper stated that the data presented was "based on work done studying lead uptake from contaminated soils." The ingestion rate was assumed to change at different ages, and was given as 0 for age group 0 to 9 months, as 1 g/day for age group 9 to 18 months, as 10 g/day for age group 1.5 to 3.5 years, as 1 g/day for age group 3.5 to 5 years, and as 0.1 g/day for a 5-year-old child. Experiments reported by Binder dealt with children between the ages of 1 to 5 years, living in East Helena, Montana, and analyzed the feces of the children for silica, aluminum, and titanium. Back-calculation of the soil ingestion rates using the analytical results showed the ingestion rate to be in the range of 0.12 to 1.8 g/day of soil. The lower numbers are based on the analysis of silica and aluminum, while the higher numbers were obtained from the titanium results. No definitive explanation was provided regarding the discrepancies in the results based on silica, aluminum, or titanium.

Dermal Contact with Soil

Deposition of contaminated soil, dirt, or dust on human skin can provide another pathway for human intake of contaminants. These contaminants can be absorbed through human skin when contaminated particulates come

248

into contact with the skin. Exposure evaluation requires an estimation of the amount of the particulates on skin, and the extent or rate of absorption (which will be dependent upon the type of chemicals involved; some chemicals are readily absorbed, while others are not).

Many factors may affect the amount of soil that can be accumulated on human skin. Such factors include the size of the exposed skin area, contact time, type of soil, soil conditions, and type of activity in which the exposed individual was engaged (e.g., the amount of soil deposited on the skin of children playing in a contaminated area is likely to differ from that deposited on on the skin of adults working in a garden).

Lepow (1975) and Roels et al. (1980), using adhesive tape, measured the amount of soil and dirt accumulated by children on exposed areas such as hands, palms, and fingers. The measured amounts of soil ranged from 0.5 to 1.5 $mg/cm^2$, with an average value of 1 $mg/cm^2$. It should be noted that this is an average value over the surface of the exposure area, and that some parts of the body may have more accumulation of soil than others. The exposed surface area of an adult is estimated to range from about 900 to 2,900 $cm^2$.

The variability of the factors mentioned above makes it difficult to arrive at an average value for the amount of soil and dirt accumulated on skin. Assumptions of (1) an average soil deposition of 1 $mg/cm^2$, and (2) an exposed surface area of about 1,000 $cm^2$, provide an average daily deposition rate of 1 g/day (within a variability range as high as a factor of two) (U.S. EPA, 1984).

Investigators at the CDC estimated the amounts of daily deposition of soil on skin according to age group. Their estimates indicate that the daily amount of soil deposited on skin is 0 for age group 0 to 9 months; 1 g for age group 9 to 18 months; 10 g for age group 1.5 to 3.5 years; 1 g for age group 3.5 15 years; and 100 mg at age 15 years (Kimbrough et al., 1984).

## Transport of Groundwater Contaminants

Contaminated leachate will impact groundwater quality. The area-source groundwater model (Hwang, 1985), among others referred to in the references, can be used to estimate the extent of a contaminant's retardation, dispersion, and dilution in the groundwater medium. The precipitation rate is needed to estimate the leachate generation rate, which becomes a source term in the groundwater fate and transport model. The retardation factor will affect the time of pollutant travel.

The groundwater transport analysis can also be used to back-calculate leachate concentrations entering groundwater below a hazardous waste facility, based on allowable concentrations at a compliance point. This functional relationship is substituted in Eq. (4) for use in estimating the permissible concentration in soil. Also, the soil-water partition coefficient is needed under the assumption that equilibrium is established between the contaminants in soil and leachate.

## Other Routes of Exposure

Exposure to contaminants may occur when people eat vegetables grown on contaminated soil, animals that have grazed in contaminated pasture or wildlands, or fish from contaminated waters; or when people breathe particulate matter on which contaminants are adsorbed. Since the concentration of a contaminant in soil eventually affects its concentration in food material, the use of Eqs. (10), (13), and (15) requires a partition coefficient that describes the chemical equilibrium between the soil and equilibrating food material. The use of Eq. (11) for estimating exposure from particulate inhalation requires information on the concentration of wind-blown particulate matter in the air. It can be noted that the Federal primary ambient air quality standard for particulate matter is 75 $\mu g/m^3$.

It is difficult to estimate the chemical equilibrium partitioning between the soil and plants, and between plants and animals, although the bioaccumulation factor can be estimated from data on the octanol-water

partition coefficient (Veith, 1980). The extent of plant uptake of a
chemical, and its distribution in animals, will depend on the type of
chemical, the type of plant, and the texture of the soil in which the
plant is grown. Plant uptake is generally small in the case of many
organic chemicals. The values of $K_{sv}$ for dieldrin, for example, range
from 0.03 to 0.67. The lower value is for cotton hay and cucumbers grown
in silt loam and Si-Cl loam, and the higher value is for soybean hay grown
in sandy loam (Nash et al., 1970).

Values for other pertinent parameters needed in the multimedia exposure assessment are listed in Table 1.

Table 1. Parameter Values Used in Exposure Evaluation

| | | |
|---|---|---|
| Body weight[a]: | Adult | 70 kg |
| | Child | 10 kg |
| Lifetime: | | 70 years |
| Soil ingestion rate: | ica, t pi | 3 g/day with p ages 1-6 0.6 g/day withouca, ages 1-6 |
| L rmal contact rate of soil: | | 1 g/day |
| Dai y consump- t n rate: | | |
| Fish | | 0.0065 kg/day |
| Vegetables | | 0.01 kg/day |
| Meat | | 0.63 kg/day |
| Air inhalation rate: | Adult | 20 m$^3$/day |
| | Child | 10 m$^3$/day |
| Water ingestion rate: | Adult | 2 L/day |
| | Child | 1 L/day |

[a]Average values.

251

## A CASE STUDY FOR PCBs CLEANUP LEVELS

The method described above is applied to develop advisory levels for polychlorinated biphenyls cleanup, in order to protect public health from short-term and long-term exposures. Exposure pathways considered in developing the advisories include drinking water, ingestion of PCB-contaminated soil by children and adults, inhalation of ambient air, and dermal contact. Other exposure routes, such as food intake and ingestion of fish which have bioaccumulated PCBs, are considered to the extent that these pathways constitute relatively important exposure routes.

In view of the high bioaccumulation factor for PCBs, the consideration of bioaccumulation is important in setting permissible PCB levels in surface water in which aquatic animals live, or its bottom sediments. If one of these routes is a controlling factor in relation to the exposure route or human intake considered, the advisory levels presented here need to be reevaluated.

The short-term acceptable intakes derived from animal studies for noncarcinogenic effects can be used to establish permissible levels of PCBs for short-term exposure. However, this AI value will not be shown for illustration in this paper. The permissible PCB concentrations in soil for each carcinogenic risk level from $10^{-5}$ to $10^{-7}$ are based on the potency factor of 4 $(mg/kg \cdot day)^{-1}$, rounded off from two separate evaluations reported for PCB-1260. As an example, the average daily dose rate corresponding to $10^{-6}$ risk can be calculated from the potency slope as $1.75 \times 10^{-5}$ mg/day.


## CHEMICAL AND PHYSICAL PROPERTIES OF PCBs EVALUATED

Commercial-grade PCBs marketed as Aroclors in the United States are mixtures of many chlorinated biphenyl compounds in various proportions. Each PCB compound may exhibit its own toxicological characteristics and physical and chemical properties. These facts complicate the exposure analysis and the subsequent estimation of the allowable concentrations in

drinking water and soil. It is likely that chemical and physical properties reported in the literature for each Aroclor designation represent an
average property for the mixture. To define the variability of safe
levels for contamination by different Aroclor designations, exposure
analyses have been performed for several Aroclors: Aroclor 1242, 1248,
1254, and 1260. Pertinent chemical and physical properties for these
Aroclors are shown in Table 2.

The widespread distribution of PCBs in the environment suggests that
PCBs are transported from treatment, storage, and disposal facilities
through the atmosphere in the vapor and adsorbed form on particulate matter. Soil contaminated with PCBs will exert vapor pressure for volatilization.

TABLE 2. Chemical and Physical Properties of Aroclors Evaluated

|  | PCB-1242 | PCB-1248 | PCB-1254 | PCB-1260 |
|---|---|---|---|---|
| Molecular weight | 266.5 | 299.5 | 328.4 | 377.5 |
| $K_{ow}$ | 380,000 | 1,300,000 | 1,070,000 | 14,000,000 |
| Specific gravity | 1.38 | 1.445 | 1.538 | 1.62 |
| Solubility in water (mg/L) | 0.24 | $5.4 \times 10^{-2}$ | $1.2 \times 10^{-2}$ $- 0.03$ | $2.7 \times 10^{-3}$ |
| Vapor pressure (mmHg) at 25°C | $4.06 \times 10^{-4}$ | $4.94 \times 10^{-4}$ | $7.71 \times 10^{-5}$ | $4.05 \times 10^{-5}$ |
| Henry's Law constant (atm·m$^3$/g mol) | $5.73 \times 10^{-4}$ | $3.51 \times 10^{-3}$ | $8.37 \times 10^{-3}$ | $7.13 \times 10^{-3}$ |

Bioaccumulation factor 31,200 L/kg

Soil-water partition coefficient 40-1000 L/kg (U.S. EPA, 1980)

Experimental data (U.S. EPA, 1980) suggest that PCBs are strongly adsorbed on earth materials, including soil. PCBs adsorbed on soil, or present in mixture with soil, will be subject to ingestion if the contaminated sites are accessible to children or adults with habitual pica. The reported values for the soil-water partition coefficient, $K_d$, range from 1000 $cm^3$ water/g soil for clay material to 40 $cm^3$ water/g soil for sandy material.

EXAMPLE CALCULATIONS

Unsteady-State Emission Rate from Contaminated Soil with No Cover

In this example, soil contaminated with PCB-1254 is assumed to be exposed to the atmosphere in the absence of a cover. The average emission rate is obtained by averaging the instantaneous emission rate over a time period, t (sec), using Eq. (33). For $D_i$ = 0.05 $cm^2$/s, $\varepsilon$ = 0.35, and $K_d$ = 1000 $cm^3$ water/g soil:

$$D_{ei} = 0.05 \ (0.35)^{1/3} = 0.0352 \ cm^2/s$$

$$\frac{H}{K_d} \text{ in } \frac{g \text{ soil}}{cm^3 \text{ air}} = \frac{8.37 \times 10^{-3}}{1000} \times 41 = 0.000343 \ \frac{mg/cm^3 \text{ air}}{mg/g \text{ soil}}$$

Hence,

$$\alpha = \frac{0.023}{0.35 + 2.65(0.65) \ \frac{1}{0.000343}} = 2.45 \times 10^{-6} \ cm^2/s$$

The average emission rate when the initial concentration in soil is $C_{SO}$ = $10^{-6}$ g/g (= 1 ppm), can be obtained by averaging over 70 years:

- 70 years (= $2.2075 \times 10^9$ sec) average:

$$\bar{N}_A = 0.00034(0.35) \ \frac{2(0.0352)(10^{-6})}{3.14(2.45 \times 10^{-6})(2.2075 \times 10^9)}$$

$$= 6.4 \times 10^{-14} \text{ g/cm}^2 \cdot \text{s}$$

Although the cleanup level is initially unknown, solution of Eq. (2) or (17) for $C_s$ does not require trial and error calculations, because the dilution factor can be calculated independent of the contaminant concentration in soil in the absence of a cover material.

Unsteady-State Emission from Contaminated Soil with Soil Cover Applied

In this example, clean soil is applied on top of the contaminated soil surface. Emissions will occur through the cover material, and will remain unsteady below the vapor saturation point. The average emission rate, under the assumption of an initial contamination depth of L=200 cm, can be obtained using Eq. (38). The average emission over a period of 70 years ($2.2075 \times 10^9$ sec) can be obtained as:

$$\bar{N}_A = 5 \times 10^{-14} \text{ g/cm}^2 \cdot \text{s}$$

When the clean cover material is applied on top of the cleaned or spill site, the solution of Eq. (2) or (17) for $C_s$ requires an initial estimate for $C_s$, and a trial-and-error calculation.

An Example of the Air Dilution Factor Calculation

This example illustrates the method of calculating the air dilution factor needed in the exposure analysis. The emission rates are used as input in applying dispersion modeling to obtain the ambient air concentration, $C_a$, downstream of the site. An average wind speed of 10 mph (=4.47 m/s) is used in this exposure evaluation. The on-site ambient air concentration is calculated using Eq. (41). For the 70-year average emission rate without cover, calculated above, the onsite air concentration for a 0.5-acre site can be obtained as follows:

$$C_a = \frac{6.4 \times 10^{-14} \times 10^6 (0.5)(4.047 \times 10^7)}{45(4.47)(0.5)(2)} = 6.4 \times 10^{-3} \ \mu g/m^3$$

The vapor phase concentration of PCB being emitted from the soil surface is assumed to be in equilibrium with the assumed soil PCB concentration (1 ppm = 1 $\mu g/g$). This vapor phase concentration, before being mixed with air, is:

$$C_{as} = \frac{H}{K_d} \ (41) \ C_s = 0.000343 \ g \ soil/cm^3 \cdot (10^6 cm^3/m^3) \cdot 1 \ \mu g/g$$

$$= 343 \ \mu g/m^3$$

Hence, the dilution factor, D, for the on-site ambient air becomes

$$D = 6.4 \times 10^{-3}/343 = 1.87 \times 10^{-5}$$

The concentration of PCBs in soil corresponding to the saturation point can be estimated from the knowledge of vapor pressure and air-soil partitioning. For example, since the vapor pressure of Aroclor-1254 (7.71 x $10^{-5}$ mmHg) corresponds to the saturation concentration, $C_{as}$, of 1,372.7 $\mu g/m^3$, the PCB concentration in soil at the saturation point is:

$$C_{sm} = \frac{1,362.7 \ \mu g/m^3}{[K_{as}(g \ soil/cm^3 \ air) \times 10^6 \ cm^3/m^3]}$$

$$= 1,362.7/(0.000343 \times 10^6) = 4 \ \mu g/g$$

for a value for $K_d$ of 1,000 $cm^3/g$ used in this example. The saturation concentration is dependent upon the value of the air-soil partition coefficient. The concentration of PCBs in saturated vapor and the corresponding PCB concentrations in soil are calculated for $K_d$=1000 and 40 $cm^3/g$. Table 3 lists the calculated values. Since the PCB concentration in vapor phase cannot exceed the saturation concentration, the values of $C_s$ in Eq. (2) or (17) cannot exceed the maximum value, $C_{sm}$.

Table 3. Concentration of PCBs in Soil at Saturation Vapor Pressure at Two $K_d$ Values

| | PCB soil concentration in equil. with saturated vapor (ppm) | | Saturated vapor concentration ($\mu g/m^3$) |
|---|---|---|---|
| | K =1000 | K =40 | |
| Aroclor-1242 | 250 | 10 | 5823.3 |
| Aroclor-1248 | 55.3 | 2.2 | 7962.8 |
| Aroclor-1254 | 4 | 0.16 | 1362.7 |
| Aroclor-1260 | 28.2 | 1.13 | 822.8 |

RESULTS OF EXAMPLE CALCULATIONS

For exposure pathways by soil ingestion, vapor inhalation, and dermal contact with soil, Eq. (17) is applied to solve for $C_s$ corresponding to the lifetime cancer risk dose at $10^{-6}$ risk:

$$C_s = \frac{1.75 \times 10^{-5} \times 1000}{[0.6(0.3)(0.323)+0.000343(1.87 \times 10^{-5})(20)(0.5)(10^6)+(1)(0.05)]0.5}$$

$$= 0.2 \text{ ppm}$$

257

This value of $C_s$ should be reused to calculate a new value for $C_s$, which is also found to be 0.2 ppm. The results of the calculations are given in Table 4. In these calculations, it is assumed that GI = 0.3, ABA = 0.5, ABS = 0.05, and SM = 0.5. For short-term exposure calculations, the value for SM = 1 can be assumed, meaning that the frequency of presence on-site for a given day is 1. The results of the long-term exposure evaluation for carcinogenic effects are also given in Table 4 for risk values ranging from $10^{-5}$ to $10^{-7}$. It should be remembered that Table 4 is prepared for PCB-1254 at $K_d$ = 1000. The permissible levels are evaluated at two different rates of soil ingestion. Calculations show that at a particulate concentration of 75 $\mu g/m^3$, the PCB intake by inhalation of particulate matter at contaminated sites is relatively unimportant. Hence, the route of exposure by inhalation of particulates in the ambient air is not considered in the example calculation for PCBs.

When all four Aroclors and the upper and lower values of $K_d$ (1000 and 40) are considered, the permissible levels can be summarized as ranges within which a permissible level of PCB cleanup should lie. These ranges reflect the different Henry's Law constant for each Aroclor in addition to the water-soil partition coefficient, and are shown in Table 5. The symbol, VS, indicates that the emission occurs at vapor saturation conditions, and hence no upper bound limit of PCB concentration for inhalation exposure can be determined. It is assumed that at a distance of 0.1 km from the site, the exposure occurs through inhalation only.

Similar calculations were performed under the scenario that the site is covered with clean soil material or its equivalent after cleanup. The normal detection range for PCB analysis is considered to be in the range of 0.05 to 0.1 ppm, below which the soil will be considered clean. The results of these calculations are shown in Table 6 for PCB-1254 at $K_d$=1000. The effect of cover material would be to retard the emission rate and to reduce the effective concentration of PCBs in the soil.

Table 4. PCB-1254 Levels in Soil Corresponding to AIs with no Cover Material (PPM)

| | Lifetime intake level | | |
| | 0.00175 | 0.0175 | 0.175 |
| | ($10^{-7}$ risk) | ($10^{-6}$ risk) | ($10^{-5}$ risk) |
|---|---|---|---|
| Soil ingestion[a]<br>Dermal contact<br>Inhalation | 0.01 | 0.1 | 1 |
| Soil ingestion[b]<br>Dermal contact<br>Inhalation | 0.02 | 0.2 | 2 |
| On-site<br>inhalation only | 0.05 | 0.5 | 7 |
| 0.1 km from site | 7 | VS | VS |

[a]Children ages 1-6, with pica, consuming 3 g soil/day.
[b]Children ages 1-6, without pica, consuming 0.6 g soil/day.
[c]Soil-air participation coefficient=$3.43 \times 10^{-4}$ g/cm$^3$ at $K_d$=1000.


Table 5. Levels of Evaluated PCBs in Soil Corresponding to AIs with no Cover (PPM)

| | Lifetime intake level | | |
| | 0.00175 | 0.0175 | 0.175 |
| | ($10^{-7}$ risk) | ($10^{-6}$ risk) | ($10^{-5}$ risk) |
|---|---|---|---|
| Soil ingestion[a]<br>Dermal contact<br>Inhalation | 0.008-0.01 | 0.08-0.1 | 0.8-2 |
| Soil ingestion[b]<br>Dermal contact<br>Inhalation | 0.01-0.06 | 0.1-0.6 | 1-6 |
| On-site<br>inhalation only | 0.01-0.2 | 0.1-2.0 | 1-20 |
| 0.1 km from site | 2-220 | 90-VS | VS |

[a]Children ages 1-6, with pica, consuming 3 g soil/day.
[b]Children ages 1-6, without pica, consuming 0.6 g soil/day.

Table 6. PCB-1254 Levels in Soil Corresponding to AIs with 10-Inch Clean Soil Cover

| | Lifetime intake level | | |
| --- | --- | --- | --- |
| | 0.00175 ($10^{-7}$ risk) | 0.0175 ($10^{-6}$ risk) | 0.175 ($10^{-5}$ risk) |
| Soil ingestion[a] Dermal contact Inhalation | 0.02 | 0.2 | 2 |
| Soil ingestion[b] Dermal contact Inhalation | 0.03 | 0.3 | 4 |
| On-site inhalation only | 0.08 | 0.8 | 14 |
| 0.1 km from site | 14 | VS | VS |

[a]Children ages 1-6, with pica, consuming 3 g soil/day.
[b]Children ages 1-6, without pica, consuming 0.6 g soil/day.
[c]Soil-air partition coefficient=$3.43 \times 10^{-4}$ g/cm$^3$ at $K_d$=1000.


DISCUSSION

Except for the calculations pertaining to estimation of emission rate and dispersion modeling, the method presented is straightforward in estimating multimedia exposures for use in determining cleanup levels of contaminants in soil.

There are uncertainties associated with input parameters in estimating exposures. Although the ranges of values for permissible cleanup levels have been obtained for PCB examples without defining uncertainty limits, site-specific evaluations can narrow the confidence intervals.

Recent experimental investigations show that the capillary effect may play a role in controlling the volatile emission rate when cover material is applied (DuPont, 1985). The data indicate that, for certain land-spread chemicals, the emission rate increases rather than decreases as time elapses, as predicted by solving for Fick's law of diffusion. The liquid waste initially spread on land and covered by soil will exert capillary tension in the dry zone of cover material, and will slowly rise toward the soil surface. This is suspected to be the cause of the increasing emission rate at a longer elapsed time than at the beginning of the initial cover-up. The increased emission will have an effect on lowering permissible levels of contaminants in soil when inhalation exposures are significant. The degree of liquid saturation to produce this "wicking" effect, if any, is unknown.

In case study examples for PCBs, it is assumed that the 10-inch clean cover material used remains undisturbed by human activities on the site. At times this assumption may be arbitrary, because an opportunity could exist that would place the contaminated soil surface in contact with the atmosphere because of inadvertent disturbances of soil surfaces, construction activities, utility installation, precipitation, or the activities of children playing on the site, to name a few. Such contingencies will require additional thickness of cover material, or, alternatively, the site should be made inaccessible to children or should be kept free of any activities that would lead to disturbances of the soil surfaces. Spills on top of the clean cover will result in a situation equivalent to the surface contamination case, requiring a more stringent concentration limit in soil. In such cases, the results given for the 10-inch thick cover material in the example calculations would not apply.


NOMENCLATURE

A      Cross-sectional area of interest in soil column ($cm^2$)

ABA    Absorption factor for inhaled contaminant

ABP    Absorption factor for a contaminant on particulate matter being inhaled

ABS    Absorption factor for dermally contacted contaminant

AI     Acceptable intake (mg/day)

BCF    Bioconcentration factor (L/kg)

BW     Body weight (kg)

| | |
|---|---|
| C | Concentration of a contaminant in the vapor phase in soil pores ($\mu g/m^3$) |
| $C_a$ | Ambient air concentration at exposure location ($\mu g/m^3$) |
| $C_{as}$ | Contaminant concentration in air at the soil surface where emission occurs ($\mu g/m^3$) |
| $C_F$ | Concentration of contaminant in fish (mg/kg) |
| $C_L$ | Concentration of contaminant in leachate (mg/L) |
| $C_m$ | Concentration of pollutant in meat (mg/kg) |
| $C_o$ | Initial concentration of contaminant in soil (mg/kg) |
| $C_p$ | Concentration of contaminant in particulates being inhaled ($\mu g/m^3$) |
| CR | Dermal contact rate of soil (g/day) |
| $C_s$ | Concentration of contaminant in soil (mg/kg [=$\mu g/g$]) |
| $C_{sm}$ | Contaminant concentration in sol corresponding to vapor saturation (mg/kg [=ppm]) |
| $C_w$ | Concentration of contaminant in water (mg/L) |
| $D_{ei}$ | Effective diffusivity ($cm^2/s$) (=$D_i \varepsilon^{1/3}$) |
| $D_i$ | Molecular diffusivity ($cm^2/s$) |
| D | Air dilution factor (=$C_a/C_{as}$) |
| F | Correction factor for changing body weight when AI is based on 70 kg BW and soil ingestion lasts only during childhood as compared to lifetime; for soil ingestion during ages 1 through 5, F = 0.323 |
| $f_g$ | Functional relationship defining fate and transport of contaminants in leachate to groundwater |
| GI | Absorption rate in gastrointestinal tract |
| H | Henry's Law constant (atm·$m^3$/mol) |
| IH | Daily inhalation of air ($m^3$/day) |
| IF | Daily fish consumption (kg/day) |
| IM | Daily meat consumption (kg/day) |
| IV | Daily vegetable consumption (kg/day) |
| IW | Daily water consumption (L/day) |
| IR | Daily ingestion rate of soil (g/day) |

k        Biodegradation constant (L/day)

$K_{as}$     Soil-air partition coefficient (g soil/$cm^3$ air [$=41H/K_d$])

$K_d$      Soil-water partition coefficient ($cm^3$ water/g soil)

$K_{LS}$     $= C_s/C_L$

$K_{sv}$     $= C_v/C_s$

$K_{vm}$     $= C_m/C_v$

LS       Width dimension of contaminated area perpendicular to wind
         direction

LT       Exposure time over a lifetime (70 years) (days)

MH       Mixing height (m)

$N_A$      Emission rate ($\mu g/cm^2 \cdot s$)

$\overline{N}_A$      Average emission rate ($\mu g/cm^2 \cdot s$)

POT      Potency slope factor $(mg/kg \cdot day)^{-1}$

Q        Total emission rate ($\mu g/s$)

SL       Safe levels for noncancer effects (mg/kg·day)

SM       Exposure frequency over a lifetime

t        Time (s)

z        Depth measured from the soil-air interface (cm)

V        Average wind speed within mixing zone = (0.5) wind speed
         at the mixing height (m/s)


Greek Letters

$\varepsilon$        Porosity

$\overline{\rho}_s$        Bulk density of soil ($g/cm^3$)

$\rho_s$        True density of soil ($g/cm^3$)


REFERENCES

Binder, S., 1985, "Estimating the Amount of Soil Ingested by Young Children
    Through Trace Elements," Report by the Centers for Disease Control.
Bruce, D.B., 1969, "Workshop of Atmospheric Dispersion Estimates," U.S.
    Department of Health, Education and Welfare, Pub. No.: 999-AP-26.
Crank, J., 1985, "The Mathematics of Diffusion," Oxford University Press,
    New York.

Dacre, J.C., Rosenblatt, D.H., and Cogley, D.R., 1980, Preliminary Pollutant limit values for human health effects, Environ. Sci. Technol.
14:778.

DuPont, R.R., (November, 1985), Evaluation of air emissions release rate
model predictions of hazardous organics from land treatment facilities, Presented at the 1985 Annual AIChE Meeting, Chicago, IL.

Farmer, W.J., Yang, M.-S., Letey, J., and Spencer, W.F., 1980, "Land Disposal of Hexachlorobenzene Wastes," EPA-600/2-80-119, U.S. Environmental Protection Agency, Cincinnati, OH.

Hwang, S.T., 1982, Toxic emissions from land disposal facilities, Environ.
Progress 1:46.

Hwang, S.T., 1985, Assessing exposure to groundwater contaminants migra-
ted from hazardous waste facilities, in: "Proceedings, Conference on
Management of Uncontrolled Hazardous Waste Sites," Cincinnati, OH.

Kimbrough, R.D., Falk, H., and Stehr, P., 1984, Health implications of
2,3,7,8-tetrachloro-dibenzodioxin (TCDD) contamination of residual
soil, J. Toxicol. Env. Health 14:47.

Lepow, M.L., 1975, Investigations in sources of lead in the environment of
urban children, Env. Res. 10:45.

Nash, R.G., Beall, M.L., Jr., and Woolson, E.A., 1970, Plant uptake of
chlorinated insecticides from soils, Agron. J. 62:369-372.

Roels, H.A., et al., 1980, Exposure to lead by the oral and the pulmonary
routes of children living in the vicinity of a primary lead smelter,
Env. Res. 22:84.

Snyder, W.S., 1975, "Report of the Task Group on Reference Man. International Commission of Radiological Protection No. 23," Pergamon Press,
New York, NY.

Thibodeaux, L.J., and Hwang, S.T., 1982, Landfarming of petroleum wastes--
modeling the air emission problem, Environ. Progress 1(1):42-46.

U.S. Environmental Protection Agency, 1980, "Attenuation of Water-Soluble
Polychlorinated Biphenyls by Earth Materials," EPA-600/2-80-027,
Washington, D.C.

U.S. Environmental Protection Agency, 1984, "Risk Analysis of TCDD-Contami-
nated Soil." EPA-600/8-84-031, Washington, D.C.

Veith, G.D., 1980, An evaluation of using partition coefficients and water
solubility to estimate bioconcentration factors for organic chemicals
in fish, J. Fish. Res. Board Can.

Wark, K., and Warner, C.F., 1981, "Air Pollution: Its Origin and Control,"
Harper and Row, New York, NY.

Wedeen, R.P., Malik, D.K., Baiuman, V., and Bogden, J.D., 1978, Geophagic
lead nephropathy: case report, Environ. Res. 17:409.

# A MULTIMEDIA STUDY OF HAZARDOUS WASTE LANDFILL GAS MIGRATION

Robert D. Stephens[*], Nancy B. Ball[**], and Danny M. Mar[**]
[*]Hazardous Materials Laboratory, California Department of Health Services
[**]California Public Health Foundation
2151 Berkeley Way, Berkeley, California 94704

## INTRODUCTION

Hazardous waste landfills pose uniquely challenging environmental problems which arise as a result of the chemical complexity of waste sites, their involvement of many environmental media, and their very size and volume. Chemical substances of every description have or could be found at a typical hazardous waste landfill. These substances would represent the full range of physical and chemical properties, from inert to reactive, from nonvolatile to highly volatile, and from ionic, water miscible to nonpolar, hydrophobic compounds. Given this range of physico-chemical properties, and that landfills are earth containment facilities, there is the probability that all environmental compartments, air, water (surface and ground), soil, and biota will be involved. The additional factor which complicates the understanding of landfills is their size and volume. Although for certain purposes landfills can be considered quasi-point sources, in reality they are generally large and often situated in several different geological and hydrogeological regimes. The actual volume or mass of material existing at a landfill makes clear definition of its chemistry impractical.

Codisposal facilities which mixed municipal solid waste with industrial waste show unique characteristics. The anaerobic degradation of the solid waste resulting in the generation of large volumes of methane and carbon dioxide significantly affects the mechanisms and pathways of the migration of contaminants from the landfill. This is particularly true for the volatile components of the industrial waste, but there is some evidence that the migrating gas affects the less volatile compounds as well. Studies have been made at a landfill of this type existing in southern California. The approach to date has been primarily directed toward the generation of chemical composition data of landfill gases, of the ambient air in the vicinity of the landfill, and of the air inside homes adjacent to the landfill. In addition, data on the composition of ground waters in the vicinity of the landfill have been obtained.

The landfill under study, which is located in southern California, has been in operation since 1963 as a municipal disposal site. In 1972, the landfill began accepting hazardous waste as well, although as the legal definition of hazardous waste was adopted in California at about that time, there was undoubtedly disposal of waste prior to 1972 which would have been classified as hazardous under the 1972 definition. It is estimated that $1.3 \times 10^7$ tons of refuse were disposed of during the

265

twenty year period from 1963 to 1983. Records indicate that during the period from 1977 to 1983 approximately 5 x 10$^8$ gallons of liquid hazardous waste were disposed. The assumption was made that leachate would be prevented by the absorption of the liquids by the trash. The problems resulting from landfill gas production and the consequent gas driven migration were not considered.

AIR QUALITY STUDIES

Earlier Studies

Two studies on the composition of the landfill gases and emission of non-methane hydrocarbons were done from 1979 to 1980 (Eutek, Inc., 1981; University of Southern California, 1981). The work by Eutek, principally an odor study, attempted through gas analysis and simple emission measurements to characterize the odors. Although the study did not succeed in determining the chemical nature of the odor problem, the report did highlight the presence of numerous volatile organic compounds in the landfill gas and in the ambient air directly over the landfill. The USC study had a similar focus on odor characterization through ambient air measurements. The results of this study were similar to those presented in the Eutek report, with the principal exception that chloroethene (vinyl chloride) was not found in the landfill gas, whereas the Eutek study found it to be a major component. This disagreement between the two studies concerning the presence of vinyl chloride probably reflects measurement difficulties evident in the USC work.

Following the discovery of vinyl chloride in the landfill gas and the establishment of a state ambient air standard for vinyl chloride in 1978 at 10 ppb, ambient air monitoring around the landfill site was begun, around 1980-81. It was subsequently found that the ambient standard was being regularly exceeded (up to ca. 70 ppb). As a result of these findings, a community air monitoring program sponsored by several governmental agencies was initiated. The purposes of the study were to:

* Evaluate the effectiveness of gas control systems designed to control the emissions of volatile organic compounds (VOC) from the landfill,

* Determine the types, concentrations, and distributions of VOCs in the neighboring community, and

* Provide a data base for an exposure assessment which would support a risk assessment.

This study was conducted through the establishment of a network of 6 monitoring stations around the landfill, and one station remote from the landfill as background. The nine compounds analysed for in the Tedlar bag whole air samples are listed in Table 1, along with their possible origin in the landfill and the range of concentrations reported. Table 2 shows the working detection limits achieved in the program. The results of this study indicated that six of the compounds measured in ambient air samples were elevated in the vicinity of the landfill relative to the background. The highest concentration was in a sector to the south of the landfill. Results from the six stations are given in Table 3. Table 4 gives results for the same seven compounds, each normalized to its concentration at the background station. With the exception of benzene, one can readily see the effect of the landfill on community ambient air. Results for benzene are not so readily interpreted, due to the relatively high levels of benzene in the ambient air of the Los Angeles basin. In addition, there is

TABLE 1. Compounds Selected for Monitoring.

| Compound (Synonym) | Uses and Probable Waste Sources | Concentration Range in Landfill Gas (ppm) |
| --- | --- | --- |
| Chloroethene (Vinyl Chloride) | Manufacture of PVC and other copolymers; organic systhesis; adhesives for plastics; refrigerant; manufacture of vinyl chloride also a significant waste source. | 83 – 12800 |
| 1,1-Dichloroethene (Vinylidene Chloride) | Copolymerized with vinyl chloride to manufacture types of saran; adhesives for synthetic fibers. | ND – 1200 |
| Trans-1,2-Dichloroethene (Acetylene Dichloride) | General solvent for organic materials, dye extraction, perfumes, lacquers, thermoplastics, organic synthesis. | ND – 800 |
| Trichloroethene (TCE) | Metal degreasing, extraction solvent for oils, fats, waxes; solvent dyeing; dry cleaning; refrigerant and heat exchange fluid; organic synthesis; fumigant; cleaning and drying electronic parts. | ND – 1000 |
| Tetrachloroethene (PCE) | Dry cleaning solvent; vapor-degreasing solvent; drying agent for metals; heat exchange medium; manufacture of fluorocarbons. | ND – 1500 |

TABLE 1 (continued). Compounds Selected for Monitoring.

| Compound (Synonym) | Uses and Probable Waste Sources | Concentration Range in Landfill Gas (ppm) |
|---|---|---|
| 1,2-Dichloroethane (Ethylene Dichloride) | Manufacture of vinyl chloride; organic synthesis; anti-knock agent in gasoline; paint, varnish, and finish removers; metal degreasing; soaps and scouring compounds; wetting and penetrating agent; ore flotation. | ND - 5000 |
| Trichloromethane (Chloroform) | Manufacture of fluorocarbon refrigerants and propellants; dyes and drugs; general solvent; fumigant; insecticides. | 5.1 0.038 (Ambient Air) |
| Benzene | Manufacture of styrene, phenol, synthetic detergents, cyclohexane for nylon, aniline, DDT, various other insecticides, fumigants; solvens, paint remover; rubber cement; anti-knock agent in gasoline. | 10 - 2000 |
| Chlorobenzene | Manufacture of chloronitrobenzene, DDT, aniline; intermediate and solvent for organic synthesis. | ND - 500 |

TABLE 2.   Detection Limits for Compounds.

| Compound | Detection Limit Range[a] (ppb) |
|---|---|
| Chloroethene (Vinyl Chloride) | 2-3 |
| Tetrachloroethene (PCE) | 0.1-0.2 |
| Trichloroethene (TCE) | 0.1-0.2 |
| 1,1-Dichloroethene (Vinylidene Chloride) | 0.1-0.3 |
| 1,2-Dichloroethane (Ethylene Dichloride) | 0.2-0.4 |
| Trichloromethane (Chloroform) | 0.02-0.1 |
| Trans-1,2-Dichloroethene | 1-3 |
| Chlorobenzene | 10[b] |
| Benzene | 2-20 |

[a]Detection limits from two laboratory data sets.

[b]Sampling by charcoal tube and analysis by one laboratory.

TABLE 3.  Summary of Ambient Air Data[a] (ppb)

| Compound | Station | | | | | |
|---|---|---|---|---|---|---|
| | A | B | C | D | E | F |
| Chloroethene (Vinyl Chloride) | 7.1–7.3 | 4.5–5.5 | 2–3 | 3.8–4.1 | 2.0–3.0 | 2–3 |
| Tetrachloroethene (PCE) | 2.1–3.7 | 1.8–2.9 | 1.4–2.4 | 2.3–3.0 | 1.5–2.1 | 1.5–3.4 |
| Trichloroethene (TCE) | 0.8–1.0 | 0.8–1.8 | 0.2–0.3 | 1.6–1.7 | 0.6–0.8 | 0.8–1.2 |
| 1,1-Dichloroethene (Vinylidene Chloride) | 1.1–1.3 | 0.7–1.0 | 0.1–0.3 | 0.7–0.8 | 0.1–0.4 | 0.3–0.4 |
| 1,2-Dichloroethane (Ethylene Dichloride) | 1.3–3.0 | 0.8–2.8 | 0.4–0.7 | 0.8–2.8 | 0.7–0.9 | 0.4–1.1 |
| Trichloromethane (Chloroform) | 0.3–0.5 | 0.3–0.6 | 0.2–0.6 | 0.6–1.0 | 0.2 | 0.2 |
| Benzene[b] | 4.8 | 4.6 | 3.6 | 4.6 | 3.0 | 3.0 |

[a]Each pair of values represents the means of the two laboratory data sets calculated for the entire study period. On those days when a compound was not detected, its concentration was assumed to be equal to the limit of detection.

[b]One set of data used. All of the second set were below the limit of detection.

TABLE 4. Ratio of Ambient Air Concentrations of Stations A-F to Control Station, Station C[a].

| | | | Station | | | |
|---|---|---|---|---|---|---|
| | A | B | C | D | E | F |
| Chloroethene (Vinyl Chloride) | 2.4-3.6 | 1.8-2.3 | 1.0 | 1.4-1.9 | 1.0 | 1.0 |
| Tetrachloroethene (PCE) | 1.5 | 1.2 | 1.0 | 1.2-1.6 | 0.9-1.1 | 1.4-2.1 |
| Trichloroethene (TCE) | 2.9-3.9 | 3.0-7.2 | 1.0 | 6.3-6.6 | 2.1-3.0 | 2.8-5.0 |
| 1,1-Dichloroethene (Vinylidene Chloride) | 3.6-11 | 2.4-9.3 | 1.0 | 2.7-6.5 | 1.0-1.3 | 1.5-3.0 |
| 1,2-Dichloroethane (Ethylene Dichloride) | 3.3-4.5 | 1.9-4.3 | 1.0 | 1.9-4.3 | 1.3-1.7 | 1.0-1.7 |
| Trichloromethane (Chloroform) | 0.4-2.7 | 1.0-1.4 | 1.0 | 1.7-3.1 | 0.4-1.1 | 0.4-1.0 |
| Benzene[b] | 1.4 | 1.3 | 1.0 | 1.3 | 0.8 | 0.8 |

[a]Ratios are calculated for each laboratory separately and expressed as a range.
[b]One laboratory data set available.

a significant similarity between the concentration profiles relative to vinyl chloride, of the community ambient air and the landfill gas. This comparison can be seen in Figure 1.

In 1984, as a result of a routine survey of ambient air in the neighborhoods surrounding the landfill, methane was found in concentrations approaching the lower explosive limit (5000 ppm) in the same southern sector which in 1982 was identified as having elevated VOC levels. Nineteen homes were evacuated as a result of elevated levels of flammable gases inside the homes. Following evacuation, a preliminary survey showed methane concentrations to 3000 ppm and vinyl chloride to 400 ppb in the homes.

As a result of these preliminary findings, a more comprehensive ambient air, landfill gas, and indoor air monitoring program was initiated. The main elements of this study were as follows:

* Whole air Tedlar bag samples taken inside 19 homes designated as priority 1; samples to be speciated for the 9 VOC's listed in Table 1, in addition to methane.

* Charcoal tube samples taken inside approximately 50 homes designated as priority 2 which were analyzed for vinyl chloride, and methane.

* Whole air ambient air samples taken in the vicinity of priority 1 and 2 homes which were analyzed for the 9 VOCs.

* Shallow subsurface (10-20cm) probes monitored for total combustible gases (portable FID).

* A variety of meteorological measurements.

Although the details of this study are to be reported elsewhere, the data as shown in Table 5 clearly indicate that vinyl chloride concentrations, at least, are elevated in the homes near the landfill relative to the ambient concentrations in the Los Angeles basin. Not shown on Table 5 are the data for background homes, which had no detectable vinyl chloride. The influence of the landfill gases on the indoor concentrations of the other VOCs is not so clear due to insufficient data on these compounds and the variety of other potential sources. Additional studies are planned on these compounds.

## Current Studies

In early 1985 a study of broader perspective was initiated following the discovery of ground water contamination in several locations by compounds similar to those reported in the landfill gas. The objectives of this study were generally to identify and evaluate the several possible pathways of migration of chemicals from the landfill and to understand the controlling mechanisms. An understanding of the mechanisms should allow identification of the key factors in contaminant migration such as hydrogeology, physical chemical equilibria, or meteorology. Description of these mechanisms would in addition provide a proper foundation for site remediation and an exposure/risk assessment.

## GROUND WATER QUALITY STUDIES

Monitoring of the ground water at the site has been performed systematically since autumn 1983. Presently, data are being collected from

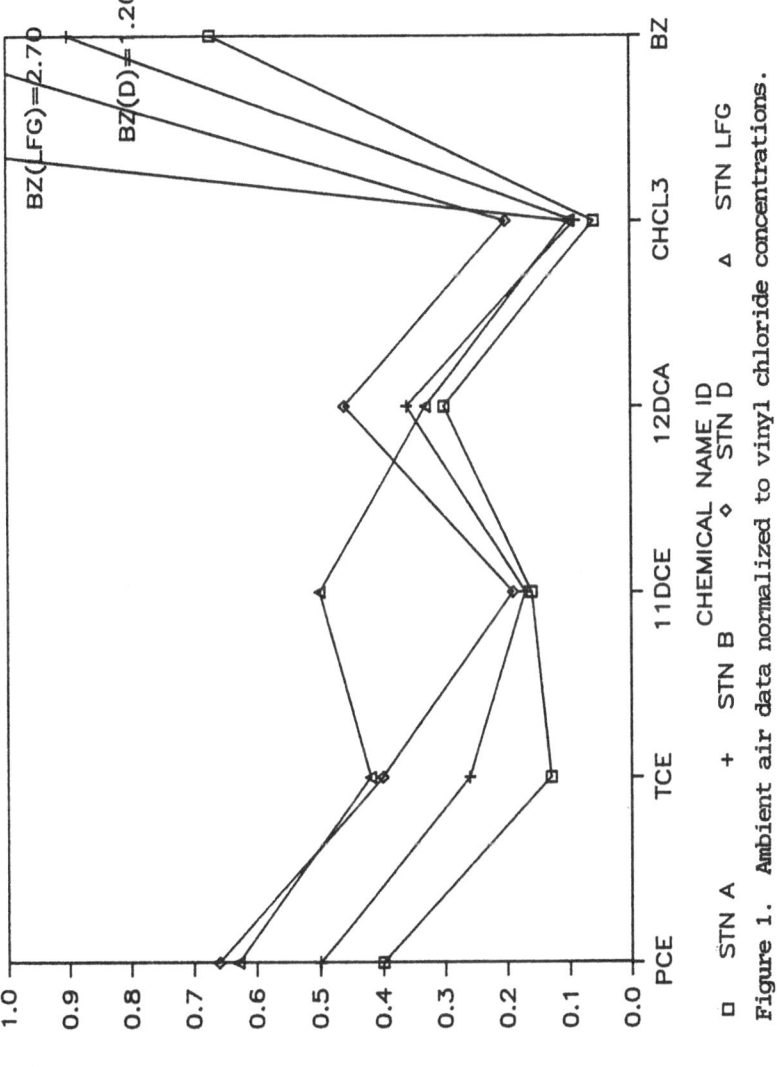

Figure 1. Ambient air data normalized to vinyl chloride concentrations.

273

TABLE 5. Air Monitoring Results.

| Analyte | Priority 1 Mean (n=108) | Priority 1 Max. | Priority 2 Mean (n=420) | Priority 2 Max. | L.A. Basin Mean (n=350) | L.A. Basin Max. |
|---|---|---|---|---|---|---|
| Vinyl Chloride | 4 | 7 | 4 | 9.3 | <0.5 | <0.5 |
| 1,1-DCE | 1 | 2 | - | - | <0.1 | <0.1 |
| $CHCl_3$ | <1 | <1 | - | - | 0.07 | 3 |
| 1,1,1-TCA | 4.7 | 14 | - | - | 1.2 | 9.7 |
| 1,2-DCA | 5.6 | 14 | - | - | 0.14 | 0.4 |
| $CCl_4$ | <1 | <1 | - | - | - | - |
| $C_6H_6$ | <1 | <1 | - | - | 5.6 | 16 |
| TCE | 2.8 | 6 | - | - | 0.4 | 2 |
| $C_2Cl_4$ | 3.4 | 6 | - | - | 1.1 | 6.3 |
| $CH_4$ | 14800 | 64900 | - | - | - | - |

Units are ppb.

wells distributed in four areas around the site, which have been designated Areas I through IV. Major organic constituents of the landfill gases have been detected in at least one well in each of the areas. Possible mechanisms for ground water contamination include landfill leachate entering ground water directly and landfill gases, driven by methane, contacting and dissolving in ground water. Consequently, the distribution of ground water contamination across the site may be governed by ground water flow, by transport of landfill gases in the vadose zone with subsequent ground water contact, or a combination of the two. The relative importance of these mechanisms in the four areas of the site has been examined, based on currently available data.

From Table 6, which lists the physical properties of the major organic contaminants, it can be seen that the vapor pressure of vinyl chloride is much greater than that of the other organic contaminants. The Henry's Law constant, which describes the partitioning of the compound between water and air, can be estimated from the ratio of the vapor pressure of the pure compound to its maximum solubility in water. This calculated Henry's Law constant is much greater for vinyl chloride than for the other compounds, primarily as a result of its high vapor pressure. One would expect that as the landfill gas contacts the ground water, vinyl chloride would become dissolved in the ground water to a lesser extent than would the less volatile organics such as tetrachloroethene, trichloro-ethene, and benzene and that vinyl chloride dissolved in the ground water would tend to volatilize out of the ground water to a greater extent. The relative proportion of vinyl chloride to the other trace organic contam-inants was used as a first estimation of the influence of landfill gases on ground water contamination.

TABLE 6. Physical Properties of Selected Volatile Organic Compounds

| Compound | Formula | Molecular Weight | Density (g/mL) | Boiling Point (°C atm) | Solubility (mg/L) | Vapor Pressure (mm Hg) | Henry's Law Constant (atm-m$^3$/mol) |
|---|---|---|---|---|---|---|---|
| Vinyl Chloride | $CH_2=CHCl$ | 62.5 | 0.91 | -13.9 | 1100 | 2660 | 6.38 |
| Trichloro-ethene | $ClCH=CCl_2$ | 131.4 | 1.46 | 86.7 | 1100 | 60 | $9.89 \times 10^{-3}$ |
| Tetrachloro-ethene | $Cl_2C=CCl_2$ | 165.8 | 1.63 | 121.4 | 150 | 14 | $1.98 \times 10^{-2}$ |
| Benzene | $C_6H_6$ | 78.1 | 0.88 | 80.1 | 1780 | 76 | $4.32 \times 10^{-3}$ |

All data except Henry's Law constants are from Verschueren (1977). Henry's Law constants are from Kavanaugh and Trussell (1980). Additional solubility values are given in Table 12; additional Henry's Law constants for vinyl chloride are given in Table 11.

TABLE 7.  Selected Ground Water Data. (Data from 10/85)

|  | VC | BZ | TCE | PCE | TOX |
|---|---|---|---|---|---|
| **Area I** | | | | | |
| (MW-29) | 227 | 370 | 1220 | 593 | 8640 |
| (P-4) | 2000 | 140 | 660 | 290 | 1910 |
| **Area II** | | | | | |
| (MW-16P2) | 2400 | 5 | 3 | 0.5 | 290 |
| **Area III** | | | | | |
| (MW-30) | 35 | 46 | 70 | 56 | 3900 |
| **Area IV** | | | | | |
| (MW24A) | 2600 | 1200 | 430 | 250 | 81300 |

Units are in ug/L

In Table 7, data from October 1985 are shown for vinyl chloride (VC) and three other solvents, benzene (BZ), tetrachloroethene (PCE), and trichloroethene (TCE), in the most contaminated well(s) in each of the areas.  Differences in the relative  proportion of vinyl chloride in each of the areas can be seen.  In Area II, the ground water contamination is clearly dominated by vinyl chloride; vinyl chloride concentrations being almost three orders of magnitude higher than benzene, trichloroethene, or tetrachloroethene.  In Area III, the concentrations of the four solvents are fairly similar, with vinyl chloride being slightly less than the other three.  Ground water samples in Areas I and IV have intermediate compositions.  In Area IV, while the vinyl chloride concentration is higher than the others, it is within the same order of magnitude.  In Area I, data from the two most contaminated wells, MW-29 and P-4, suggest different trends.  The contamination in the sample from P-4 is dominated by vinyl chloride, similarly to Area IV, while the vinyl chloride in the MW-29 sample is slightly less concentrated than the other three, similarly to Area III.

While the relative compositions shown in Table 7 appear to be characteristic of the ground water in these areas, variability in the data caused by analytical uncertainty and seasonal and spatial variations obscure the conclusions somewhat.  Analytical uncertainty was examined through data from split samples which were analyzed by four laboratories. The results of these analyses are shown in Table 8.  While large analytical uncertainties occur in all analyses, vinyl chloride shows the greatest variability among laboratories.  Because of the high volatility of vinyl chloride, this component is more likely to volatilize out of the samples during collection, transport, and sample preparation.  In addition, reliable analytical standard solutions are difficult to prepare and maintain.

The variation over six months in measured concentrations of vinyl chloride, benzene, tetrachloroethene, and trichloroethene in ground water samples from MW-29 is shown in Figure 2.  While the concentrations of each constituent vary greatly over time, the relative composition of the ground

TABLE 8.  Split Sample Results for P-4.

|  | LAB 1 | LAB 2 | LAB 3 | LAB 4 |
|---|---|---|---|---|
| Vinyl Chloride | 3255 | 785 | 880 | 2000 |
| Tetrachloroethene | 980 | 315 | 340 | 290 |
| Trichloroethene | 1265 | 595 | 650 | 660 |
| Trichloroethane | 30 | N/A | 18 | 11 |
| t-Dichloroethene | 1115 | 810 | 620 | 1100 |
| 1,2-Dichloroethane | N/A | 205 | 210 | 340 |
| Chloroform | 425 | 265 | N/A | 370 |
| Methylene Chloride | 285 | 300 | 380 | 220 |
| Benzene | N/A | 101 | 115 | 140 |

Units are ug/L
N/A is Not Analyzed

water remains fairly constant. Figure 3 shows these same analyses, normalized to vinyl chloride.  In general, the ratios remain fairly constant indicating a similar composition with the analytical uncertainty.  It is not clear whether the low concentrations measured in June and July reflect low concentrations in the ground water or increased volatilization from samples collected in warm weather.

Samples of landfill gas, collected from gas probes along the southeast perimeter of the site in Area I, are described in Table 9.  The gas composition is dominated by vinyl chloride, with the concentration of vinyl chloride being one to two orders of magnitude greater than those of the other three gases.

At first examination, it would appear that the same wastes are resulting in both ground water and gas contamination, however, the explanation becomes more complex when the Henry's Law constants are considered. Using the Henry's Law constants listed in Table 5, one can calculate the concentrations of solvents in equilibrium with the landfill gas sample from probe 215.65, closest to MW-29, and compare them to measured concentrations in the corresponding ground water sample from MW-29 (Table 10). If the landfill gas and the ground water are in communication with each other, a calculation of the partitioning using the Henry's Law constants can be used to estimate whether the two phases are in equilibrium and, if not, how close to equilibrium they are.  The calculation itself does not indicate whether the two are in communication; that must be ascertained independently through hydrogeologic investigations.

For tetrachloroethene, trichloroethene, and benzene, the measured ground water concentrations are in MW-29 (Table 10) are within an order of magnitude of the calculated equilibrium concentrations, within the uncertainty of the Henry's Law calculations, however, the vinyl chloride concentration measured in the ground water is over three orders of magnitude greater than the calculated concentration contributed by the landfill gas. The calculation instead estimates that the ground water would be in equilibrium with landfill gas containing almost 38,000 ppm(v/v) vinyl chloride, far greater than has ever been detected at the site (see Table 1).  Although the range of calculated Henry's Law constants for vinyl chloride in the literature is relatively small (Table 11), because of the wide range of solubilities and vapor pressures from

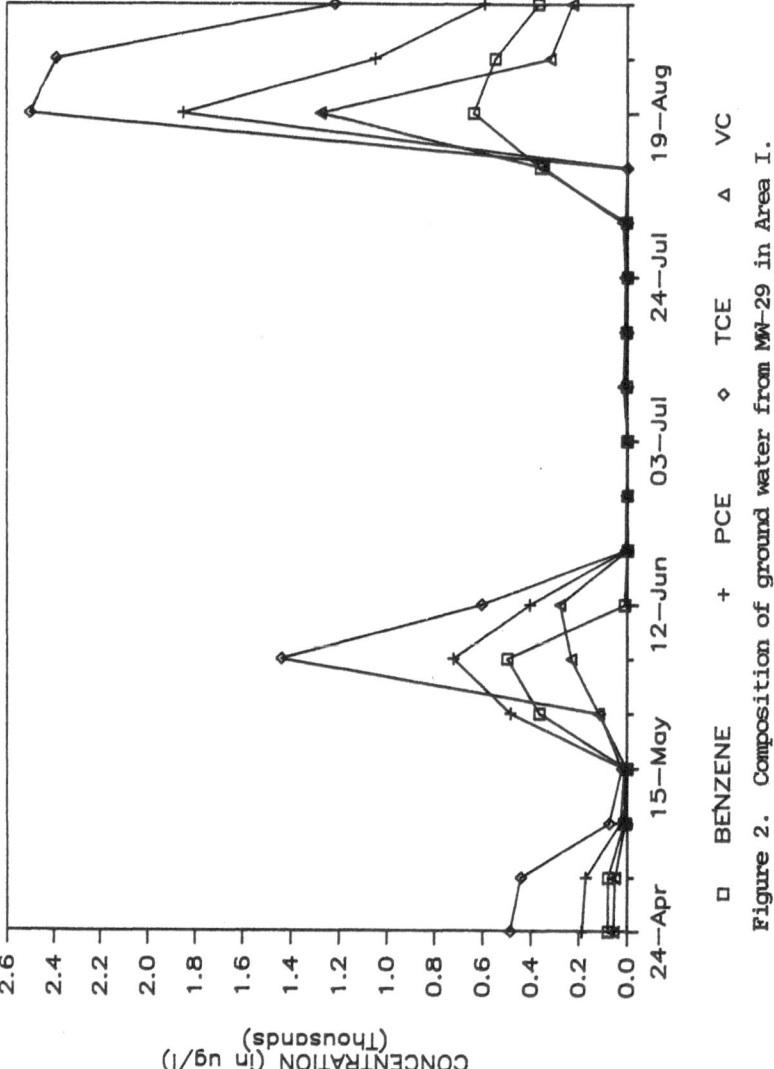

Figure 2. Composition of ground water from MW-29 in Area I.

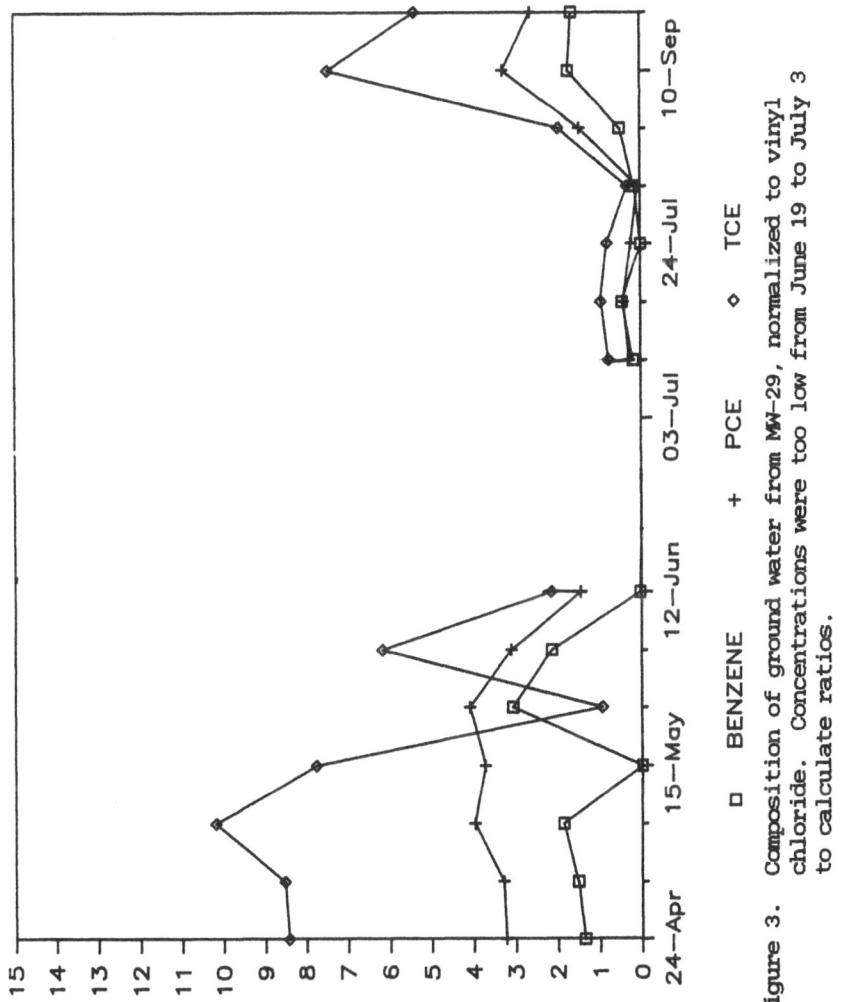

**Figure 3.** Composition of ground water from MW-29, normalized to vinyl chloride. Concentrations were too low from June 19 to July 3 to calculate ratios.

□ BENZENE   + PCE   ◇ TCE

TABLE 9.  Landfill Gas Composition. (Data from 12/84)

Area I Perimeter Probes

| Probe | VC | BZ | TCE | PCE | 1,2DCA | Methane |
|-------|-----|-----|------|------|--------|---------|
| 112.85 | 200 | 33 | 23 | 39 | 100 | 33% |
| 215.65 | 92 | 12 | 6 | 8 | 42 | 0.1% |
| 228.5 | 250 | <1 | 3.5 | 15 | 25 | 19% |
| 302.5 | 40 | 2 | 3 | 5 | 2 | 10% |
| OS-16 | 190 | <1 | 3 | 8 | 1 | 16% |
| OS-17 | 3 | <1 | <1 | <1 | <1 | 0.3% |

Entire Site Composite Samples

| | | | | | | |
|-------|-----|-----|------|------|--------|---------|
| B-2 Outlet | 300 | 140 | 32 | 140 | 210 | 27% |
| B-3 Outlet | 160 | 41 | 17 | 38 | 280 | 17% |

All units are in ppm (v/v) unless otherwise noted.

TABLE 10.  Gas/Ground Water Equilibrium.

| Compound | H $\frac{atm-m^3}{mol}$ | Gas Probe Concentration[a] (ppm v/v) | Calculated[b] Equil. conc. in g.w. (ug/L) | Measured Concentration[c] in g.w. (ug/L) | $\frac{Measured}{Calculated}$ |
|----------|------|------|------|------|------|
| VC | 6.38 | 92 | 0.90 | 370 | 411 |
| BZ | $4.3 \times 10^{-3}$ | 12 | 217 | 84 | 0.4 |
| PCE | $2.0 \times 10^{-2}$ | 8 | 67 | 370 | 6 |
| TCE | $9.9 \times 10^{-3}$ | 6 | 80 | 550 | 7 |

a
 Measured 12/84 from probe near MW-29
b
 Calculation assumes the total pressure is 1 atm.  Higher subsurface gas pressures will result in proportionately higher calculated ground water concentrations.
c
 Measured 12/84 from MW-29

TABLE 11.  Effect of Vinyl Chloride Henry's Law Constant on Calculated
           Ground Water Composition.

| Reference | $H(atm\text{-}m^3/mol)$ | Measured Conc. in Gas (ppm v/v) | Calculated Conc. in Water (ug/L) |
|---|---|---|---|
| Kavanaugh & Trussell (1980) | 6.38* | 92 | 0.90 |
| Lyman et al. (1982) | 2.4* | 92 | 2.40 |
| EPA (1980) | 7.3 | 92 | 0.79 |

* Constants were converted to reported units.

which they were calculated, there could be significant uncertainty in the
value of the constant.  In order to explain the observed concentrations in
the ground water and the landfill gas, the Henry's Law constant for vinyl
chloride would have to be approximately $1.5 \times 10^{-2}$, much lower  than the
reported 6.4.

Possible sources of vinyl chloride in the ground water include
dissolved vinyl chloride leached from the waste and transported in the
ground water as a contaminant plume, dissolution of vinyl chloride from
landfill gases, and production of vinyl chloride as a breakdown product of
other chlorinated hydrocarbons in the ground water.  Henry's Law calculations argue against dissolution from landfill gases. Vinyl chloride from
the landfill leachate and the breakdown of other contaminants remain
possible sources.

Concentrations of total organic halogens, TOX, were high in many of
the ground water samples, much higher than can be accounted for by
identified chlorinated hydrocarbons.  Jekel and Roberts (1970) found TOX
to be a measure of the more polar, nonvolatile, higher molecular weight
halogenated compounds which are not amenable to identification by gas
chromatography.  These compounds have been observed to travel through
aquifers rapidly with little change in concentration.  They may, therefore, be good indicators of contaminated ground water flow.  While these
compounds are not expected to volatilize from the ground water, they may
be precursors to more volatile halogenated compounds.

Decreases in concentrations of organic contaminants in ground water
may result from volatilization into the vadose zone, adsorption onto soil
materials, dilution during ground water transport, and microbial transformation into other compounds.  Vinyl chloride does not appear to be in
equilibrium with the landfill gas and therefore volatilization may be
hindered hydrogeologically or through poor diffusion away from the water
table.  These barriers to diffusion would be expected to prevent communication of the other contaminants with the landfill gases as well, with the
coincidence of measured and calculated concentrations then having no
physical meaning.  On the other hand, if the coincidence of measured and
calculated concentrations of PCE, TCE, and benzene are taken as evidence
of equilibrium between these landfill gases and the ground water, then
either doubt is cast on the accuracy of the Henry's Law constant for vinyl
chloride or evidence is given for in situ production of vinyl chloride.

Adsorption of the organic contaminants onto soil surfaces has been estimated by either octanol-water partition coefficients or by solubility parameters (Karickhoff et al., 1979; Chiou et al., 1979). A regression equation for the partitioning of chlorinated hydrocarbons between ground water and soil organic matter was reported by Chiou et al. (1979) as:

$$\log K_{OC} = -0.557 \log S + 4.040$$

where

$$K_{OC} = \frac{\text{ug hydrocarbon adsorbed / g soil organic carbon}}{\text{ug hydrocarbon in solution / ml ground water}}$$

Both the Henry's Law constant and the estimation of $\log K_{OC}$ depend upon solubility measurements which, for compounds as volatile as vinyl chloride, vary greatly with the experimental conditions such as temperature and pressure. In Table 12, the concentrations of adsorbed contaminants in equilibrium with ground water from MW-29 are calculated using different literature values for solubility. As the regression equation was derived for chlorinated hydrocarbons, benzene was omitted from the calculations.

Based on these calculations, the adsorption of TCE and PCE is expected to be greater than that of vinyl chloride regardless of the vinyl chloride solubility value used. Greater adsorption would, in turn, increase the retardation of these compounds, thus decreasing their velocity in the ground water. Leachate from the buried waste could become relatively enriched in vinyl chloride through adsorption of PCE, TCE, and dichlorinated alkanes onto soil surfaces as well as through breakdown of these compounds into vinyl chloride. This scenario requires that communication between ground water and the landfill gas be minimal to prevent the preferential loss of vinyl chloride to the gas phase.

Summarization of the relationship of the gas and ground water phases includes the following possibilities:

* The gas and ground water phases are not in equilibrium or communication; vinyl chloride becomes relatively enriched in the ground water phase through depletion of PCE and TCE resulting from adsorption reactions and through degradation into vinyl chloride.

* The gas and ground water phases are in communication; the discrepancy between the measured and calculated equilibrium concentrations in the ground water are explained as follows:

  - the gas and ground water phases are in equilibrium with each other; the Henry's Law constant for vinyl chloride is in error;

  - vinyl chloride is volatilizing from the ground water, but is not yet in equilibrium with the gas phase, either because of continuing production at a faster rate than that of volatilization or because the initial vinyl chloride concentrations in the leachate have not volatilized to equilibrium.

By coupling the known history and setting of the different areas of the site with the possible mechanisms of transport, reaction, and degradation, one can begin to explain the distribution and mode of dissemination

TABLE 12.  Effect of Solubility Uncertainty on Estimates of Adsorption.

| | Solubility (mg/L) | Ref. | log S (umol/L) | log $K_{oc}$[*] | Measured conc. in water (ug/L) | Calculated conc. in soil (ug/g OC) |
|---|---|---|---|---|---|---|
| VC | 2700 | (1) | 4.64 | 1.46 | 227 | 6.5 |
| | 90 | (2) | 3.16 | 2.28 | | 43 |
| | 8840 | (3) | 5.15 | 1.17 | | 34 |
| | 2770 | (4) | 4.65 | 1.45 | | 64 |
| | 60 | (5) | 2.98 | 2.38 | | 54 |
| TCE | 1100 | (1) | 3.92 | 1.86 | 1220 | 87 |
| | 1000 | (2) | 3.88 | 1.88 | | 92 |
| | 1080 | (6) | 3.92 | 1.86 | | 88 |
| PCE | 140 | (1) | 2.93 | 2.41 | 593 | 150 |
| | 400 | (2) | 3.38 | 2.16 | | 85 |
| | 150 | (6) | 2.96 | 2.39 | | 150 |
| | 200 | (7) | 3.08 | 2.32 | | 120 |

(1)  AWWA, 1983. Measured at 20°C.
(2)  Lyman et al., 1982. Measured at 20°C.
(3)  Horvath, 1982.  Measured at 25°C, saturated vapor pressure.
(4)  Horvath, 1982.  Measured at 25°C, 1 atm pressure.
(5)  EPA, 1980.  Measured at 25°C.
(6)  Horvath, 1982.  Measured at 20°C, saturated vapor pressure.
(7)  Estimated from Chiou et al., 1979. Measured at 20°C.

[*]$\log K_{oc}$ = -0.557 log S + 4.040 (Chiou et al., 1979)

where $K_{oc}$ = $\dfrac{\text{ug ads/g OC}}{\text{ug/mL solution}}$

of contaminants at the site and suggest further data needs and sampling procedures to increase that understanding.

In Area 1, ground water from P-4 is dominated by vinyl chloride whereas ground water from MW-29 shows more influence of the less volatile organics. There is historical evidence of a leaky sump which contained condensate from the gas collection system. Although no data are available concerning the chemical composition of the contents of this unlined sump, samples have been taken from condensate from the present gas collection system. Table 13 lists gas condensate data for two samples of condensate from the gas collection system. As expected, the gas condensate is depleted in the more volatile vinyl chloride and relatively enriched in the other three solvents. Leachate from this unlined sump would contain a lower proportion of vinyl chloride than would ground water contaminated solely from landfill gases in the vadose zone. This gas condensate leachate plume may explain, in part, the differences between samples from P-4 and MW-29. In other words, the differences may be explainable through source differences. On the other hand, P-4, being further from the source, may reflect increased degradation coupled with a diffusion barrier decreasing or preventing volatilization of the vinyl chloride produced.

Ground water data from Area I are arranged in Table 14 to show the distribution of vinyl chloride with depth in the various wells sampled. Only the wells which are perforated near the ground water surface contained vinyl chloride. This is consistent with the idea that diffusion into the vadose zone is hindered either through the formation of diffusion layers or through physical barriers to gas flow.

While the extremely high proportion of vinyl chloride in the ground water in Area II may be the result of degradation of chlorinated compounds into vinyl chloride which did not degrade further to a significant degree, an alternative explanation is that these high concentrations are the result of the leaching of vinyl chloride-containing wastes. Historical records state that aqueous solutions of vinyl chloride were injected into the Class I area at various times during the site's active history.

TABLE 13.  Gas Condensate Data

| Compound | Sump P84 4/84 | Sump J 9/85 |
|---|---|---|
| Vinyl chloride | 910 | 70 |
| Methylene chloride | 12,000 | 2,430 |
| 1,1-Dichloroethane | 210 | 4,300 |
| t-1,2-Dichloroethene | 3,600 | 50 |
| 1,2-Dichloroethane | 130,000 | 2,770 |
| 1,1,1-Trichloroethane | | 1,730 |
| 1,1,2-Trichloroethane | 14,000 | 600 |
| 1,1,2,2-Tetrachloroethene | 1,400 | 1,340 |
| Toluene | 8,000 | |
| Chlorobenzene | 3,200 | ·1,790 |
| Xylenes (total) | 2,690 | |

Units are ug/L

TABLE 14.  Relationship of Screened Interval to Vinyl Chloride Concentration
in Area I.  (Data from October 1985)

| Well | Screened Interval | Depth to Water | Water Elevation | Height of Water Above Screened Interval | Vinyl Chloride (ug/L) |
|------|-------------------|----------------|-----------------|------------------------------------------|------------------------|
| MIR-1 | 10-60 | 42 | 748 | 0 | 140 |
| MIR-3 | 10-50 | 19 | 758 | 0 | 170 |
| MW-29 | 99-178 | 153 | 653 | 0 | 227 |
| P-4 | 86-98 | 85 | N/A | 1 | 639 |
| EPAW-2A | 35-80 | 25 | N/A | 10 | ND |
| EPAW-3 | 60-100 | 39 | N/A | 21 | ND |
| P-5 | 149-161 | 84 | N/A | 65 | ND |
| EPAW-2B | 185-240 | 79 | N/A | 106 | ND |
| P-6 | 196-210 | 80 | N/A | 116 | ND |
| MW-31 | 181-220 | 11 | 688 | 170 | ND |

N/A = Surface elevation data not available.
ND = Below detection limit

Ground water contamination in Area III, which is fairly evenly distributed among the four compounds investigated, may reflect ground water
flow of essentially unaltered waste leachate, degradation of compounds
into vinyl chloride at a low rate, or degradation at a higher rate coupled
with moderate volatilization into the gas phase.

Contamination in Area IV appears to involve ground water flow.  Data
from the many wells in this area support the idea of a contaminant plume
extending from behind the barrier westward towards P-11.  Known hydrogeology of the basin suggests that this canyon contains the natural
discharge path for the site.  The high concentration of vinyl chloride
suggests that if this is the result of degradation, volatilization cannot
occur at the rate of vinyl chloride production.

FUTURE STUDIES

Before the mode of contaminant dispersion in the subsurface can be
understood, additional sampling will be required.  Future sampling will
include soil gas samples taken at different depths in conjunction with
ground water samples at the water table.  The hydrogeology will be investigated further to better map the ground water flow patterns.  Diffusion
barriers will be suggested through model-deduced diffusion layers or
through detection of physical, hydrogeological strata. Henry's Law
constants applicable to the site can be evaluated through careful analysis
of ground water samples and headspace in equilibrium with those samples.
This would eliminate possible effects of mixed compounds on individual
constants.

Ground water and gas composition will be analyzed with distance along
the respective flow paths, looking for evidence of transformation of the
higher chlorinated species to vinyl chloride. The biotransformation of

chlorinated hydrocarbons to vinyl chloride, which has been suggested by several workers (Barrio-Lage et al., 1986), appears to be consistent with many of the findings of this study. We have actually noted during the analysis of ground waters, that samples held for periods of four to six weeks show a dramatic increase in vinyl chloride concentration. Further work is being carried out to elucidate this process.

The other major area of uncertainty of the current data is TOX. In some of the areas of the landfill TOX levels exceed that of identified contaminants by orders of magnitude. TOX is a generic measurement without specific chemical identification, however it is assumed to be indicative of higher molecular weight, polar contaminants of low volatility. The presence of significant concentrations of halogenated organics of low volatility is evidence of contaminated ground water flow. The evidence would, however, be much stronger if at least some of the compounds giving rise to the high TOX could be characterized. To this end, an effort utilizing LC/MS and LC/FTIR is being made to characterize this complex, low volatility mix.

The effects of the production and transport of methane throughout the landfill are not yet fully understood. The concentration of methane at the facility perimeter is in the range of 10,000 - 20,000 ppm and, within the facility, in the range of 50-100%. In this environment, the equilibrium between gas and liquid phases may be significantly affected. We have been unable to find literature references to this phenomenon. There may also be a mass dilution effect in the gas phase resulting from methane generation, particularly if much of the generation takes place in the unsaturated zones. Such a dilution of the volatile organics would account for concentrations of vinyl chloride in the gas phase which are lower than those calculated at equilibrium. Equilibrium calculations have assumed 1 atmosphere pressure in the landfill. Pressures may exceed that value significantly in localized areas. Pressures may be reduced in other areas as a result of pumping by the gas collection system.

By knowing not only the distribution of contaminants, but also the mechanisms involved in their transport, a more effective program can be undertaken to stop dissemination and begin remediation.

REFERENCES

AWWA, 1983, "Occurrence and removal of volatile organic chemicals from drinking water", AWWA Research Foundation, Denver.

Barrio-Lage, G., Parsons, F. Z., Nassar, R. S., and Lorenzo, P. A., 1986, Sequential dehalogenation of chlorinated ethenes. Environ. Sci. Technol., 20:96.

Chiou, C. T., Peters, L. J., and Freed, V. H., 1979, A physical concept of soil-water equilibria for nonionic organic compounds, Science, 206:831.

Eutek, Inc., "BKK landfill odor study final report", prepared for the City of West Covina, February 27, 1981.

Horvath, A. L., 1982, "Halogenated Halocarbons", Marcel Dekker, Inc., New York.

Jekel, M. R. and Roberts, P. V., 1980, Total organic halogen as a parameter for the characterization of reclaimed waters: measurement, occurrence, formation, and removal, Environ. Sci. and Technol., 14:970.

Karickhoff, S. W., Brown, D. S., and Scott, T. A., 1979, Sorption of hydrophobic pollutants on natural sediments, Water Research, 13:241.

Kavanaugh, M. C. and Trussell, R. R., 1980, Design of aeration towers to strip volatile contaminants from drinking water, Jour. AWWA, 72:684.

Lyman, W. J., Reehl, W. F., and Rosenblatt, D. H., 1982, "Handbook of chemical property estimation methods", McGraw-Hill Book Company, New York.

United States Environmental Protection Agency, 1980, Innovative and Alternative Technology Assessment Manual, February 1980, EPA-430/9-78/009, U.S.E.P.A. Municipal Construction Division, Washington, D.C.

University of Southern California, "Investigation of odorous and volatile compounds for BKK class I landfill site in the city of West Covina", prepared for BKK Corporation, July 1981.

Verschueren, K., 1977, "Handbook of Environmental Data on Organic Chemicals", Van Nostrand Reinhold Company, New York.

THE TOTAL EXPOSURE ASSESSMENT METHODOLOGY (TEAM) STUDY:
DIRECT MEASUREMENT OF PERSONAL EXPOSURES THROUGH AIR
AND WATER FOR 600 RESIDENTS OF SEVERAL U.S. CITIES

Lance Wallace
U.S. EPA
401 M Street, SW
Washington, DC  20460

Edo Pellizzari, Linda Sheldon, Ty Hartwell,
    Charles Sparacino, and Harvey Zelon
Research Triangle Institute
Research Triangle Park, NC 27709

INTRODUCTION

We are exposed daily to toxic or carcinogenic chemicals in our air, water, and food.  However, the extent of our exposure is not known in many cases due to the lack of sufficiently sensitive measuring instruments (Wallace 1982) or to the lack of sufficiently extensive population exposure studies.  Without knowing the extent of our exposure, governmental agencies such as the Environmental Protection Agency (EPA) cannot quantify relative risks of different chemicals and thereby prioritize efforts to reduce public exposure, nor can the major sources of exposure be determined.

We report here on the results of a five-year EPA study of personal exposures of urban populations to a number of organic chemicals in air and drinking water in several U.S. cities.

Goals

The TEAM Study was planned in 1979 (Pellizzari, 1980) and completed in 1985.  The goals of this study were: (1) to develop methods to measure individual total exposure (exposure through air, food, and water) and resulting body burden (concentrations in breath, blood, and urine) of toxic and carcinogenic chemicals, and (2) to apply these methods to estimate the exposures and body burdens of urban populations in several U.S. cities.  To achieve these goals, the following approach was adopted:

1. A small personal sampler was developed to measure personal exposure to airborne toxic chemicals.

2. A specially-designed spirometer was developed to measure the same chemicals in exhaled breath.

3. A survey design involving a 3-stage stratified probability selection approach was adopted to insure inclusion of potentially highly exposed groups.

## Pilot Study

A pilot study (Entz, 1982; Wallace, 1984a; Pellizzari, 1982; Sparacino, 1982a,b) was conducted between July and December 1980 to test 30 sampling and analytical protocols for four groups of chemicals potentially present in air, water, food, house dust, blood, breath, urine, and human hair.

The four groups of chemicals were:

1. Volatile organics (15 target chemicals including benzene, vinyl chloride, chloroform, and tetrachloroethylene)

2. Semivolatile organics (8 target pesticides and PCBs)

3. Metals (lead, cadmium, arsenic)

4. Polyaromatic hydrocarbons (6 compounds including benzo-a-pyrene)

The results of the pilot study indicated that the TEAM goals could be met at present for only one group of compounds: the volatile organics. Adequate methods existed to determine their concentrations in personal air, ambient air, exhaled breath, and drinking water. They were seldom present in food (with the exception of chloroform in beverages) so that food could safely be ignored.

Each of the other three groups of chemicals had major measurement method problems in one or another medium. Both metals and pesticides have a major route of exposure in solid foods--yet the sampling and analytical protocols for measuring individual meals do not exist. For the PAHs, no personal monitor capable of collecting sufficient amounts to analyze existed.

Thus, the main TEAM Study concentrated on the volatile organics. This group of some hundreds of compounds includes a dozen or so known or suspected human carcinogens, including many organics contained in the list of 37 Hazardous Air Pollutants that the Agency must decide whether to regulate or not; several solvents of interest to the Office of Toxic Substances; and compounds that the Office of Drinking Water may regulate soon.

## Main Study

The main TEAM Study measured the personal exposures of 600 people to a number of toxic or carcinogenic chemicals in air and drinking water. The subjects were selected to represent a total population of 650,000 residents of cities in New Jersey, North Carolina, North Dakota, and California. These areas included several cities with intensive petroleum refining and chemical manufacturing activity (Bayonne and Elizabeth, NJ; Los Angeles, CA; Antioch and Pittsburg, CA). Smaller studies were undertaken in an urban area without chemical activity (Greensboro, NC), and a rural town expected to have near-background concentrations (Devils Lake, ND).

## New Jersey, North Carolina, and North Dakota

The first phase of the main study began in the fall of 1981

(Hartwell, 1984; Pellizzari, 1981, 1984a,b, 1985a; Wallace 1985a,b).
Three hundred and fifty-five volunteers from the New Jersey cities were
selected from over 10,000 residents screened by a probability sampling
technique to represent 128,000 persons (over the age of seven) who live
in the two neighboring cities. One hundred and eight geographic areas
throughout the two cities were selected for monitoring. Each
participant carried a personal sampler with him during his normal daily
activities for two consecutive 12-hour periods. (One resident in each
of the 108 sampling segments had an identical sampler operating in the
backyard for the same two 12-hour periods.) All participants also
collected two drinking water samples. At the end of the 24-hour
sampling period, all participants gave a sample of exhaled breath, which
was analyzed for the same compounds. All participants also completed a
questionnaire on their age, sex, occupations, and activities during the
sampling period. An extensive quality assurance program was carried out
on all sampling/analysis activities.

A return visit was made to 157 of the original participants in the
summer of 1982, and a final visit was made to 49 of these 157 persons in
January-February of 1983.

A small comparison study was undertaken in Greensboro, North
Carolina in May 1982. Greensboro was selected because it has a similar
population to the Bayonne-Elizabeth area and has similar small
industries, but no chemical manufacturing or petroleum refining
operations. A group of 25 subjects was selected by a 3-step sampling
scheme to represent 131,000 Greensboro residents.

A second comparison site was selected to investigate whether a
rural and agricultural population in a small town far from any industry
might show clear differences in personal exposure. Once again, 25
subjects, representing 7000 Devils Lake, North Dakota residents, were
selected by a stratified probability design.

## Response Rates

The response rates to the household screening stage ranged from 85%
in New Jersey to 95% in North Carolina and 96% in North Dakota.

In the selection of individuals to be monitored the response rate
ranged from a low of 51% in New Jersey (first visit) to 67% in North
Dakota and 80% in North Carolina. The return visits to the New Jersey
respondents showed successively higher response rates of 79% and 91%.

The overall response rate is a product of the rates at each stage.
Thus, the New Jersey rate (first visit) is 85% x 51% = 43%. The North
Carolina overall response rate is 76% and the North Dakota rate is 64%.

## Measurement Methods

Personal and outdoor air samplers employed a glass cartridge
containing the solid granular sorbent Tenax-GC®. A small Dupont pump
drew air at ~30 mL/min through the cartridge for ~12 hrs to collect a
target volume of ~20 L. A sampling vest was designed to hold the pump
and the cartridge close to breathing level while leaving the
participant's hands free for normal activities.

Breath samples were collected using a specially-designed spirometer
mounted in a van. The subject provided the breath sample at his home in

the evening (6-9 pm) at the end of the 24-hour sampling period.

Water samples were collected by each participant from the tap or other sources at work or at home. Samples were collected in 2-oz. glass jars containing sodium thiosulfate to quench residual chlorine reactions.

Air and breath samples were analyzed by capillary gas chromatography mass spectrometry (GC-MS) techniques followed by a combination of manual and automated analyses of spectra. Water samples were analyzed by a purge and trap GC method utilizing a Hall Electroconductivity detector for halogenated compounds and a flame ionization detector for aromatics.

A total of nearly 5000 air, breath, and drinking water samples were collected for 400 respondents (600 person-days) in the New Jersey, North Carolina, and North Dakota sites. This represented about 95% of all samples originally scheduled.

## Quality of the Data

An extensive quality assurance (QA) program was carried out. About 30% of all samples were either blanks, spikes, or duplicates. Every type of analysis was repeated for 10% of samples in external QA laboratories (IIT Research Institute and the University of Miami Medical School). Audits of all laboratory activities were undertaken by EPA's Environmental Monitoring Systems Laboratory at Research Triangle Park, North Carolina (EMSL-RTP) and spiked samples were supplied by EMSL-RTP (air) and EPA's Environmental Monitoring and Support Laboratory in Cincinnati (water). A separate QA report was written by an independent laboratory (Northrop Corporation) concluding that no significant analytical differences could be found among the three air monitoring laboratories (Research Triangle Institute, IIT Research Institute, and EMSL-RTP).

Precision. Results of the duplicate analyses for prevalent target compounds in air and breath samples indicate that median coefficients of variance (CV) ranged from 20-40% during most visits. In general, precision for water samples was excellent ( <10% CV).

For New Jersey, the target chemicals may be sorted into several categories based on the percent of samples exceeding the quantifiable limit, which for most compounds was about 1 $\mu g/m_3$ (Table 1).

The first class, ubiquitous chemicals that were usually found in 80-100% of air and breath samples, includes two common solvents (1,1,1-trichloroethane and tetrachloroethylene); several aromatic components of gasoline, paints, and other petrochemical products (benzene, the xylene isomers, and ethylbenzene); and two isomers of dichlorobenzene, used in moth crystals and deodorizers.

The second class, compounds often but not always found in all sample types, includes one additional solvent (trichloroethylene); a compound mainly found in drinking water (chloroform); and a common component of consumer products (styrene, used in insulation, carpets, and rubber products). An indication that the probable source of styrene and the dichlorobenzenes was in the home can be discerned in the much greater frequencies of measurable amounts in personal air samples (70-80%) compared to outdoor air samples (10-30%).

Table 1.

TARGET COMPOUNDS SORTED BY PERCENT MEASURABLE IN BREATH
AND AIR SAMPLES -- ALL THREE SEASONS

|  | Range of % Measurable |
|---|---|
| **Ubiquitous Compounds** | |
| Benzene | 55 - 100 |
| Tetrachloroethylene | 66 - 100 |
| Ethylbenzene | 62 - 100 |
| o-Xylene | 58 - 100 |
| m,p-Xylene | 68 - 100 |
| m,p-Dichlorobenzene | 44 - 100 |
| 1,1,1-Trichloroethane | 33 - 99 |
| | |
| **Often Found** | |
| Chloroform | 4 - 92 |
| Trichloroethylene | 33 - 79 |
| Styrene | 46 - 91 |
| | |
| **Occasionally Found** | |
| Vinylidene Chloride | 0 - 95 |
| 1,2-Dichloroethane | 0 - 22 |
| Carbon Tetrachloride | 0 - 53 |
| Chlorobenzene | 2 - 40 |
| o-Dichlorobenzene | 1 - 34 |
| Bromodichloromethane | 0 - 24 |
| Dibromochloromethane | 0 - 1 |
| Bromoform | 0 - 1 |
| Dibromochloropropane | 0 - 1 |

The third class of substances were only occasionally found (<10% measurable in most sample types). This class includes ethylene dichloride, vinylidene chloride, carbon tetrachloride, bromodichloromethane, chlorobenzene, and o-dichlorobenzene.

Finally, three brominated substances almost never found in air or breath included bromoform, dibromochloromethane, and dibromochloropropane.

Fewer target chemicals were found in drinking water in New Jersey (Table 2). Only the three trihalomethanes were ubiquitous. A second group of three solvents appeared at low concentrations in nearly all tap water samples collected in Elizabeth but in hardly any of the Bayonne samples.

For the two comparison sites in Greensboro, North Carolina and Devils Lake, North Dakota, most of the prevalent chemicals in New Jersey air and breath samples continued to be found (Table 3). Only carbon tetrachloride appeared considerably less often than in New Jersey. In water samples, the same chemicals (trihalomethanes) made their appearance as in New Jersey (Table 4).

## Concentrations--New Jersey First Season (Fall 1981)

Frequency distributions for the combined Bayonne-Elizabeth target population of 128,000 persons are shown for all personal air, outdoor air, and breath samples of several prevalent chemicals (Figures 1-3). Notable are the great range of exposures ( < 1 $\mu g/m3$ to > 100 $\mu g/m^3$); the greater personal exposures than outdoor concentrations; and the greater breath concentrations than outdoor concentrations in most cases.

Personal exposures were invariably greater than outdoor concentrations for all prevalent target chemicals. The arithmetic means of the daytime and nighttime (i.e., indoor) personal air exposures are several times the outdoor mean concentrations. Since equilibrium concentrations in exhaled breath are often only 20-40% of inspired concentrations, the remainder being metabolized or excreted through other pathways, the fact that breath levels often exceeded outdoor levels is further indication that exposures are higher than would be expected from observed outdoor concentrations.

The median value for chloroform in drinking water was 67 $\mu g/L$; in air, 3.2 $\mu g/m^3$. Drinking water accounted for most exposure to bromodichloromethane, since the chemical was detected in only 3% of the personal air samples.

The decreasing importance of outdoor levels in contributing to the higher exposures was illustrated by comparing indoor nighttime exposures (when persons were almost invariably inside their homes) to outdoor nighttime concentrations for the 75th percentile and the 99th percentile of each distribution. The ratios increased from 2-5 at the 75th percentile up to 10-20 at the 99th percentile for most of the target chemicals. The daytime personal air exposures were usually the highest, as expected since this time period included the commuting and occupational activities. However, the nighttime personal air exposures, when people were normally sleeping, were nearly as high. In fact, all eleven prevalent chemicals had much higher maximum nighttime indoor concentrations than maximum nighttime outdoor concentrations, sometimes 100 times higher.

Table 2.

TARGET COMPOUNDS SORTED BY PERCENT MEASURABLE
IN WATER SAMPLES -- NJ-- ALL THREE SEASONS

|  | Range of % Measurable |
| --- | --- |
| Ubiquitous Compounds | |
| Chloroform | 99 - 100 |
| Bromodichloromethane | 99 - 100 |
| Dibromochloromethane | 93 - 100 |
| | |
| Often Found | |
| 1,1,1-Trichloroethane | 46 - 50 |
| Trichloroethylene | 44 - 51 |
| Tetrachloroethylene | 43 - 53 |
| | |
| Occasionally Found | |
| Vinylidene Chloride | 26 - 43 |
| 1,2-Dichloroethane | 1 |
| Benzene | 1 - 25 |
| Carbon Tetrachloride | 6 - 18 |
| Bromoform | 2 - 6 |
| Chlorobenzene | 0 - 1 |
| Dichlorobenzene isomers | 0 - 3 |
| | |
| Never Found | |
| Ethylbenzene | 0 |
| Styrene | 0 |
| Xylene isomers | 0 |

Table 3.

TARGET COMPOUNDS SORTED BY PERCENT MEASURABLE
IN AIR AND BREATH SAMPLES--NC AND ND

| Category and Compound | Range of % Measurable | |
| | NC | ND |
|---|---|---|
| Ubiquitous Compounds | | |
| 1,1,1-Trichloroethane | 72-76 | 80-91 |
| Tetrachloroethylene | 50-100 | 73-95 |
| m,p-Dichlorobenzene | 71-80 | 56-89 |
| Ethylbenzene | 90-100 | 60-80 |
| o-Xylene | 90-100 | 66-91 |
| m,p-Xylene | 85-100 | 80-97 |
| Benzene | * | * |
| | | |
| Often Found | | |
| Chloroform | 47-68 | 22-65 |
| Trichloroethylene | 8-68 | 33-52 |
| Styrene | 41-64 | 59 |
| | | |
| Occasionally Found | | |
| 1,2-Dichloroethane | 4-14 | 5-17 |
| Carbon tetrachloride | 4- 6 | 8-14 |
| Bromodichloromethane | 0 | 14 |
| Chlorobenzene | 0-16 | 7-44 |
| o-Dichlorobenzene | 0- 2 | 0-10 |
| Bromoform | 0- 4 | 0 |
| | | |
| Never Found | | |
| Dibromochloromethane | 0 | 0 |
| Dibromochloropropane | 0 | 0 |

* Benzene was ubiquitous, but high background contamination
  prevented quantifying the results.

Table 4.

TARGET COMPOUNDS SORTED BY PERCENT MEASURABLE
IN DRINKING WATER SAMPLES--NC AND ND

| | Range of % Measurable | |
|---|---|---|
| Category and Compound | NC | ND |
| **Ubiquitous Compounds** | | |
| Chloroform | 93 | 100 |
| Bromodichloromethane | 93 | 73 |
| **Often Found** | | |
| Dibromochloromethane | 93 | 18 |
| 1,1,1-Trichloroethane | 24 | 42 |
| **Occasionally Found** | | |
| Tetrachloroethylene | 74 | 0 |
| Vinylidene Chloride | 10 | 0 |
| Carbon Tetrachloride | 3 | 0 |
| Trichloroethylene | 5 | 5 |
| Toluene | NM[a] | 30 |
| 1,2,-Dichloroethane | 0 | 2 |
| Chlorobenzene | 0 | 2 |
| Bromoform | 0 | 8 |
| Dichlorobenzene isomers | 0 | 2 |
| **Never Found** | | |
| Benzene | NM | 0 |
| Styrene | NM | 0 |
| Ethylbenzene | NM | 0 |
| Xylene isomers | NM | 0 |

[a]Not Measured

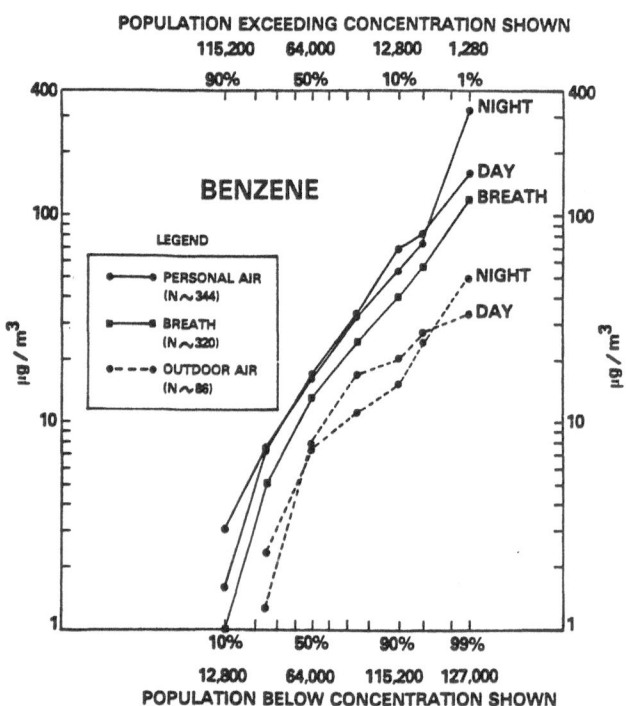

Figure 1. Benzene: Estimated frequency distributions of personal air exposures, outdoor air concentrations, and exhaled breath values for the combined Elizabeth-Bayonne target population (128,000). All air values are 12-hour integrated samples. The breath value was taken following the daytime air sample (6:00 am - 6:00 pm). All outdoor air samples were taken in the vicinity of the participants' homes.

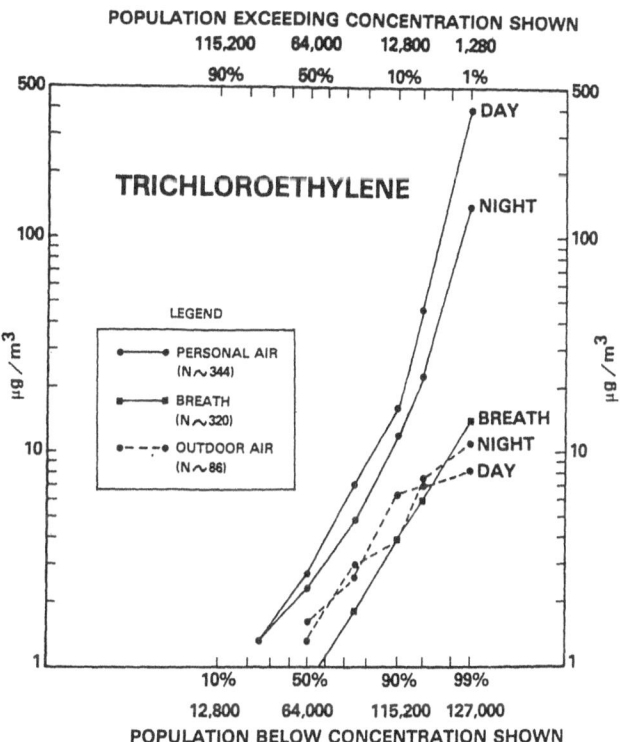

Figure 2. Trichloroethylene: Estimated frequency distributions of personal air exposures, outdoor air concentrations, and exhaled breath values for the combined Elizabeth-Bayonne target population (128,000). All air values are 12-hour integrated samples. The breath value was taken following the daytime air sample (6:00 am - 6:00 pm). All outdoor air samples were taken in the vicinity of the participants' homes.

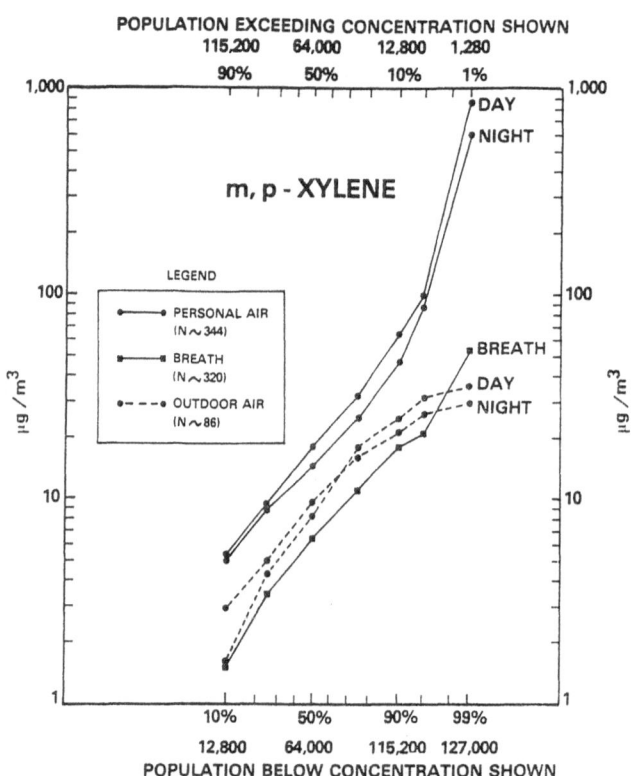

Figure 3. m,p-Xylene: Estimated frequency distributions of personal air exposures, outdoor air concentrations, and exhaled breath values for the combined Elizabeth-Bayonne target population (128,000). All air values are 12-hour integrated samples. The breath value was taken following the daytime air sample (6:00 am - 6:00 pm). All outdoor air samples were taken in the vicinity of the participants' homes.

## Comparison of the Three Seasons in New Jersey

Air, breath, and water concentrations are compared for all three seasons in New Jersey in Tables 5 and 6. In 28 of 30 cases, the mean personal air exposures exceeded the outdoor air concentrations, usually by factors of 2-10 (Table 5). The most extreme example was the combined m- and p-dichlorobenzene isomers, with arithmetic means indoors of about 50 $\mu g/m^3$ compared to outdoor values of less than 2 $\mu g/m^3$. The maximum personal air values for all chemicals were consistently in the hundreds or thousands of $\mu g/m^3$, while the maximum outdoor concentrations were usually less than 100 $\mu g/m^3$. Even breath maximum values normally exceeded the outdoor air maxima. Finally, the comparison of drinking water values across the three seasons (Table 6) shows clearly that only the three trihalomethanes had nonnegligible concentrations in the tap water samples. Also clear is the sharp decline in the winter levels.

The observation of higher indoor than outdoor values was corroborated in the second and third season. An increase in the indoor/outdoor ratios of the median and 90th percentile values was observed from summer to fall to winter, consistent with a reduced air exchange rate in winter in the presence of indoor sources.

However, the great range of indoor concentrations observed cannot be explained by differences in air exchange rates. Homes may differ by a factor of 100 in concentrations of one or more of their chemicals, whereas air exchange rates do not vary by more than a factor of 5 in most cases.

## Comparison of Greensboro with Devils Lake

Although the two smaller studies in Greensboro, North Carolina and Devils Lake, North Dakota were carried out in different seasons, a limited comparison indicates that the same chemicals with few exceptions were prevalent in air, breath, and water in the two cities. Moreover, personal air and breath levels were similar in both cities.

Greensboro. A total of 242 samples were collected, of which 110 were quality control or quality assurance samples. Blank values were very high for 1,1,1-trichloroethane and benzene. Thus, the benzene and 1,1,1-trichloroethane data should be viewed with caution. Precision was very good for air duplicates and acceptable for breath duplicates.

Personal air exposures were again greater than outdoor air exposures for most of the target chemicals (Table 7). (The low number of outdoor air samples makes this only a tentative conclusion.) A large range in personal air exposures and breath concentrations is again evident, although peak daytime personal air values are somewhat below those observed in the winter season in New Jersey. Correlations between breath and daytime personal air exposures were significant for only three of eight prevalent chemicals.

Devils Lake. A total of 237 air, water, and breath samples were collected, of which 108 were QA/QC samples. As with the Greensboro samples, high and variable background concentrations of benzene and 1,1,1-trichloroethane occurred. Median coefficients of variance of duplicate samples were in the usual ranges of 10-30% for air, but rather high levels of 30-70% for breath samples.

Table 5.

ARITHMETIC MEANS ($\mu g/m^3$) FOR AIR AND BREATH CONCENTRATIONS
OF ORGANIC COMPOUNDS IN NEW JERSEY

| Chemical | Fall 1981 (128,000)[a] | | | Summer 1982 (109,000) | | | Winter 1983 (94,000) | | |
|---|---|---|---|---|---|---|---|---|---|
| | Overnight Air | | Breath | Overnight Air | | Breath | Overnight Air | | Breath |
| | Personal | Outdoor | | Personal | Outdoor | | Personal | Outdoor | |
| Chlorinated Compounds | | | | | | | | | |
| 1,1,1-Trichloroethane | 110 | 5.4 | 15 | 21 | 10 | 15 | 31 | 1.4 | 4.0 |
| m,p-Dichlorobenzene | 56 | 1.5 | 8.1 | 49 | 1.4 | 6.3 | 54 | 1.2 | 6.2 |
| Tetrachloroethylene | 11 | 3.7 | 13 | 9.0 | 4.0 | 10 | 13 | 1.9 | 11 |
| Trichloroethylene | 7.3 | 2.1 | 1.8 | 4.8 | 7.8 | 5.9 | 3.0 | 0.2 | 0.6 |
| Carbon Tetrachloride | 14 | 1.2 | 1.3 | 1.2 | 1.0 | 0.4 | nd[c] | nd | 0.6 |
| Chloroform | 8.7 | 1.2 | 3.1 | 4.6 | 12 | 6.3 | 4.0 | 0.1 | 0.3 |
| Aromatics | | | | | | | | | |
| Benzene | 30 | 8.6 | 19 | nc[b] | nc | nc | nc | nc | nc |
| m,p-Xylene | 55 | 11 | 9.0 | 19 | 11 | 10 | 29 | 8.5 | 4.7 |
| o-Xylene | 16 | 4.0 | 3.4 | 8.0 | 4.3 | 5.4 | 9.8 | 3.1 | 1.6 |
| Ethylbenzene | 13 | 3.8 | 4.6 | 7.8 | 3.5 | 4.7 | 11 | 3.4 | 2.1 |
| Styrene | 2.7 | 0.9 | 1.2 | 2.0 | 0.6 | 1.6 | 2.2 | 0.6 | 0.7 |

[a]Population of Elizabeth and Bayonne for which estimates apply

[b]Not calculated--cartridges contaminated

[c]Not detected (most samples)

Table 6.

ARITHMETIC MEANS AND MAXIMA ($\mu g/m^3$) OF ORGANIC COMPOUNDS
IN NEW JERSEY DRINKING WATER

| Chemical | Fall 1981 (128,000)[a] Mean | Max | Summer 1982 (109,000)[b] Mean | Max | Winter 1983 (94,000)[c] Mean | Max |
|---|---|---|---|---|---|---|
| Chloroform | 70 | 170 | 61 | 130 | 17 | 33 |
| Bromodichloromethane | 14 | 23 | 14 | 54 | 5.4 | 16 |
| Dibromochloromethane | 2.4 | 8.4 | 2.1 | 7.2 | 1.4 | 3 |
| 1,1,1-Trichloroethane | 0.6 | 5.3 | 0.2 | 2.6 | 0.2 | 1.6 |
| Trichloroethylene | 0.6 | 4.2 | 0.4 | 8.3 | 0.4 | 3.4 |
| Tetrachloroethylene | 0.4 | 3.3 | 0.4 | 9.3 | 0.4 | 5.0 |
| Toluene | 0.4 | 2.7 | -- | -- | -- | -- |
| Vinylidene Chloride | 0.2 | 2.4 | 0.1 | 2.5 | 0.2 | 0.9 |
| Benzene | -- | -- | 0.7 | 4.8 | -- | -- |

[a,b,c] Population of Bayonne and Elizabeth to which estimates apply.

Table 7.

INDOOR/OUTDOOR RATIOS IN GREENSBORO,NC

| Ratio | Median Values Indoor[a] | Outdoor[b] | Ratio (I/O) | Maximum Values Indoor | Outdoor | (I/O) |
|---|---|---|---|---|---|---|
| Chloroform | 2.3[c] | 0.14[c] | 15 | 5.5[c] | 1.3[c] | 4 |
| 1,1,1-Trichloroethane | 26 | 60 | 0.5 | 110 | 275.0 | 0.4 |
| Benzene | 11 | 0.4 | 20 | 43 | 82.0 | 0.5 |
| Carbon Tetrachloride | 1.3 | 0.1 | 10 | 3.6 | 0.45 | 8 |
| Trichloroethylene | 1.0 | 0.2 | 5 | 8.7 | 2.4 | 3 |
| Tetrachloroethylene | 2.8 | 0.7 | 4 | 57 | 1.7 | 30 |
| Styrene | 0.8 | 0.1 | 8 | 3.1 | 0.31 | 10 |
| m,p-Dichlorobenzene | 3.4 | 0.4 | 8 | 72 | 1.7 | 40 |
| Ethylbenzene | 2.2 | 0.3 | 7 | 20 | 3.3 | 6 |
| o-Xylene | 3.7 | 0.6 | 6 | 26 | 3.8 | 7 |
| m,p-Xylene | 6.4 | 1.5 | 4 | 62 | 11.0 | 6 |

[a] N = 24 (overnight personal air samples)
[b] N = 6
[c] $\mu g/m^3$

Personal air exposures again exceeded outdoor air concentrations for all target compounds, although caution is indicated since the number of outdoor air samples was extremely small (Table 8). Most chemicals were not measurable in outdoor air, but indoor levels remained comparable to those observed in Greensboro. Drinking water concentrations of chloroform were exceedingly low (< 1 µg/L).

## Correlations of Personal Air Exposures with Breath Concentrations.

Ten of the eleven prevalent chemicals in the breath of the 355 New Jersey residents were significantly correlated (most at probabilities p <.0001) with the previous 12-hour average air exposures (Table 9). (The 11th chemical, chloroform, showed a significant correlation between breath and drinking water concentrations.) Since many of these chemicals are metabolized, excreted through other pathways than breath, and stored in different body compartments for different characteristic residence times, and since their concentration in breath depends partially on the previous blood concentration at the beginning of the monitoring period and also on the time history of air concentrations over the 12-hour monitoring period, strong correlations with a single 12-hour integrated concentration cannot be expected. Thus, the existence of significant correlations with previous exposures in air or water for every one of the eleven prevalent chemicals is strong evidence for the trustworthiness of the observed air and breath concentrations.

## CALIFORNIA

The final phase of the TEAM Study took place in California (Pellizzari, 1985b; Wallace, 1985c). Two areas were selected for study: Los Angeles and Contra Costa County. Ten new chemicals were added to the California study: decane, octane, undecane, and dodecane, (three of which are common co-carcinogens, i.e, promoters), 1,1,1,2-tetrachloroethane, 1,1,2,2-tetrachloroethane, 1,2-dibromoethane, 1,4-dioxane, α-pinene, and bromoform (the latter because of groundwater supplies in the California communities).

An initial survey of the two areas took place in September 1983 to collect ambient air and water samples. Ninety-seven ambient air samples were collected and analyzed by GC/MS. From these, 22 samples were selected for comprehensive qualitative identification of the 26 target chemicals. Ten drinking water samples were also collected from five municipal water plants and analyzed for four trihalomethanes.

## SURVEY RESULTS

## Los Angeles—First Season

Between February 3 and March 2, 1984, 117 residents of the South Bay section of Los Angeles (Torrance, Carson, Hermosa Beach, Redondo Beach, Manhattan Beach, Lawndale, El Nido, Lomita, Harbor City, Walteria, Hollywood Riviera, and Victoria Park) participated in the study. As in New Jersey, they collected two consecutive 12-hour personal air samples and gave a breath sample at the end of the 24-hour monitoring period (usually between 6 pm and 9 pm). The technicians collected a tap water sample on their final two visits to each home. These were analyzed separately and averaged. Participants also filled out the household questionnaire and a 24-hour activity recall diary.

Table 8.

INDOOR/OUTDOOR RATIOS IN DEVILS LAKE, ND

| Ratio | Median Values | | Ratio (I/O) | Maximum Values | | (I/O) |
|---|---|---|---|---|---|---|
| | Indoor[a] | Outdoor[b] | | Indoor | Outdoor | |
| Chloroform | 0.14 | 0.05[c] | 1 | 2.8 | 0.78 | 3 |
| 1,1,1-Trichloroethane | 37 | 0.05 | 100 | 1100 | 5.0 | 200 |
| Benzene | – –[d] | – – | –– | – – | – – | –– |
| Carbon Tetrachloride | 0.8 | 0.46[c] | 1 | 10 | 0.84 | 12 |
| Trichloroethylene | 0.7 | 0.08[c] | 3 | 32 | 1.1 | 30 |
| Tetrachloroethylene | 4.4 | 0.69 | 3 | 45 | 3.4 | 13 |
| Styrene | – – | – – | –– | – – | – – | –– |
| m,p-Dichlorobenzene | 1.7 | 0.07[c] | 10 | 230 | 2.0 | 110 |
| Ethylbenzene | 2.8 | 0.03[c] | 10 | 11 | 1.8 | 6 |
| o-Xylene | 3.5 | 0.05[c] | 10 | 19 | 1.0 | 19 |
| m,p-Xylene | 8.4 | 0.05[c] | 10 | 40 | 2.2 | 18 |

[a] N = 23 (overnight personal air samples)
[b] N = 5
[c] Not detectable – value equals 1/2 the limit of detection
[d] Data uncertain based on quality assurance results


Table 9.

SPEARMAN CORRELATIONS BETWEEN BREATH CONCENTRATIONS AND PRECEDING
DAYTIME 12-HOUR PERSONAL EXPOSURES TO ELEVEN COMPOUNDS
IN NEW JERSEY, NORTH CAROLINA, AND NORTH DAKOTA

| | NJ1[a] (N=330) | NJ2[b] (N=130) | NJ3[c] (N=47) | ND[d] (N=23) | NC[e] (N=23) |
|---|---|---|---|---|---|
| Chloroform | .07 | -.11 | -.03 | -.01 | .45* |
| 1,1,1-Trichloroethane | .28* | .28* | .32* | .71* | –– |
| Benzene | .21* | ––[f] | –– | –– | .22 |
| Carbon Tetrachloride | .24* | -.01 | –– | -.23 | -.53* |
| Trichloroethylene | .38* | .10 | .35* | .26 | .38 |
| Tetrachloroethylene | .46* | .23* | .37* | .53* | .58* |
| Styrene | .19* | .20* | .19 | –– | .32 |
| m,p-Dichlorobenzene | .54* | .38* | .61* | .63* | .68* |
| Ethylbenzene | .33* | .22* | .44* | .12 | -.01 |
| o-Xylene | .26* | .22* | .45* | .21 | .28 |
| m,p-Xylene | .32* | .27* | .48* | .19 | .08 |

[a] Fall 1981
[b] Summer 1982
[c] Winter 1983
[d] Fall 1982
[e] Spring 1982
[f] Data uncertain based on quality assurance results
* Significant at $p < .05$ level

## Response Rates

A total of 1966 homes were screened (1260 in Los Angeles, 604 in Contra Costa County) with an 88% completion rate. From the information collected on more than 5000 residents of these homes, a total of 311 were selected to participate, with 188 (60%) completing the study. Thus, the overall response rate was 53% (88% x 60%).

## Samples Collected

During the two visits to Los Angeles and the single trip to Contra Costa about 600 air samples, 500 water samples, and 200 breath samples were collected together with an additional 500 quality control samples, for a total of 1845 samples. This represented 98% of all samples scheduled.

## QUALITY CONTROL RESULTS

40 blank cartridges for air and breath samples normally contained less than 10 ng (the equivalent of 0.5 $\mu g/m^3$) of all chemicals except benzene (15-36 ng), chloroform (2-58 ng), and 1,1,1-trichloroethane (6-36 ng).

Recoveries on 41 control cartridges ranged between 70-130% for most chemicals, with the exception of the four trihalomethanes (42-200%). Cartridges loaded with deuterated benzene, chlorobenzene, and p-xylene gave recoveries ranging between 70-95%, indicating acceptable operating losses.

Blanks for the water samples were very clean. However, recoveries were generally low: 50-90%.

86 duplicate air and breath samples displayed median precisions of about 10-20%. Only chloroform and chlorobenzene were always worse than 20%. 48 duplicate water samples gave excellent precisions of 1-13% for the trihalomethanes.

## MONITORING RESULTS

### Frequency of Detection

All 26 target chemicals were found in at least a few air or water samples. Many were present in nearly every air or breath sample (Table 10). The 11 prevalent airborne chemicals in New Jersey were also prevalent in California; in addition, 6 of the 10 new target chemicals were also present much of the time.

In drinking water (Table 11) bromoform appeared in 70-90% of the samples, compared to almost none in the surface-water supplies of New Jersey. Once again the common solvents (trichloroethylene, tetrachloroethylene, and 1,1,1-trichloroethane) were present but at very low levels.

### Concentrations in Air and Breath
Los Angeles--February 1984. The 117 participants represented a

Table 10.

TARGET COMPOUNDS SORTED BY PERCENT MEASURABLE
IN AIR AND BREATH SAMPLES

Range of Percent Measurable

| | Los Angeles, CA 1st Season | 2nd Season | Antioch/Pittsburg, CA |
|---|---|---|---|
| **Ubiquitous** | | | |
| 1,1,1-Trichloroethane | 99-100 | 89-100 | 49-100 |
| Benzene | 95-100 | 79-100 | 82-100 |
| Tetrachloroethylene | 97-100 | 99-100 | 58-100 |
| Ethylbenzene | 82-100 | 70-100 | 64-100 |
| o-Xylene | 91-100 | 57-100 | 58-100 |
| m,p-Xylene | 100 | 100 | 84-100 |
| **Often Found** | | | |
| n-Octane | 81-99 | 59-94 | 29-96 |
| n-Decane | 53-96 | 25-81 | 48-100 |
| m,p-Dichlorobenzene | 79-100 | 61-87 | 0-75 |
| Styrene | 47-100 | 37-94 | 56-91 |
| Carbon tetrachloride | 12-100 | 11-100 | 14-96 |
| α-Pinene | 62-98 | 47-92 | 0-85 |
| Chloroform | 36-99 | 31-80 | 12-79 |
| **Occasionally Found** | | | |
| Trichloroethylene | 50-97 | 4-66 | 0-72 |
| n-Undecane | 56-99 | 48-74 | 8-88 |
| n-Dodecane | 30-96 | 17-45 | 0-77 |
| 1,2-Dichloroethane | 4-68 | 0-23 | 0-30 |
| o-Dichlorobenzene | 13-59 | 0-19 | 0-19 |
| 1,4-Dioxane | 8-70 | 3-21 | 5-25 |
| Chlorobenzene | 1-12 | 0-8 | 0-18 |
| 1,2,-Dibromoethane | 0-4 | 0-13 | 0-2 |
| 1,1,1,2-Tetrachloroethane | 0-3 | 0-12 | 0-18 |
| 1,1,2,2-Tetrachloroethane | 0-10 | 0-18 | 0-18 |

Table 11.

TARGET COMPOUNDS SORTED BY WEIGHTED PERCENT MEASURABLE
IN DRINKING WATER SAMPLES

| | Range of Percent Measurable | | |
| | Los Angeles | | Antioch/Pittsburgh |
| | Jan-Feb 1984 | May 1984 | |
| --- | --- | --- | --- |
| Ubiquitous | | | |
| Chloroform | 94 | 86 | 94 |
| Bromodichloromethane | 93 | 96 | 96 |
| Dibromochloromethane | 89 | 85 | 85 |
| | | | |
| Often Found | | | |
| Bromoform | 69 | 90 | 69 |
| | | | |
| Occasionally Found | | | |
| 1,1,1-Trichloroethane | 48 | 14 | 10 |
| Tetrachloroethylene | 22 | 19 | 94 |
| Trichloroethylene | 8 | 12 | 66 |
| Chlorobenzene | 13 | 5 | 6 |

Table 12.

ESTIMATES OF AIR AND BREATH EXPOSURES FOR 360,000 LOS ANGELES
RESIDENTS (February 1984)

| | Personal Air (N=225) | Outdoor Air (N=48) | Breath (N=110) |
| --- | --- | --- | --- |
| 1,1,1-Trichloroethane | 96[a] | 34[a] | 39[a] |
| m,p-Xylene | 28 | 24 | 3.5 |
| m,p-Dichlorobenzene | 18 | 2.2 | 5.0 |
| Benzene | 18 | 16 | 8.0 |
| Tetrachloroethylene | 16 | 10 | 12 |
| o-Xylene | 13 | 11 | 1.0 |
| Ethylbenzene | 11 | 9.7 | 1.5 |
| Trichloroethylene | 7.8 | 0.8 | 1.6 |
| n-Octane | 5.8 | 3.9 | 1.0 |
| n-Decane | 5.8 | 3.0 | 0.8 |
| n-Undecane | 5.2 | 2.2 | 0.6 |
| n-Dodecane | 2.5 | 0.7 | 0.2 |
| α-Pinene | 4.1 | 0.8 | 1.5 |
| Styrene | 3.6 | 3.8 | 0.9 |
| Chloroform | 1.9 | 0.7 | 0.6 |
| Carbon tetrachloride | 1.0 | 0.6 | 0.2 |
| 1,2-Dichloroethane | 0.5 | 0.2 | 0.1 |
| p-Dioxane | 0.5 | 0.4 | 0.2 |
| o-Dichlorobenzene | 0.4 | 0.2 | 0.1 |
| TOTAL (19 Compounds) | 240 | 120 | 80 |

[a] Average of arithmetic means of day and night 12-hour samples ($\mu g/m^3$)
[b] Arithmetic mean

total of 360,000 residents of the South Bay section of Los Angeles. The highest personal exposures (Table 12) are to 1,1,1-trichloroethane, p-xylene, p-dichlorobenzene, benzene, and tetrachlorothylene. Outdoor concentrations, particularly at night, are unusually high, exceeding daytime outdoor levels by 50% or more. Breath means range from 10-30% of personal exposures for most chemicals except tetrachlorothylene (75%) and benzene (45%). The four straight-chain hydrocarbons added for the California study maintained consistent relationships among themselves in both outdoor and indoor air, with octane and undecane the highest, docecane the lowest.

Los Angeles--May 1984. The second trip to 50 of the original participants resulted in estimates of exposures for 330,000 Los Angeles residents. Concentrations were considerably reduced, both personal and outdoor. However, the same chemicals appeared in roughly the same order. Outdoor overnight values no longer exceeded daytime levels, and personal exposures nearly always exceeded outdoor concentrations (Table 13).

Contra Costa--June 1984. 71 residents of Antioch and Pittsburg, CA represented a target population of 91,000 persons. Air and breath exposures were lower than in Los Angeles (Table 14), but again the same five chemicals were responsible for the highest exposures. The relative concentrations of the straight-chain hydrocarbons were different in Contra Costa, with decane highest outdoors.

## Concentrations in Drinking Water

Again, only trihalomethanes were found in significant quantities in drinking water (Table 15). However, California supplies, which included some groundwater sources, had relatively higher concentrations of brominated compounds.

## Correlations between Breath and Environmental Concentrations

Correlations between breath concentrations and preceding personal air exposures were significant for many chemicals (Table 16): correlations with outdoor air concentrations were almost never significant. Only chloroform in drinking water showed occasional significant correlations with breath concentrations.

## SUMMARY

The major findings of the TEAM Study may be summarized as follows:

1. Personal air exposures to essentially every one of the prevalent target chemicals were greater than outdoor concentrations.

2. A major reason for these higher personal exposures appears to be elevated indoor air levels at work and at home.

3. The elevated indoor air levels appear to be due to a variety of sources, including consumer products, building materials, and personal activities.

4. The breath levels correlated significantly with personal air exposures to nearly all chemicals but did not correlate with outdoor air levels. This is further corroboration of the

Table 13.

ESTIMATES OF AIR AND BREATH EXPOSURES FOR 330,000 LOS ANGELES
RESIDENTS (May 1984)

| | Personal Air (N=100) | Outdoor Air (N=47) | Breath (N=51) |
|---|---|---|---|
| 1,1,1-Trichloroethane | 44a | 5.9a | 23a |
| m,p-Xylene | 24 | 9.4 | 2.8 |
| Tetrachloroethylene | 15 | 2.0 | 9.1 |
| m,p-Dichlorobenzene | 12 | 0.8 | 2.9 |
| Benzene | 9.2 | 3.6 | 8.8 |
| Ethylbenzene | 7.4 | 3.0 | 1.1 |
| o-Xylene | 7.2 | 2.7 | 0.7 |
| α-Pinene | 6.5 | 0.5 | 1.7 |
| Trichloroethylene | 6.4 | 0.1 | 1.0 |
| n-Octane | 4.3 | 0.7 | 1.2 |
| n-Decane | 3.5 | 0.7 | 0.5 |
| n-Undecane | 4.2 | 1.0 | 0.7 |
| n-Dodecane | 2.1 | 0.7 | 0.4 |
| Styrene | 1.8 | 0.9 | 0.9 |
| p-Dioxane | 1.8 | 0.2 | 0.05 |
| Chloroform | 1.1 | 0.3 | 0.8 |
| Carbon tetrachloride | 0.8 | 0.7 | 0.2 |
| o-Dichlorobenzene | 0.3 | 0.1 | 0.04 |
| 1,2-Dichloroethane | 0.1 | 0.06 | 0.05 |
| TOTAL (19 Compounds) | 150 | 33 | 56 |

[a] Average of arithmetic means of day and night 12-hour samples ($\mu g/m^3$)
[b] Arithmetic mean

Table 14.

ESTIMATES OF AIR AND BREATH EXPOSURES FOR 91,000 RESIDENTS OF
CONTRA COSTA COUNTY (Antioch/Pittsburg--June 1984)

| | Personal Air (N=136) | Outdoor Air (N=20) | Breath (N=67) |
|---|---|---|---|
| 1,1,1-Trichloroethane | 16[a] | 2.8[a] | 16[a] |
| m,p-Xylene | 11 | 2.2 | 2.5 |
| Benzene | 7.5 | 1.9 | 7.0 |
| m,p-Dichlorobenzene | 5.5 | 0.3 | 3.7[*] |
| Tetrachloroethylene | 5.6 | 0.6 | 8.6 |
| o-Xylene | 4.4 | 0.7 | 0.6 |
| Ethylbenzene | 3.7 | 0.9 | 1.2 |
| Trichloroethylene | 3.8 | 0.1 | 0.6 |
| n-Octane | 2.3 | 0.5 | 0.6 |
| n-Decane | 2.0 | 3.8 | 1.3 |
| n-Undecane | 2.7 | 0.4 | 1.2 |
| n-Dodecane | 2.1 | 0.2 | 0.4 |
| α-Pinene | 2.1 | 0.1 | 1.3 |
| Carbon tetrachloride | 1.3 | 0.4 | 0.2 |
| Styrene | 1.0 | 0.4 | 0.7 |
| Chloroform | 0.6 | 0.3 | 0.4 |
| o-Dichlorobenzene | 0.6 | 0.07 | 0.08 |
| p-Dioxane | 0.2 | 0.1 | 0.2 |
| 1,2-Dichloroethane | 0.1 | 0.05 | 0.04 |
| TOTAL (19 Compounds) | 72 | 16 | 62 |

[a] Average of arithmetic means of day and night 12-hour samples ($\mu g/m^3$)
[*] One very high value removed


Table 15.

ESTIMATES OF DRINKING WATER CONCENTRATIONS FOR CALIFORNIA RESIDENTS

| Chemical | Los Angeles (N=117) Feb. 1984 Arith. Mean | SE | Los Angeles (N=52) May 1984 Arith. Mean | SE | Contra Costa (N=71) June 1984 Arith. Mean | SE |
|---|---|---|---|---|---|---|
| Chloroform | 14[a] | 1.41[a] | 29[a] | 3.4[a] | 42[a] | 3.1[a] |
| Bromodichloromethane | 11 | 0.84 | 20 | 2.3 | 21 | 1.4 |
| Dibromochloromethane | 9.4 | 0.91 | 28 | 3.1 | 8 | 0.56 |
| Bromoform | 0.8 | 0.14 | 8 | 2.4 | 0.8 | 0.09 |
| 1,1,1-Trichloroethane | 0.15 | 0.04 | 0.08 | 0.02 | 0.09 | 0.04 |
| Trichloroethylene | 0.08 | 0.01 | 0.07 | 0.02 | 0.06 | 0.01 |
| Tetrachloroethylene | 0.07 | 0.01 | 0.04 | 0.02 | 0.10 | 0.09 |

[a] $\mu g/L$

311

Table 16.

Table 16.

SPEARMAN CORRELATIONS BETWEEN BREATH AND PRECEDING AIR CONCENTRATIONS
(Measurable Amounts Only)

| | Breath vs. Daytime Personal Air | | | Breath vs. Daytime Outdoor Air | | |
|---|---|---|---|---|---|---|
| | LA1[a] (N=11-112) | LA2[b] (13-49) | CC[c] (10-58) | LA1[a] (N=8-24) | LA2[b] (7-24) | CC[c] (7) |
| Trichloroethylene | 0.74* | 0.84* | 0.72* | −0.05 | NC[d] | NC |
| m,p-Dichlorobenzene | 0.71* | 0.40* | 0.46* | 0.54* | 0.60* | NC |
| Tetrachloroethylene | 0.32* | 0.36* | 0.44* | 0.11 | −0.09 | NC |
| 1,1,1-Trichloroethane | 0.57* | 0.62* | 0.11 | −0.17 | 0.19 | NC |
| Ethylbenzene | 0.31* | 0.45* | 0.13 | −0.12 | 0.29 | NC |
| o-Xylene | 0.39* | 0.51* | 0.03 | 0.14 | −0.22 | NC |
| m,p-Xylene | 0.42* | 0.44* | 0.16 | 0.02 | 0.14 | −0.29 |
| Benzene | 0.25* | 0.25 | 0.07 | −0.04 | 0.11 | NC |
| Styrene | 0.31* | 0.12 | 0.06 | 0.08 | 0.23 | NC |
| n-Octane | 0.31* | 0.38* | 0.25 | −0.07 | 0.53 | NC |
| n-Decane | 0.22 | 0.63* | 0.01 | −0.09 | NC | NC |
| n-Undecane | 0.10 | 0.34 | 0.09 | 0.22 | 0.56 | NC |
| n-Dodecane | 0.23 | 0.66 | NC | 0.33 | NC | NC |
| Chloroform | −0.06 | 0.17 | NC | NC | NC | NC |
| Carbon tetrachloride | −0.32 | NC | 0.05 | NC | NC | NC |
| α-Pinene | 0.21* | 0.10 | 0.10 | −0.15 | −0.12 | NC |

[a] Los Angeles--First trip--February 1984
[b] Los Angeles--Second trip--May 1984
[c] Contra Costa (Antioch/Pittsburg)--June 1984
[d] Not calculated--N $\leq$ 5
* Significant at p < 0.05

relative importance of indoor air compared to outdoor air.

5. A number of specific sources of exposure were identified including:
    a. Smoking (benzene, xylenes, ethylbenzene, styrene)
    b. Passive smoking (same chemicals)
    c. Visiting dry cleaners (tetrachloroethylene)
    d. Auto exhaust (benzene)
    e. Various occupations, including: chemicals and plastics, scientific laboratories, garage or repair work, metal work, printing, etc. (most target chemicals)

6. Other sources were hypothesized, including:
    a. Hot showers (chloroform in air)
    b. Room air fresheners or moth crystals (p-dichlorobenzene)
    c. Opaquing fluid (trichloroethylene)
    d. Insulation, carpets, cushions (styrene)
    e. Aerosol sprays (1,1,1-trichloroethane)

7. In most cases, these sources far outweighed the impact of traditional "major" point sources (chemical plants, petroleum refineries, petrochemical plants) and area sources (dry cleaners and service stations) on personal exposure.

8. For all chemicals except the trihalomethanes, the air route provided >99% of the exposure. Water provided nearly all of the exposure to the three brominated trihalomethanes, and a significant portion of personal exposure to chloroform.

9. Measurement of exhaled breath proved to be a very sensitive and non-invasive way to determine body burden.

## CONCLUSIONS

As a first effort in directly measuring exposures in air and water, the TEAM Study was able to estimate such exposures and also to discover potential sources of exposure. The inclusion of a body burden measurement (exhaled breath) was extremely useful in providing corroboration of the exposure measurements. For this set of compounds, breath was greatly superior to blood as a sample medium: it is more sensitive, allowing identification of up to 200 compounds, and it is noninvasive, an important aid to obtaining acceptable response rates from volunteers.

The TEAM Pilot Study indicated that new sampling and analytical protocols for measuring individually cooked meals would have to be developed before an adequate multimedia study of exposure could be carried out on metals or pesticides.

## REFERENCES

Entz, R., Thomas, K., and Diachenko, G. (1982) Residues of volatile halocarbons in food using headspace gas chromatography. J. Agric. Food Chem. 30:846-849.

Hartwell, T., Perrit, R., Zelon, H., Whitmore, R., Pellizzari, E., and Wallace, L. (1984) Comparison of indoor and outdoor levels for air volatiles in New Jersey, in <u>Indoor Air, v. 4, Chemical characterization and Personal Exposure,</u> Swedish Council for Building Research, Stockholm, pp. 81–86.

Pellizzari, E.D., Erickson, J.D., Giguere, M.T., Hartwell, T.D., Williams, S.R., Sparacino, C.M., Zelon, H., and Waddell, R.D. (1980) <u>Preliminary Study on Toxic Chemicals in Environmental and Human Samples: Work Plan, Vols. I and II, (Phase I), U.S. EPA,</u> Washington, DC.

Pellizzari, E.D., Hartwell, T., Zelon, H., Leininger, C., Erickson, M., Cooper, S., Whittaker, D., and Wallace, L. (1982) <u>Total Exposure Assessment Methodology (TEAM) Prepilot Study--Northern New Jersey,</u> U.S. EPA, Washington, DC.

Pellizzari, E., Sparacino, C., Sheldon, L., Leininger, C., Zelon, H., Hartwell, T., and Wallace, L. (1984a) Sampling and analysis for volatile organics in indoor and outdoor air in New Jersey, in <u>Indoor Air, v. 4, Chemical characterization and Personal Exposure,</u> Swedish Council for Building Research, Stockholm, pp. 221–226.

Pellizzari, E., Hartwell, T., Sparacino, C., Sheldon, L., Whitmore, R., Leininger, C., and Zelon, H. (1984b) Total Exposure Assessment Methodology (TEAM) Study, First Season, Northern New Jersey, Interim Report, Contract #68-02-3679, U.S. EPA, Washington, DC.

Pellizzari, E.D., Perritt, K., Hartwell, T.D., Michael, L.C., Whitmore, R., Handy, R.W., Smith, D., and Zelon, H. (1985a) <u>Total Exposure Assessment Methodology (TEAM) Study: Elizabeth and Bayonne, New Jersey; Devils Lake, North Dakota; and Greensboro, North Carolina: Volume II, Final Report,</u> Contract #68-02-3679, U.S. EPA, Washington, DC.

Pellizzari, E.D., Perritt, K., Hartwell, T.D., Michael, L.C., Whitmore, R., Handy, R.W., Smith, D., and Zelon, H. (1985b) <u>Total Exposure Assessment Methodology (TEAM) Study: Selected Communities in Northern and Southern California: Volume III, Final Report,</u> Contract #68-02-3679, U.S. EPA, Washington, DC.

Sparacino, C., Leininger, C., Zelon, H., Hartwell, T., Erickson, M., and Pellizzari, E. (1982) <u>Sampling and Analysis for the Total Exposure Assessment Methodology (TEAM) Prepilot Study,</u> U.S. EPA, Washington, DC.

Wallace, L.A. (1982) Direct measurement of individual human exposures and body burden: research needs, <u>J. Env. Sci. and Health</u> 4:531–540.

Wallace, L.A., Pellizzari, E., Hartwell, T., Rosenzweig, M., Erickson, M., Sparacino, C., and Zelon, H. (1984a) Personal exposure to volatile organic compounds: I. direct measurement in breathing-zone air, drinking water, food, and exhaled breath, <u>Env. Res.</u> 35:293–319.

Wallace, L., Pellizzari, E., Hartwell, T., Zelon, H., Sparacino, C., and Whitmore, R. (1984b) Analysis of exhaled breath of 355 urban residents for volatile organic compounds, in <u>Indoor Air, v. 4, Chemical characterization and Personal Exposure,</u> Swedish Council for Building Research, Stockholm, pp. 15–20.

Wallace, L., Pellizzari, E., Hartwell, T., Sparacino, C., Sheldon, L., and Zelon, H. (1985a) Personal exposures, indoor-outdoor relationships and breath levels of toxic air polklutants measured for 355 persons in New Jersey, _Atmos. Env._ 19:1651-1661.

Wallace, L., Pellizzari, E., Hartwell, T., Zelon, H., Sparacino, C., and Whitmore, R. (1985b) Concentrations of 20 volatile compounds in the air and drinking water of 350 residents of New Jersey compared to concentrations in their exhaled breath, accepted by _J. Occ. Med._

Wallace, L., Pellizzari, E., Perritt, K., Hartwell, T., Michael, L., Whitmore, R., Handy, R., Smith, D., and Zelon, H. (1985c) _Total Exposure Assessment Methodology (TEAM) Study: Summary and Conclusions, Volume I_, Final Report, Contract #68-02-3679, U.S. EPA, Washington, DC.

MULTIMEDIA DESIGN PRINCIPLES APPLIED TO THE DEVELOPMENT OF THE GLOBAL

BASELINE INTEGRATED MONITORING NETWORK

G. B. Wiersma and
M. D. Otis

Idaho National Engineering Laboratory
Idaho Falls, Idaho

## ABSTRACT

A global background monitoring network for making estimates of
pollutant levels and estimates of compounds that have natural as well
as man-made reference levels is being developed.  This program is being
developed through use of a multimedia systems approach.  This paper
will deal with the design of the project, a description of the current
pilot project, and a discussion of the future implementation of the
entire global project.  Emphasis will be on the value of the use of
a systems approach in designing and evaluating an environmental monitoring program on a global scale.

## INTRODUCTION

Monitoring systems have traditionally concentrated on a single
medium for sampling.  Partly, this was the result of the way many of
the environmental regulations such as the Clean Water Act and the Clean
Air Act, were written.  It was also partly a result of a lack of the
application to monitoring systems of the appreciation of the interactive
nature of natural ecosystems, and finally, it was the result of the
failure to use available mathematical techniques to help relate movement
of pollutants from one environmental media to another.

However in the 1970s, researchers began to question the efficacy
of using single media monitoring programs to make environmental assessments.
Shuck and Morgan (1975) and Behar (1979) discussed the need for what
was called an "integrated monitoring network".  Essentially, this was
designing monitoring networks in a multimedia fashion.  In these networks,
a pollutant source was linked to an identified critical receptor.
The procedures, however for designing and analyzing a multimedia monitoring
network were not fully developed by these authors.

Mathematical procedures applicable to designing and analyzing
multimedia environmental monitoring also began to be investigated in
the 1970s.  Lindell (1978) suggested that the commitment concept previously
used in radiation exposure estimates be applied to non-radioactive

pollutants.  Burton (1981) further expanded upon the use of commitment
equations to help relate sources to receptors and to understand how
those sources interacted between various environmental media.  The
commitment approach is simple to use and can provide estimates of the
relative pathways of exposure to a critical receptor, but the approach
depends upon the ratio of the steady state or integrated exposure level
of the pollutant in the sending compartment to the similarly derived
pollutant level in the receiving compartment.  Therefore, the approach
does not help the investigator focus on the processes affecting transfers
between compartments.  Also, feedback loops are difficult to deal with
by this method.

Munn (1973) and Eberhardt (1976) both proposed that simple kinetic
models describing pollutant movement and distribution among various
environmental compartments would be a useful technique for designing
monitoring programs.  The estimate for simple first order reactions
tended to yield poor fit to field data but the strength of the approach
was in helping to conceptualize complex systems by means of a schematic.
Wiersma et al. (1984) expanded on the concept of kinetic models in
designing environmental monitoring systems and presented examples of
field monitoring programs in which this procedure had been used.  Cohen
et al. (1984) further defined the approach and expanded this application
to organic and environmentally degradable compounds as opposed to those
compounds which were more stable in the environment.

Our experience has shown that kinetic models are extremely useful
in helping to plan and design monitoring systems.  This paper describes
the application of this technique to design a global integrated background
monitoring network.  The necessity for a global background monitoring
network has been described and called for by a number of researchers
for at least 20 years.  As early as 1965, a U.S. Presidential Advisory
Committee suggested that background ecosystem studies worldwide should
be carried out (Wenger et al., 1970).  Lundholm (1968) recommended
the establishment of a global monitoring system for remote background
areas.  He visualized that such a system would provide basic environmental
information on which man could assess changes in more impacted areas,
and upon this, base recommendations for action or have advance warning
on the effects of man's activities on natural habitat.  Concurrent
with the ecosystem studies, Lundholm described the need for coordinated
pollutant measurements.

Since then, scientists throughout the world have reemphasized
the need for a global monitoring system for the establishment of a
global network of ecological baseline stations (Ecological Research
Committee, 1970; Ad Hoc Task Force, 1970; Jenkins, 1971; Sokolov, 1980;
and Izrael, 1982).

In the late 1970s at a series of expert meetings hosted by the
World Meteorological Organization, the United Nations Environment Program's
Global Environmental Monitoring Sytem, and the Ecological Sciences
Division of UNESCO, the structure for such a global background monitoring
network was laid out and the funds identified for the initiation of
a three station pilot network, with one site in the United States,
a second site in Chile, and eventually, a third site in the Soviet
Union.

Currently, this network is underway with two sites established,
one at Olympic National Park in the U.S., and the second at Torres
del Paine National Park in southern Chile.

The objectives of this network are to establish reference levels for pollutants that have potential for global contamination; to serve as an early warning system for detecting global spread and trends of pollutants; establish background levels for selected ecosystem parameters against which data from more impacted areas can be compared; and contribute to the study of the biogeochemical cycle.

As early as 1973, Munn, in describing the need for a Global Environmental Monitoring Sytem recommended that the simplest approach to design of such a system would be the use of box models or simple kinetic models. Munn said that the global Environmental Monitoring System (GEMS) should be designed in such a way that interactions between media can be studied permitting delineation of the pathways of biogeochemical cycling. Such procedures are also needed in the development of the GEMS subset, which we are calling the Integrated Global Background Monitoring Network. Because of the nature of this integrated background monitoring project, i.e., pollutants entering a remote area primarily through the atmospheric route of exposure and moving and being distributed through a number of environmental media and finally leaving the system by a different pathway, a multimedia design is demanded. Also, the attempt to coordinate pollutant measurements with selected ecosystem parameters in a remote site demands that a multimedia sampling program be undertaken.

Our first effort at applying kinetic models to remote area studies was in the Great Smoky Mountains National Park in the United States. Wiersma (1979) used simple kinetic models to help design an environmental monitoring program and to help assess potential modifications to that monitoring program. Similar studies were carried out at Olympic National Park and Glacier National Park. Studies in these locations have provided extremely valuable experience and data bases which we have used in the development of the global integrated background monitoring system.

METHODS

This paper describes the results of our efforts at Glacier National Park and the application of kinetic models to design and analyze a multimedia environmental monitoring system in a remote area. To plan our initial sampling for Glacier National Park, a simplified schematic was devised based upon our experience sampling in Great Smoky Mountains National Park and Olympic National Park (Figure 1). Based upon this, sampling was conducted for atmospheric concentrations of trace metals, sulfates and nitrates, and samplings of water, forest litter, mosses, forest trees, plants, and soil. Dry deposition plates were used with mixed results, and hydrologic parameters were measured. Two sites were chosen, the first within three kilometers of the Logan Pass Visitors Center, and the second, a very remote site located 29 kilometers from the nearest trailhead and centered in the most inaccessible part of Glacier National Park. Details on the first two years of this study are given in Wiersma et al. (1984). The location of the sites is shown in Figure 2.

Based upon the results of this initial sampling, a more detailed and representative schematic of the system that we studied was developed (Figure 3). This schematic is specifically designed to represent a high mountain lake ecosystem. As such, it represents directly the Martha's Basin site which was the most remote site. However, with modification of the input parameters, the schematic can be readily changed to represent the second site which was an exposed site. The discussion in this paper is to show only the results for the remote site, Martha's Basin.

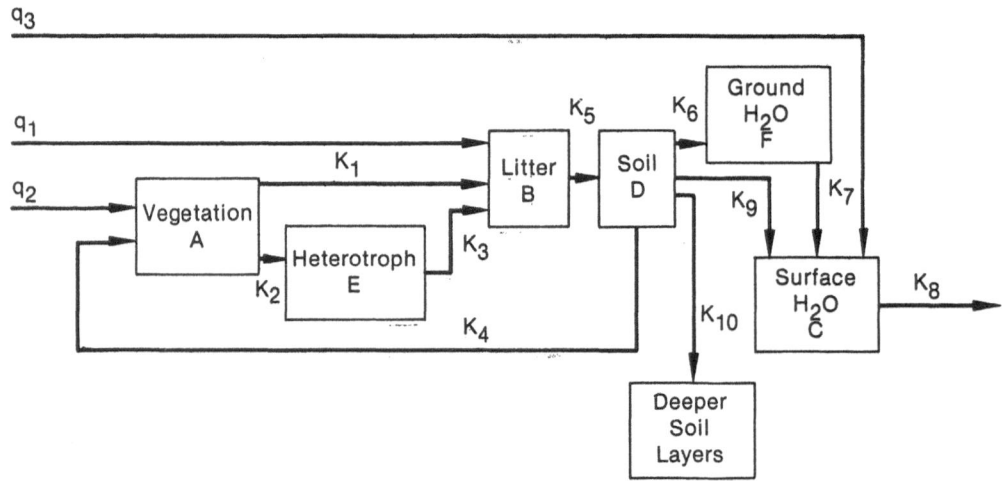

Figure 1.  Schematic of generalized system for preliminary sampling at Glacier National Park.

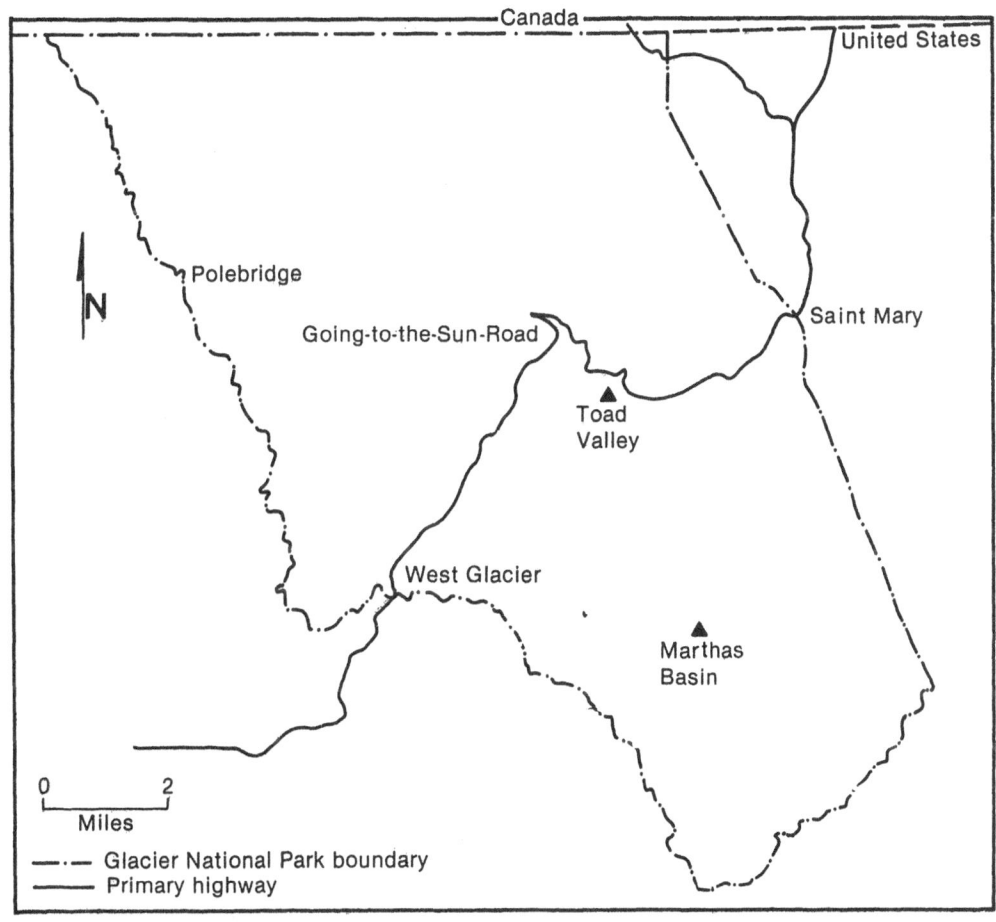

Figure 2.   Location of study sites in Glacier National Park,
Martha's Basin (remote site); Toad Valley (exposed site)

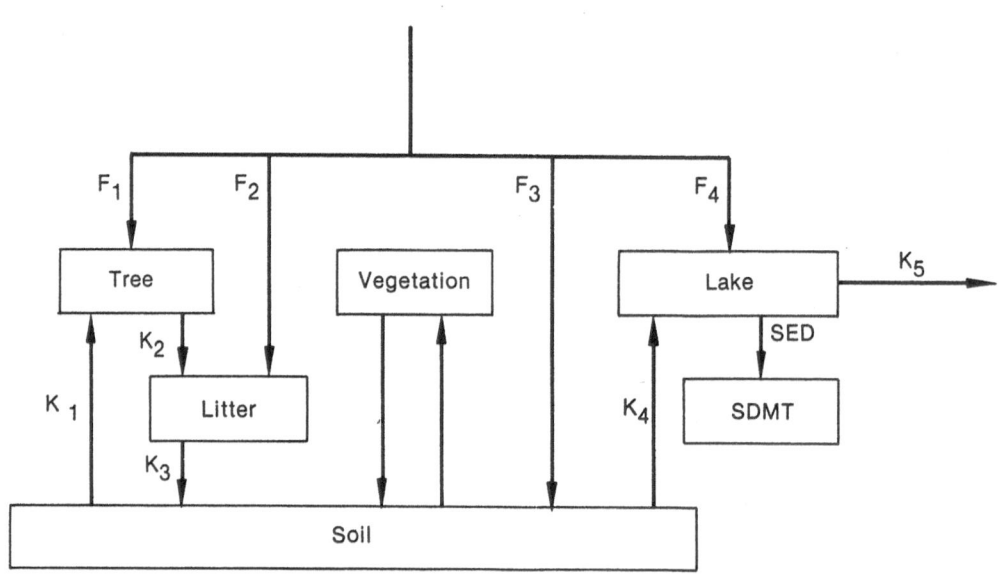

Figure 3.   Schematic of Glacier National Park Monitoring Project.

The parameters used in the model are shown in Table 1. The derivation and modification of the input parameters as necessary is described in Table 2. The results of the initial simulation are shown in Table 3 compared to field values of samples collected in the 1981-82 field season. Model simulations were taken out to equilibrium with a constant atmospheric concentration of lead equal to 5 $ng/m^3$ (Davidson et al., 1985).

RESULTS AND DISCUSSION

As a test case, we used lead as the element of concern, although in the data sets for Glacier National Park, up to 26 trace elements have been analyzed in vegetation, litter, water, and soil and up to 10 elements have been analyzed in the atmosphere.

Our inputs to the model were primarily first order estimates and we have assumed the Glacier National Park to be at equilibrium with a constant lead deposition rate. Considering these limitations, the results in Table 3 are of sufficient accuracy to help us reanalyze the system and replan future monitoring programs.

This initial analysis indicated some key pieces of information which should be collected under field conditions in future studies. These are: 1) a better definition of atmospheric deposition velocities from measurements made at the site; 2) an estimate of the leaf area index of the dominant forest type; 3) an estimate of the bulk density of the soil in the area; 4) estimates in the field of the standing biomass of the trees and the fraction of that biomass consisting of foliage (needles); 5) estimates of the mass of the litter that was sampled; 6) an estimate of the volume of the lake; 7) estimates of sedimentation rates in the lake; 8) annual increments of litterfall; and 9) estimate of annual litter turnover rates. In our initial sampling, we did attempt to gauge the streams so that we had an estimate of the output of the system, but we did not estimate the volume of the lake.

In reality, most of these measurements would not be difficult to obtain, and indeed in the pilot networks of the Global Environmental Monitoring System referred to earlier in this paper, several of these parameters are being collected simultaneously with pollutant measurements.

In addition to the above data needs, certain measurements that were needed in the model were relatively easy to obtain through aerial photographs. These included the area of the cirque basin in trees, low vegetation, bare rock and permanent snow fields.

Even after the above measurements were made, some additional information needs to be obtained. This information is: 1) The proportion of the particular element in question, in this case, lead, that was retained by needles as opposed to that which was washed off throughout the year. This could be obtained under certain types of field studies, or could be obtained from laboratory studies or from the literature; 2) Estimates of rain and snow scavenging would also be important, but probably could be approximated from literature values; and 3) Measurements of the distribution constants (Kd's) might be made in the field or laboratory. These would be the relationship of the concentration of the compound in question in the sediment divided by the concentration of the same compound in water, and the ratio between the concentration of the compound in question in vegetation and similar concentration in soil.

## TABLE 1. PARAMETERS FOR GLACIER NATIONAL PARK MODEL - MARTHA'S BASIN

Model Output Units
$\mu$g/g (solid media)
mg/L (water)

| Parameter | Value (Conventional) / Value (Model) | Units (Conventional) / Units (Model) |
|---|---|---|
| AIR | 5 / $5 \times 10^{-3}$ | $ng/m^3$ / $\mu g/m^3$ |
| DV | 0.2 / $6.3 \times 10^4$ | cm/s / m/y |
| SNOW | $3 \times 10$ / $3 \times 10^0$ | m / m |
| SCAV | $2 \times 10^4$ | $y^{-1}$ |
| ATREE | 0.20 | --- |
| AVEG | 0.20 | --- |
| ALAKE | 0.10 | --- |
| ABARE1 | 0.25 | --- |
| ABARE2 | 0.25 | --- |
| LAI | 10 | --- |
| RETAIN | 0.30 | --- |
| RHOSED | 2 / $2 \times 10^3$ | $g/cm^3$ / $kg/m^3$ |
| SEDRT | $3 \times 10^0$ / $3 \times 10^{-3}$ | mm/y / m/y |

AIR = air concentration  
SNOW= annual snowfall  
ATREE=fraction of area forested  
ALAKE=fraction of area w/surface water cover  
ABARE2=fraction of area w/bare rock draining into lake  
RETAIN=fraction of initial deposit retained in needle canopy  
SEDRT=sedimentation rate  

DV = deposition velocity  
SCAV=particle scavenging rate for snowfall  
AVEG=fraction of area w/vegetative cover  
ABARE1=fraction of area w/bare rock draining into soil  
LAI=leaf area index of forest  
RHOSED=density of sediment

TABLE 1. PARAMETERS FOR GLACIER NATIONAL PARK MODEL -
MARTHA'S BASIN
(Cont'd)

Model Output Units
$\mu g/g$ (solid media)
$mg/L$ (water)

| Parameter | Value (Conventional) / Value (Model) | Units (Conventional) / Units (Model) |
|-----------|--------------------------------------|--------------------------------------|
| KD | 700 / 700 | $\dfrac{\mu g}{g\ soil} \Big/ \dfrac{\mu g}{ml\ water}$ / L/kg |
| CF | 0.07 | $\dfrac{\mu g\ Pb}{kg\ wet\ wt\ veg} \Big/ \dfrac{\mu g\ Pb}{kg\ soil}$ |
| AMTREE | $3.5 \times 10^3$ / $3.5 \times 10^{-1}$ | kg/ha / kg/m$^2$ |
| AMLITT | $1.2 \times 10^3$ / $1.2 \times 10^{-1}$ | kg/ha / kg/m$^2$ |
| AMSOIL | $7.5 \times 10^1$ | kg/m$^2$ |
| AMLAKE | $3.3 \times 10^3$ | kg/m$^2$ |
| K1 | $1 \times 10^{-5}$ | $y^{-1}$ |
| K2 | .20 | $y^{-1}$ |
| K3 | .04 | $y^{-1}$ |
| K4 | .01 | $y^{-1}$ |
| K5 | .20 | $y^{-1}$ |

KD=distribution constant;sediment/water
AMTREE=mass of conifer needles in
    forested areas
AMSOIL=mass of soil in forested
    and vegetated areas
K1 through K5=first order rate
    constants

CF=concentration factor;veg
  (wet wt)/soil
AMLITT=mass of litter in
    forested areas
AMLAKE=average mass of water per
    surface area of lake

TABLE 2.  INPUT PARAMETER MODIFICATION
PRIOR TO SIMULATION

DRYDEP = AIR * DV

WETDEP = AIR * SNOW * SCAV

F1 = (ATREE * LAI * RETAIN) * DRYDEP

F2 = (ATREE * LAI * [1-RETAIN]) * DRYDEP

F3 = (AVEG + ABARE1) * DRYDEP + (AVEG + ABARE1 + ATREE) * WETDEP

F4 = (ALAKE + ABARE2) * (DRYDEP + WETDEP)

SED = RHOSED * ALAKE * SEDRT * KD

MSOIL = AMSOIL * (ATREE + AVEG)

MTREE = AMTREE * ATREE

MLITTER = AMLITT * ATREE

MLAKE = AMLAKE * ALAKE

## DIFFERENTIAL EQUATIONS
SOLVED DURING SIMULATION

$$\frac{d}{dt} \text{ TREE} = (\text{F1} + \text{K1} * \text{MSOIL} * \text{SOIL} - \text{K2} * \text{MTREE} * \text{TREE})/\text{MTREE}$$

$$\frac{d}{dt} \text{ LITT} = (\text{F2} + \text{K2} * \text{MTREE} * \text{TREE} - \text{K3} * \text{MLITT} * \text{LITT})/\text{MLITT}$$

$$\frac{d}{dt} \text{ SOIL} = (\text{F3} + \text{K3} * \text{MLITT} * \text{LITT} - \text{K4} * \text{MSOIL} * \text{SOIL})/\text{MSOIL}$$

$$\frac{d}{dt} \text{ VEG} = \text{CF} * \frac{d}{dt} \text{ SOIL}$$

$$\frac{d}{dt} \text{ LAKE} = (\text{F4} + \text{K4} * \text{MSOIL} * \text{SOIL} - \text{SED} * \text{LAKE} - \text{K5} * \text{MLAKE} * \text{LAKE})/\text{MLAKE}$$

$$\frac{d}{dt} \text{ SDMT} = \text{KD} * \frac{d}{dt} \text{ LAKE}$$

|  | Model Results | Field Data |
|---|---|---|
| Tree Needles | 18 µg/g | 5 µg/g (Abies Lasiocarpa) |
| Forest Litter | 60 µg/g | 82 µg/g |
| Soil | 17 µg/g | 8.5 µg/g |
| Vegetation | 1.2 µg/g | Not collected |
| Lake Water | .002 mg/L | $\leq$ .01 mg/L |
| Lake Sediment | 835 µg/g | Not sampled |

TABLE 4. MODEL PREDICTIONS FOR
RESPONSE TIME OF MAJOR ECOSYSTEM COMPONENTS

Percent of Equilibrium Pb Concentration

| Years of Simulation Time | Tree Needles | Forest Litter | Soil | Vegetation | Lake Water | Lake Sediment |
|---|---|---|---|---|---|---|
| 10 | .86 | .30 | .25 | .25 | .42 | .25 |
| 20 | .98 | .53 | .45 | .45 | .57 | .45 |
| 30 | 1.00 | .69 | .60 | .60 | .69 | .60 |
| 40 | 1.00 | .80 | .72 | .72 | .78 | .72 |
| 50 | 1.00 | .87 | .81 | .81 | .85 | .81 |
| 60 | 1.00 | .92 | .87 | .87 | .90 | .87 |
| 70 | 1.00 | .95 | .92 | .92 | .94 | .92 |

The refinement of the parameters mentioned above would result in a more accurate model. Equally important, however, it would give us a better understanding of the system in which we are measuring pollutant distribution. Furthermore, many of the parameters which are necessary for the improvement of the model performance are also potential indicators of ecosystem functioning. In particular, biomass, litterfall, and litter turnover rates, are potentially indicators of the health of the ecosystem. In addition, the relationship between pollutant distribution and movement and ecosystem functioning becomes much clearer when a logical systems approach is used.

From the analysis in Table 3, one can also obtain an estimate of where it would be most important to sample in future studies, i.e., where the highest concentrations of lead are expected to be found. In this case, while we sampled most of the important compartments, one compartment that should be sampled on any new sampling effort at Martha's Basin is the lake sediment.

The Glacier National Park experience has lead to significant changes in how we approached the design and analysis of the integrated global background monitoring network. For example, in the global background monitoring network, vegetation sampling was restricted to mosses and lichens (except for nutrient analyses) to partially avoid the wash off problem associated with tree leaves. Also, certain ecosystem parameters are being measured and several of these are also input parameters for a system model of the site in question. These include living and dead biomass and litter input rates. Other measurements discussed above will be made as more funds become available.

In addition to identifying important ecosystem parameters and equilibrium distributions among ecosystem compartments, kinetic models have the capability of estimating the response time of the system to changing conditions. One measure of response time is the time required to come into equilibrium with a constant input rate of pollutant. Table 4 presents the percent of equilibrium lead concentration for each 10 years of simulated time in the major ecosystem components modeled. Tree needles have the fastest response and lake sediment the slowest. In this example, tree needles would be the sampling medium of choice for response to the average lead deposition rate of recent years, after due consideration for possible wash off problems. Soil or sediment samples should reflect the long-term average.

Estimates of this kind are of value in planning monitoring programs for ecosystems subjected to changing impacts. No sampling plan should be initiated without some attempt at estimating the degree of non-equilibrium in the ecosystem. We recognize that the dynamics of kinetic models are often more sensitive to accurate parameter values than equilibrium concentrations are, however, the pattern of response times among compartments forces careful examination of equilibrium assumptions.

SUMMARY AND CONCLUSIONS

In summary, there are a number of advantages to using a systems approach to help design environmental monitoring programs. These are:

(1) It forces a consideration of the system as a whole rather than a series of distinct environmental components.

328

(2) It forces a consideration of the physical-chemical and biological factors effecting pollutant transport in the system.
(3) It forces examination of the dynamics of ecosystems subject to changing conditions.
(4) It sets up an analytical procedure for data analysis at the time the monitoring system is designed.
(5) It helps show the functional relationship between pollutant levels in different environmental media.
(6) It identifies gaps in current knowledge of physical-chemical and biological factors influencing transfer of pollutants and provides guidance to controlled studies addressing pollutant kinetics.
(7) It is highly compatible with monitoring programs designed to measure key ecosystem functions.

There are aso some disadvantages. These are:

(1) A very large data base is required to calculate the fractional transfers and compartment masses. Therefore, the approach loses some of its utility as the pollutants become more exotic. From an environmental view, lead is relatively well known compared to such compounds as benzene and trichloroethylene.
(2) Extensive field sampling is required in the beginning of the project, but it can be rationally reduced afterwards.
(3) Unwary users of this approach could be lulled into believing the series of equations give predictive answers. It must be remembered the fractional transfers are at best, approximations representing average conditions. At worst, they are extrapolated from marginally appropriate conditions and are literally in the first approximation category.
(4) One improvement that this analysis will not result in is the optimization of spatial placement of sampling points.

As stated earlier in the paper, monitoring has traditionally been relegated to a single media, or if it has been multimedia, it has been relegated to making pollutant measurements only. Future monitoring efforts must consider an analogous set of parameters as described in this paper. Without getting into a dialogue on the semantical differences between research and monitoring, it should be obvious that the only major difference between a research project on ecosystem functioning and a monitoring project on pollutant movement and distribution in ecosystems and potential impacts on those ecosystems is that the monitoring project would tend to have a repetitive nature, as opposed to the non-repetitive nature of the research project. Both monitoring and research projects would start with hypotheses, but the monitoring project would also proceed beyond that to the collection of data against which future hypotheses would be collected.

We believe such an integrated multimedia approach to sampling of ecosystems is extremely important. This approach is applicable not only to studies in remote areas and natural ecosystems, but also are applicable to studies on human exposures to a number of pollutants in both suburban and urban areas.

REFERENCES

Ad Hoc Task Force on GNEM. 1970. A global network for environmental monitoring. A Report to the Executive Committee, U.S. National Committee for the International Biological Program.

Behar, J. V., Shuck, E. A., Stanley, R. E. and Morgan, G. B. 1979. Integrated exposure assessment monitoring. Environmental Science and Technology 13:34-39.

Burton, N. G. 1981. The exposure commitment method in environmental pollutant assessment. Environmental Monitoring and Assessment 1:21-36.

Cohen, Y. and P. A. Ryan. 1984. Multimedia modeling of environmental transport: trichloroethylene test case. Submitted to Environmental Science and Technology.

Davidson, C. E., W. D. Gould, T. P. Mathison, G. B. Wiersma, K. W. Brown, and M. T. Reilly. 1985. Airborne trace elements in Great Smoky Mountains, Olympic and Glacier National Parks. Environmental Science and Technology 19(1):27-35.

Eberhardt, L. L., Gilbert, R. O., Hollister, H. L. and Thomas, J. M. 1976. Sampling for contaminants in ecological systems. Environmental Science and Technology 10:917-925.

Ecological Research Committee. 1970. Global Environmental Monitoring System. Swedish Natural Science Research Council, Wenner Gren Center, Stockholm, Sweden.

Izrael, Y. A. 1982. Background monitoring and its role in global estimation and forecast of the state of the biosphere. Environmental Monitoring and Assessment 2(4):369-378.

Jenkins, D. W. 1971. Global biological monitoring: Chapter in "Man's Impact on Terrestrial and Oceanic Ecosystems". Ed. by W. M. Matthews et al. MIT Press. pp. 351-370.

Lindell, B. 1978. Source-related detriment and the commitment concept: applying the principles of radiation protection to non-radioactive pollutants. Ambio 7(5-1):250-259.

Lundholm, B. 1968. Global baseline stations. Swedish Ecological Research Committee, Natural Science Research Council, Wenner Gren Center, Stockholm, Sweden.

Munn, R. E. 1978. Background discussion paper IES workshop on environmental monitoring. Institute of Environmental Studies (IES), University of Toronto, Toronto, Canada.

Schuck, E. A. and Morgan, G. B. 1975. Design of pollutant oriented integrated monitoring systems. In: Proceedings of International Conference on Environmental Sensing and Assessment. Las Vegas, Nevada. pp. 20-26.

Sokolov, V. 1981. The biosphere reserve concept in the USSR. Ambio 10(2-3):97-101.

Wenger, K. F., C. E. Ostrom, P. R. Larson and T. D. Rudolph. 1970. Potential effects of global atmospheric conditions on forest ecosystems. Summer study on critical environmental problems. July 1-31, 1970. Williams College, Williamstown, Mass.

Wiersma, G. B. 1979. Kinetic and exposure commitment analyses of lead behavior in a biosphere reserve. MARC Report 15. Monitoring and Assessment Research Center. Chelsea College, University of London, London.

Wiersma, G. B., C. I. Davidson, S. A. Mizell, R. P. Breckenridge, R. E. Binda, L. C. Hull and R. Herrmann. 1983. Integrated monitoring in mixed forest biosphere reserves. First International Biosphere Reserve Congress, Minsk, USSR.

Wiersma, G. B., C. W. Frank. M. J. Case, and A. B. Crockett. 1984. The use of simple kinetic models to help design environmental monitoring systems. Environmental Monitoring and Assessment 4(3):233-255.

Ad Hoc Task Force on GNEM. 1970. A global network for environmental monitoring. A Report to the Executive Committee, U.S. National Committee for the International Biological Program.

Behar, J. V., Shuck, E. A., Stanley, R. E. and Morgan, G. B. 1979. Integrated exposure assessment monitoring. Environmental Science and Technology 13:34-39.

Burton, N. G. 1981. The exposure commitment method in environmental pollutant assessment. Environmental Monitoring and Assessment 1:21-36.

Cohen, Y. and P. A. Ryan. 1984. Multimedia modeling of environmental transport: trichloroethylene test case. Submitted to Environmental Science and Technology.

Davidson, C. E., W. D. Gould, T. P. Mathison, G. B. Wiersma, K. W. Brown, and M. T. Reilly. 1985. Airborne trace elements in Great Smoky Mountains, Olympic and Glacier National Parks. Environmental Science and Technology 19(1):27-35.

Eberhardt, L. L., Gilbert, R. O., Hollister, H. L. and Thomas, J. M. 1976. Sampling for contaminants in ecological systems. Environmental Science and Technology 10:917-925.

Ecological Research Committee. 1970. Global Environmental Monitoring System. Swedish Natural Science Research Council, Wenner Gren Center, Stockholm, Sweden.

Izrael, Y. A. 1982. Background monitoring and its role in global estimation and forecast of the state of the biosphere. Environmental Monitoring and Assessment 2(4):369-378.

Jenkins, D. W. 1971. Global biological monitoring: Chapter in "Man's Impact on Terrestrial and Oceanic Ecosystems". Ed. by W. M. Matthews et al. MIT Press. pp. 351-370.

Lindell, B. 1978. Source-related detriment and the commitment concept: applying the principles of radiation protection to non-radioactive pollutants. Ambio 7(5-1):250-259.

Lundholm, B. 1968. Global baseline stations. Swedish Ecological Research Committee, Natural Science Research Council, Wenner Gren Center, Stockholm, Sweden.

Munn, R. E. 1973. Global Environmental Monitoring Systems (GEMS), Action Plan for Phase I. SCOPE Report 3, International Council of Scientific Unions, Scientific Committee on Problems of the Environment. Toronto, Canada.

Munn, R. E. 1978. Background discussion paper IES workshop on environmental monitoring. Institute of Environmental Studies (IES), University of Toronto, Toronto, Canada.

Schuck, E. A. and Morgan, G. B. 1975. Design of pollutant oriented integrated monitoring systems. In: Proceedings of International Conference on Environmental Sensing and Assessment. Las Vegas, Nevada. pp. 20-26.

Sokolov, V. 1981. The biosphere reserve concept in the USSR. Ambio 10(2-3):97-101.

Wenger, K. F., C. E. Ostrom, P. R. Larson and T. D. Rudolph. 1970. Potential effects of global atmospheric conditions on forest ecosystems. Summer study on critical environmental problems. July 1-31, 1970. Williams College, Williamstown, Mass.

Wiersma, G. B. 1979. Kinetic and exposure commitment analyses of lead behavior in a biosphere reserve. MARC Report 15. Monitoring and Assessment Research Center. Chelsea College, University of London, London.

Wiersma, G. B., C. I. Davidson, S. A. Mizell, R. P. Breckenridge, R. E. Binda, L. C. Hull and R. Herrmann. 1983. Integrated monitoring in mixed forest biosphere reserves. First International Biosphere Reserve Congress, Minsk, USSR.

Wiersma, G. B., C. W. Frank. M. J. Case, and A. B. Crockett. 1984. The use of simple kinetic models to help design environmental monitoring systems. <u>Environmental Monitoring and Assessment</u> <u>4</u>(3):233-255.

**SUMMARY**


In nature, pollutants do not stay where they originate but migrate
through physical, chemical and biological processes across environmental
media boundaries. Consequently, exposure and risk analyses require
knowledge of pollutant transport and accumulation in the multimedia
environment. Multimedia models of pollutant transport, described in the
first section of this book, range from the very simple to the very complex.
Irrespective of the mathematical complexity of the model, a true
forecasting capability can only be achieved through an accurate description
of the physical, chemical and biological intermedia and transformation
processes. In particular, the utilization of multimedia models, as
described in the section on multimedia analyses of exposure and risk, will
require a quantitative evaluation of modeling uncertainties. These
uncertainties are due to a combination of parameter uncertainties and a
lack of understanding of the physicochemical processes which govern
pollutant transport and transformations. If multimedia models are to
become useful in risk-assessment, an effective methodology will have to be
developed to quantify the effect of these model uncertainties.

Presently, the available multimedia models consist of either spatial
models that link individual-medium modules or models that make use of well-
mixed compartmental systems. The spatial models require a considerable
meteorological and hydrological input data base, while, comparatively, the
well-mixed compartmental models have modest parameter input requirements.
It is anticipated that a class of specialized models will continue to
emerge over the next several years. The selection of a multimedia model
for a particular application calls for a careful consideration of the
required temporal and spatial scale, the complexity of the model in
relation to available model input parameters, model interpretation, and
validation by laboratory and field measurements.

Multimedia field monitoring of pollutant concentrations and of human
exposure are scarce, which makes the validation of predictive models very
difficult. Consequently, there is a need to continue and expand current

efforts to improve our understanding of the fundamentals of intermedia transport and transformations in the environment. Another area of recent concern is that of indoor pollution. Data which are now becoming available indicate that in some instances exposure to certain chemicals in the home environment (indoors) may be more severe than exposure to pollutants outdoors. This area of research is in its embryonic stage and will require careful attention in the forthcoming years.

The papers in the second and third sections of the book reveal that the determination of human exposure and risk requires information on pollutant concentrations in the various environmental media and knowledge of pollutant intake by the human receptor. Clearly, the human body is not a sink for chemical contaminants. These contaminants may undergo various metabolic reactions in the different body organs. Additionally, there may be thermodynamic constraints that dictate the partitioning of a given toxic chemical between the body and the environment with which the body comes in contact. Therefore, the precise determination of pollutant intake and concentrations in the various body organs will require detailed models of pollutant intake, similar to the models which have been extensively employed in pharmacokinetics.

In closing, any multimedia program, whether theoretical or experimental, must first consider the role of each environmental compartment in an overall multimedia scheme. In addition, the coordination of both multimedia monitoring and modeling activities is most essential for progress toward assessing the environmental impact of emerging new technologies, and hazardous waste treatment and disposal practices. In addition to the intensive modeling efforts, there is still a need to further explore the connection between the multimedia aspects of environmental pollution and the resulting effect on multimedia pollution control strategies.

CONTRIBUTORS

Nancy B. Ball
California Public Health Foundation
2151 Berkeley Way
Berkeley, CA 94704


Yoram Cohen
Department of Chemical Engineering
5531 Boelter Hall
University of California
Los Angeles, CA 90024


J. Clarence Davies
Executive Vice President
The Conservation Foundation
1255 Twenty-third Street, NW
Washington, DC 20037


J.G. Droppo
Pacific Northwest Laboratory
P.O. Box 999
Richland, WA 99352


James W. Falco
Exposure Assessment Group
U.S. Environmental Protection Agency
401 M Street SW
Washington, DC 20460


C.S. Fang
University of Southwestern Louisiana
Lafayette, LA


Ty Hartwell
Research Triangle Institute
Research Triangle Park, NC 27709


Seong Hwang
Exposure Assessment Group (RD-689)
Office of Research and Development

U.S. Environmental Protection Agency
401 M Street SW
Washington, DC 20460


Robert H. Kadlec
Department of Chemical Engineering
Herbert H. Dow Building
University of Michigan
Ann Arbor, MI 48109-2136


William E. Kastenberg
Mechanical, Aerospace and Nuclear
    Engineering Department
University of California
Los Angeles, CA 90024


Thomas McKone
Environmental Sciences Division
Lawrence Livermore
    National Laboratory
P.O. Box 5507   L-453
Livermore, CA 94550


Donald Mackay
Department of Chemical Engineering
    and Applied Chemistry
University of Toronto
200 College Street   #217
Toronto Ontario CANADA M5S 1A4


Danny M. Mar
California Public Health Foundation
2151 Berkeley Way
Berkeley, CA 94704


Brock Neely
Dow Chemical Company
P.O. Box 1706
Midland, MI 48640

George R. Oliver
The Dow Chemical Company
P.O. Box 1706
Midland, MI 48640

M.D. Otis
Idaho National Engineering Laboratory
P.O. Box 1625
Idaho Falls, ID 83415

Sally Paterson
Department of Chemical Engineering
    and Applied Chemistry
University of Toronto
200 College Street   #217
Toronto Ontario CANADA M5S 1A4

Malcolm R. Patterson
Computing and Telecommunications
    Division
P.O. Box X
Building 4500 North   Room B226
Oak Ridge National Laboratory
Oak Ridge, TN 37831

Edo Pelizzari
Research Triangle Institute
Research Triangle Park, NC 27709

D.D. Reible
Louisiana State University
Baton Rouge, LA 70803

Linda Sheldon
Research Triangle Institute
Research Triangle Park, NC 27709

Charles Sparacino
Research Triangle Institute
Research Triangle Park, NC 27709

B.L. Steelman
Pacific Northwest Laboratory
P.O. Box 999
Richland, WA 99325

Robert Stephens
Chief, Hazardous Materials Laboratory
California Department of
    Health Services
2151 Berkeley Way   Room 237
Berkeley, CA 94704

D.L. Strenge
Pacific Northwest Laboratory
P.O. Box 999
Richland, WA 99325

Louis J. Thibodeaux
Director, Hazardous Waste
    Research Center
2418 CEBA Building
Louisiana State University
Baton Rouge, LA 70803

Lance Wallace
U.S. Environmental Protection Agency
401 M Street, Sw
Washington, DC 20460

Gene Whelan
Pacific Northwest Laboratory
P.O. Box 999
Richland, WA 99325

G. Bruce Wiersma
E.G. & G. Idaho Inc.
Idaho National Engineering Laboratory
P.OL Box 1625
Idaho Falls, ID 83415

K. Jack Yost
Director, Division of
    Sponsored Programs
School of Health Sciences
Purdue University
303 Hovde Hall
West Lafayette, IN 47907

Harvey Zelon
Research Triangle Institute
Research Triangle Park, NC 27709

# INDEX